ROUTLEDGE LIBRARY EDITIONS:
ECOLOGY

Volume 3

T0373647

TROPICAL RESOURCES

TROPICAL RESOURCES

Ecology and Development

Edited by
JOSÉ I. FURTADO, WILLIAM B. MORGAN,
JAMES R. PFAFFLIN AND KENNETH RUDDLE

LONDON AND NEW YORK

First published in 1990 by Harwood Academic Publishers

This edition first published in 2020
by Routledge
2 Park Square, Milton Park, Abingdon, Oxon OX14 4RN

and by Routledge
52 Vanderbilt Avenue, New York, NY 10017

Routledge is an imprint of the Taylor & Francis Group, an informa business

© 1990 Harwood Academic Publishers GmbH

All rights reserved. No part of this book may be reprinted or reproduced or utilised in any form or by any electronic, mechanical, or other means, now known or hereafter invented, including photocopying and recording, or in any information storage or retrieval system, without permission in writing from the publishers.

Trademark notice: Product or corporate names may be trademarks or registered trademarks, and are used only for identification and explanation without intent to infringe.

British Library Cataloguing in Publication Data
A catalogue record for this book is available from the British Library

ISBN: 978-0-367-36640-7 (Set)
ISBN: 978-0-429-35088-7 (Set) (ebk)
ISBN: 978-0-367-35349-0 (Volume 3) (hbk)
ISBN: 978-0-367-35356-8 (Volume 3) (pbk)
ISBN: 978-0-429-33091-9 (Volume 3) (ebk)

Publisher's Note
The publisher has gone to great lengths to ensure the quality of this reprint but points out that some imperfections in the original copies may be apparent.

Disclaimer
The publisher has made every effort to trace copyright holders and would welcome correspondence from those they have been unable to trace.

Tropical Resources

Ecology and Development

Edited by

JOSÉ I. FURTADO
WILLIAM B. MORGAN
JAMES R. PFAFFLIN
KENNETH RUDDLE

harwood academic publishers
chur london paris new york melbourne

© 1990 by Harwood Academic Publishers GmbH, Poststrasse
22, 7000 Chur, Switzerland. All rights reserved.

Harwood Academic Publishers

Post Office Box 197
London WC2E 9PX
United Kingdom

58, rue Lhomond
75005 Paris
France

Post Office Box 786
Cooper Station
New York, New York 10276
United States of America

Private Bag 8
Camberwell, Victoria 3124
Australia

This material originally appeared in Volume 7 of the journal *Resource Management and Optimization*.

Library of Congress Cataloging-in-Publication Data

Tropical resources : ecology and development / edited by José
I. Furtado . . . [et al.].
p. cm.
ISBN 3-7186-0514-7
1. Agricultural resources—Tropics—Management. 2. Natural
resources—Tropics—Management. 3. Agricultural ecology—Tropics.
4. Ecology—Tropics. I. Furtado, J. I.
S604.64.T76T76 1990 89-71532
333.95'11'0913—dc20 CIP

No part of this book may be reproduced or utilized in any form or by any means, electronic or mechanical, including photocopying and recording, or by any information storage or retrieval system, without permission in writing from the publisher. Printed in the United States of America.

Contents

Resource Management and Optimization
1990, Volume 7(1–4), pp. 1–37
Reprints available directly from the publisher.
Photocopying permitted by license only.
© 1990 Harwood Academic Publishers GmbH
Printed in the United States of America

ENVIRONMENTS, PRODUCTION AND RESOURCES

R.D. HILL
Department of Geography and Geology, University of Hong Kong, Hong Kong

CONTENTS

Tropical forests are probably less known than any natural environment except the ocean floor. Environments in the tropics are quite diverse, both in the biological and physical senses. Discussion is presented of the environments, resources and present and future products of what are commonly known as the tropics.

1. DEFINING THE TROPICS

Strictly speaking, the tropics are simply those regions between the Tropics of Capricorn and Cancer but environmental realities are quite otherwise. Whether delimited by climatic or by vegetational criteria the tropics spill over these artificial boundaries while containing within them enclaves which some consider to the temperate. Just how differently one section, the humid tropics, may be to a climatologist and to a student of vegetation is shown in Figure 1. In the great deserts centred upon the northern and southern tropics climatic and vegetational affinities are with the regions polewards of the Tropics rather than with regions nearer the equator. Indeed, for many a layman the image conjured up by the word 'tropical' is one of great heat, humidity and vegetational richness.

The scientist's vision of tropical lands is scarcely more refined as a comparison of maps of vegetation will quickly show. Compare the so-called world climatic map of Koeppen (who, to complicate matters even further gives vegetational names to his climatic zones) with that of Preston James[1] or the map included here (Figure 2). Suffice to say that in tropical lands, environments for the most part are extremely diverse reflecting not only significant variations in the factors of the physical environment but also highly varied impacts of man. The latter range from the minimal impacts of hunting and gathering peoples such as the Vedda of Sri Lanka or the Aeta of the Philippines to the wholesale transformation of forest to vegetations dominated by a single species such as rice and rubber, or to the structurally and floristically very different forests which remain after selective logging of rainforest. Many of these distinctly different habitats cannot be shown on world-scale maps because of their spatial characteristics. For example, the mangrove characteristic of most low wave-energy coasts except the driest, forms an elongated belt often no more than a few tens of metres wide, except in deltas where it may broaden to 50 km or more as along the Musi river in Sumatra. To take another case—the extremely varied secondary forests and scrublands which follow shifting cultivation. Here a vegetation transect only a kilometre long may cross fields of hill rice, patches of manioc and other tuber crops, a grove of bananas half-choked with rapidly-growing scrub, patches of 'forest', at various stages of regeneration, each representing a former clearing for cultivation (true jungle these), each becoming more diversified in terms of species and acquiring greater phytomass.

Thus, with the exception of the desert margins and still to a degree in the great rainforests of Amazonia, the general picture is of an often fine-grained mosaic of differing landscapes the origins of which are quite diverse. A possible indicator of diversity in general may be biological variety. While this use raises difficult questions of cause and effect, it is true that tropical areas, even quite dry ones, contain an

[1]See Figure 10.

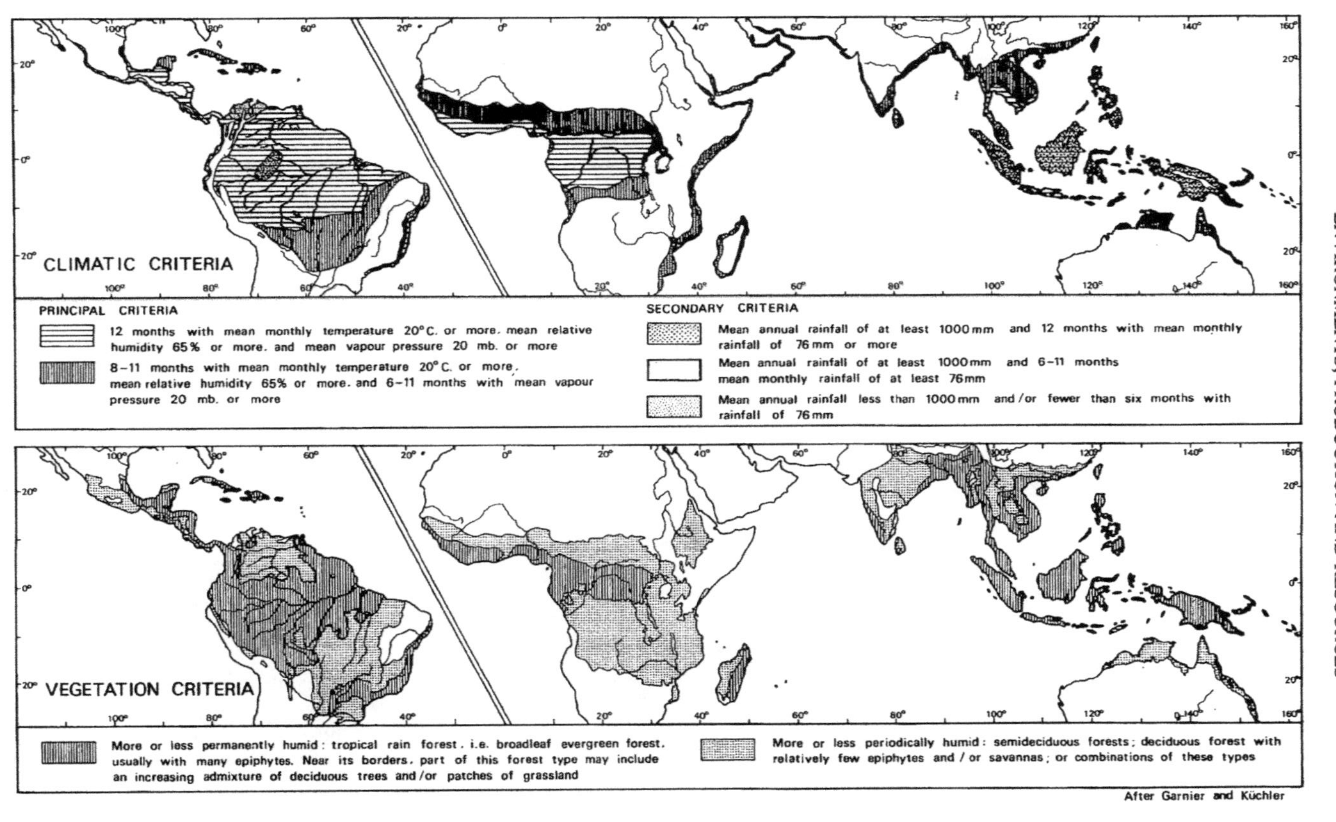

FIGURE 1. The delimitation of the humid tropics according to climatic and vegetational criteria. Compare Küchler's vegetation map with that by Soviet geographers (Figure 2) and a simplified version of that by Preston E. James (Figure 10).

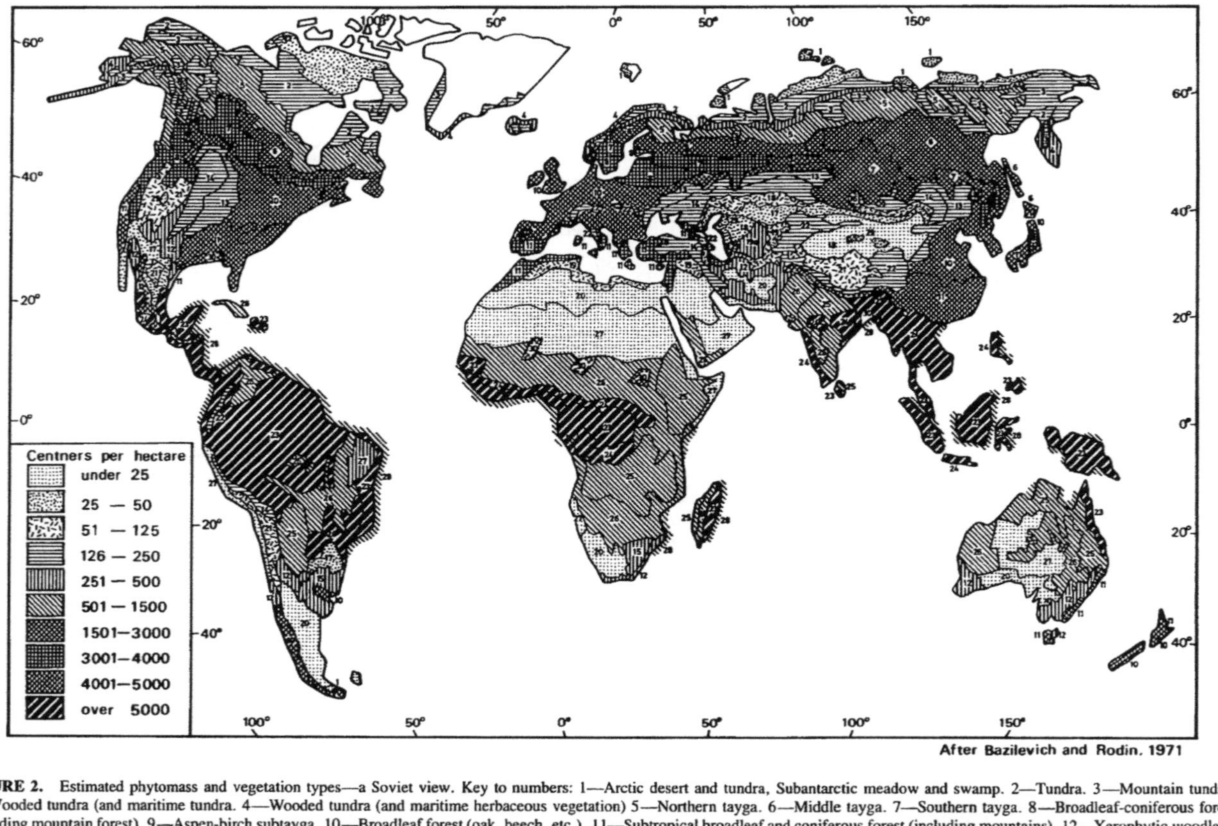

FIGURE 2. Estimated phytomass and vegetation types—a Soviet view. Key to numbers: 1—Arctic desert and tundra, Subantarctic meadow and swamp. 2—Tundra. 3—Mountain tundra. 4—Wooded tundra (and maritime tundra. 4—Wooded tundra (and maritime herbaceous vegetation) 5—Northern tayga. 6—Middle tayga. 7—Southern tayga. 8—Broadleaf-coniferous forest (including mountain forest). 9—Aspen-birch subtayga. 10—Broadleaf forest (oak, beech, etc.). 11—Subtropical broadleaf and coniferous forest (including mountains). 12—Xerophytic woodland and shrub; mountain xerophytes. 13—Wooded steppe (meadow steppe). 14—Moderately arid and arid steppe (including mountains). 15—Pampa and herbaceous savanna. 16—Dry steppe. 17—Subboreal wormwood desert. 18—Subboreal saltwort desert. 19—Subtropical semidesert. 19a—Saksaul growths. 20—Subtropical desert. 21—Highland desert. 22—Alpine and subalpine meadow. 23—Tropical rain forest. 24—Intermittently humid deciduous tropical forest. 25—Tropical xerophytic woodland. 26—Tropical savanna. 27—Tropical desert. 28—Mangrove swamp. 29—Solonchaks. 30—Subtropical and tropical herbaceous and woody growth of the fringing-forest type.

enormous range of organisms. In the rainforests of the Malay Peninsula, for example, Hsuan Keng (1974) has indicated the presence of almost 6 800 species of vascular plants. Amongst breeding land birds in the Americas, the Isthmian region contains close to 600 species (MacArthur, 1972, 212). Such diversity in the tropics may be much greater than is yet known, for some groups, the insects is one, are as yet little-studied compared with temperate regions. Unquestionably a considerable factor in explaining this diversity is the fact that biological evolution in the land areas of the tropics has not suffered the effects of Pleistocene glaciations to the same degree as temperate lands the more northerly of which were a veritable *tabla rasa* as recently as 10 000 years ago (Flenly, 1979). Speciation has thus had a longer period in which to operate in the tropics.

Too much ought not to be made of biological diversity as an indicator of environmental diversity, for some tropical environments, though physically diverse, and are not biologically diverse, not even where man has had little impact. One example is the mangrove in which there are roughly two degrees of magnitude fewer species of vascular plants than in nearby forests even though the environment contains many variations in grain-size and organic content of the substrate, rates of sediment deposition, and salinity not to mention climate. Human activities also tend to reduce biological diversity, particularly those related to agriculture, and in addition usually result in substantial reduction in biological production.

2. PHYTOMASS AND PRODUCTIVITY

Given the fine grain of many tropical environments, their great variety over short distances, it is necessary to find some measure which will reflect the gross differences within the region. One such measure is the phytomass, the weight (dry) of plants per hectare. An example is shown in Figure 2 from which it is immediately obvious that there is an enormous range, from values 0.25 tonnes per hectare in the deserts up to more than 500 tonnes per hectare in the ever-humid equatorial zones. Since temperature is only rarely a constraint to plant growth in the tropics, this pattern largely reflects differences in water supply which in turn derive from the global pattern of air mass circulation.

Production also reflects other physical factors including the phytomass, for it is to some degree true that the greater the phytomass the greater is its productivity though this is influenced in part by the speed with which biological cycling takes place (see Figure 3). Cycling is influenced to some degree by temperature though the breakdown of the litter proceeds more rapidly when temperature is high, other things being equal. (See Figure 4).

The humid tropics are thus characterized by high values for phytomass, (except where woody vegetation is replaced by grasses which have values of only 5-10 tonnes per hectare), by high annual production, again except where grasses occur, and by high

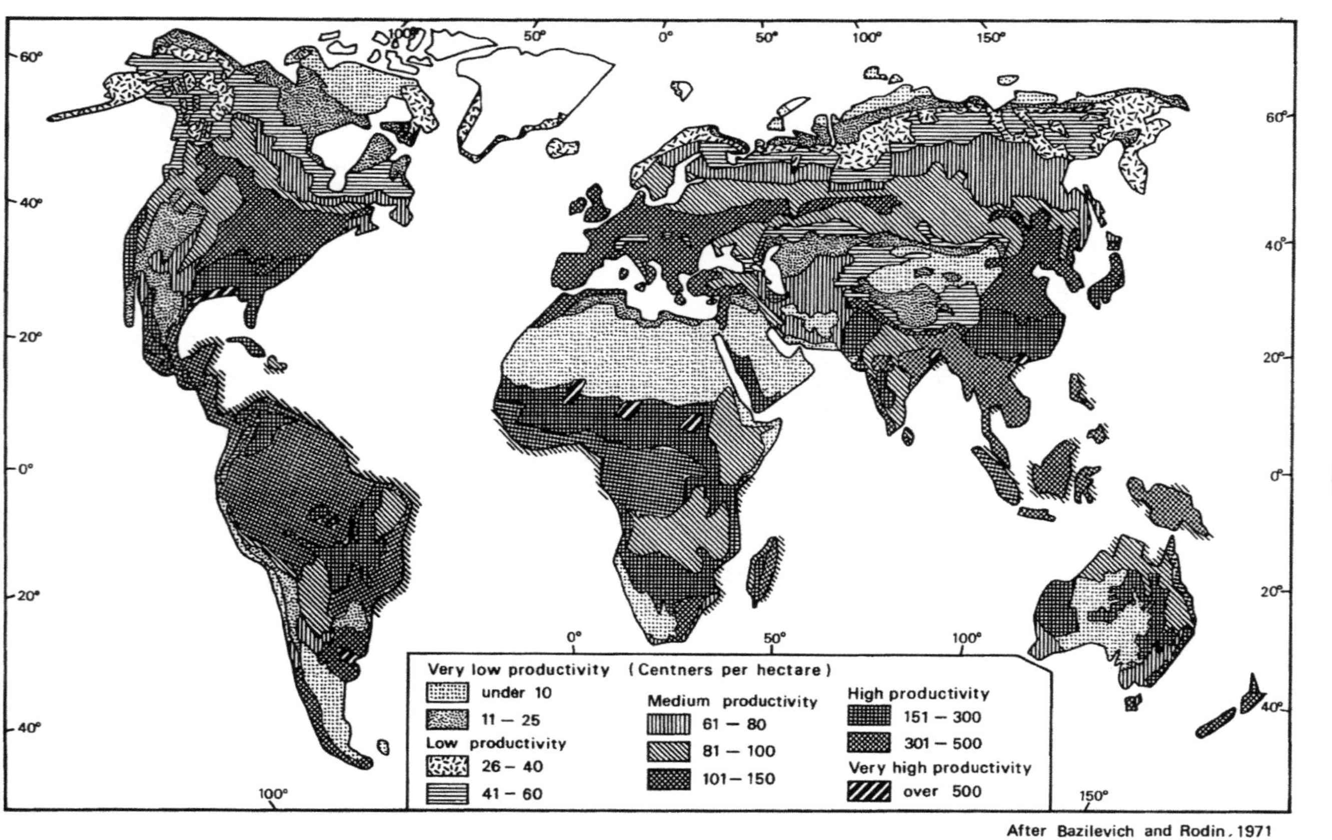

FIGURE 3. Estimated annual production of plant cover.

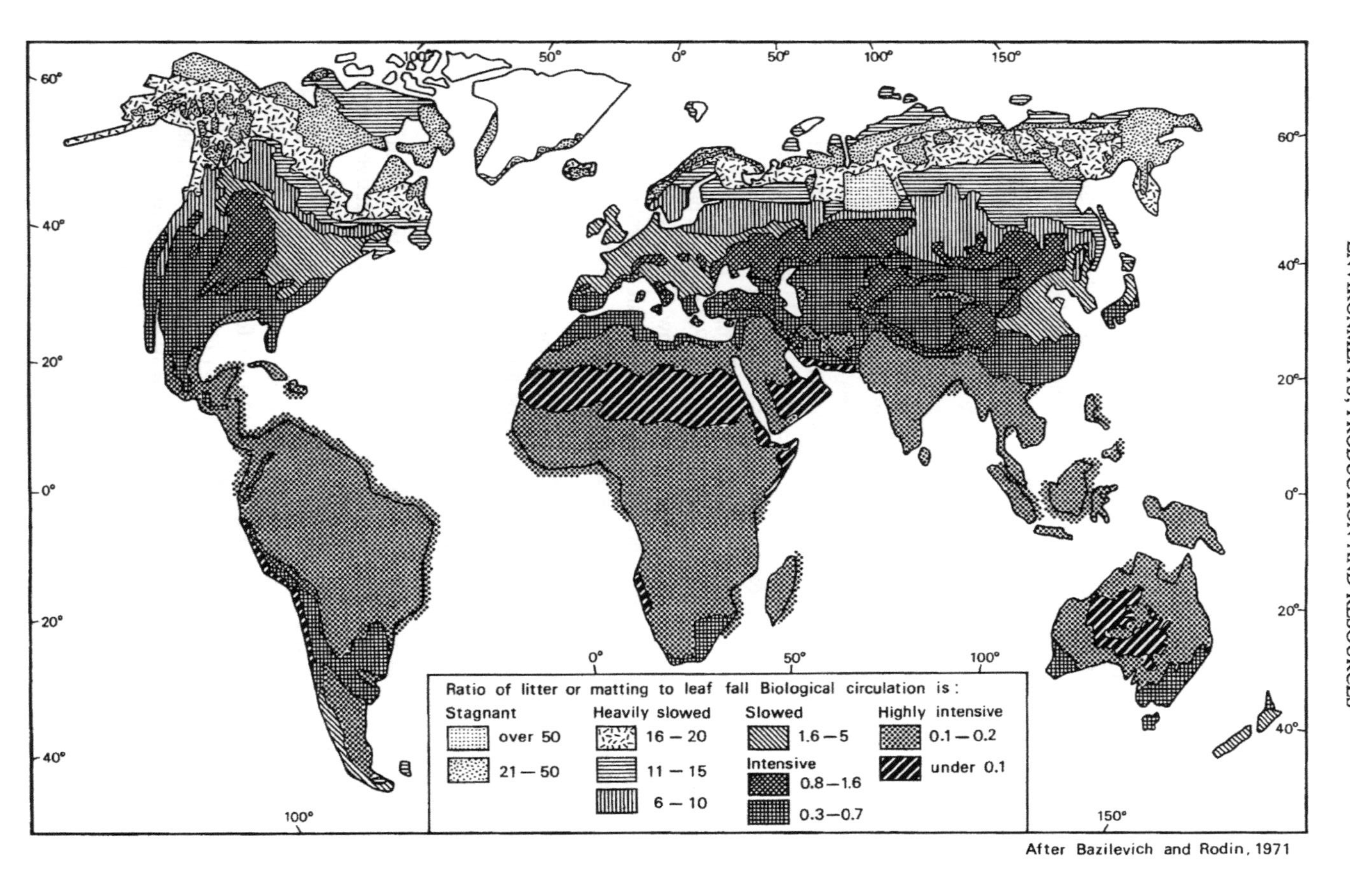

FIGURE 4. Ratio of litter to leaf fall as a measure of the rate of biological cycling.

levels of biological circulation except where soils are permanently or seasonally saturated with water, as in lowland swamp forests. In the last case, given the predominance of anaerobic decay of plant litter, circulation may be sufficiently slow to allow peat to accumulate despite high temperatures. The only other significant exception is in the cool but often extremely wet environment of tropical mountains where moderate temperatures and much rain may also lead to the accumulation of peat, sometimes forming a blanket up to 1.5 m thick even on slopes.

At the other end of the scale, in tropical arid regions phytomass and annual production are very low, because of dryness, though biological circulation remains high, reflecting rapid decay in the prevailing heat. The major exception to this pattern occurs in oases and on the flood plains of rivers, the so-called gallery vegetations. Detailed estimates of phytomass and annual production are given in Table I.

The mean values given in Table I can be compared with those for temperate lands. In the great needle-leaf forests of Eurasia and North America the phytomass averages 189 t/ha with an annual production at only seven tons per hectare while the mid-latitude broadleaf forests show average values of 342 and 13 tons per hectare respectively. Biological circulation is much slower in these temperate forests because of lower temperatures. One major consequence of such rapid biological circulation is that in most tropical areas, except alluvial lowlands and regions of basic volcanics where soils are inherently of moderate to high fertility, the bulk of the store of nutrients in the environment is locked up in the biomass. For the most part the soil litter is thin and the content of organic matter in the soil is low, the only major exception being in locations which are so continually wet as to permit anaerobic conditions to predominate. Provided that water is not a limiting factor, growth can be extremely rapid.

Water supply is an obvious factor reflecting in differences in phytomass and annual production both from major biome to major biome and within the semi-arid and arid biomes. For example though floodplain locations in semi-arid and arid zones show lower values for phytomass and production than those in the humid zone, the differences are not large basically because the perenially waterlogged conditions of humid tropical floodplains depress yields compared to those from more freely-drained sites. In contrast, mountain savanas in the semi-arid zone may be more productive than many of the lowland formations.

Elevation also serves to depress the values for phytomass and production to some degree, but not in the humid zone where it is only in the rather limited areas of montane forest, above about 1000m near the Equator, that values for phytomass are lower than nearer sea-level. In Peninsular Malaysia, for example, height and girth of forest trees when compared with the lowlands, show considerable depression of both height and girth with increasing elevation. Figure 5 shows the marked stunting of trees at higher elevations and the absence of trees of large girth. (These data are presented in place of phytomass estimates which are lacking).

A further factor in accounting for phytomass and production differences within

TABLE I
Estimates of Area, Phytomass and Annual Production of Major Tropical Vegetations

Tropical Humid Types	Area (000 Km^2)	Phytomass (t/ha)	Annual Production (t/ha)
Humid evergreen forest on red/yellow ferralitic soils	8630	650	30
Humid evergreen forest on dark red (oxidized) soils	197	600	27
Seasonally-humid evergreen & tall-grass savana on red/ yellow ferralitic soils	8786	200	16
Seasonally-humid evergreen & tall-grass savana on black tropical soils	141	80	15
Humid evergreen swamp forest on ferralitic gley soils	2244	500	25
Bog formations	664	300	150
Floodplain formations	1175	250	70
Mangrove	478	130	10
Humid tropical/sub-montane forest	2419	700	35
Seasonal tropical sub-montane forest	1175	450	22
Total	26501	Mean 440	Mean 29
Tropical Semi-arid Types			
Xerophytic forest on ferralitized brownish-red soils	4627	250	17
Grass & shrub savanna on ferralitic red-brown soils	6547	40	12
Grass & shrub savanna on tropical black soils	2050	30	11
Grass & shrub savanna on tropical solonets soils	235	20	7
Meadow & swamp savanna on ferralitized red & meadow soils	894	60	14
Floodplain formations	221	200	60
Xerophytic mountain forest on brownish-red mountain soils	497	200	15
Mountain savanna on red-brown mountain soils	938	40	12
	16009	Mean 107	Mean 14
Tropical Arid Types			
Desertlike savanna on reddish-brown soils	4290	15	4
Desert on tropical desert soils	4673	1.5	1
Psammophytic formations on sand	2820	1.0	0.1
Desert on tropical coalesced soils	216	1.0	0.2
Halophytic formations on solonchak	115	1.0	0.1
Floodplain formations	100	150	40
Mountain desert on tropical mountain desert soils	626	1.0	0.1
	12840	Mean 7.0	Mean 2.0

After Brazilevich, Rodin, Ye & Rozov, 1971.

each major biome is soil type. Amongst the humid tropical vegetation types, for example, values are lower on oxidized red soils than on the more common red-yellow ferralitic soils (soils containing substantial Fe and Al). In the seasonally-humid

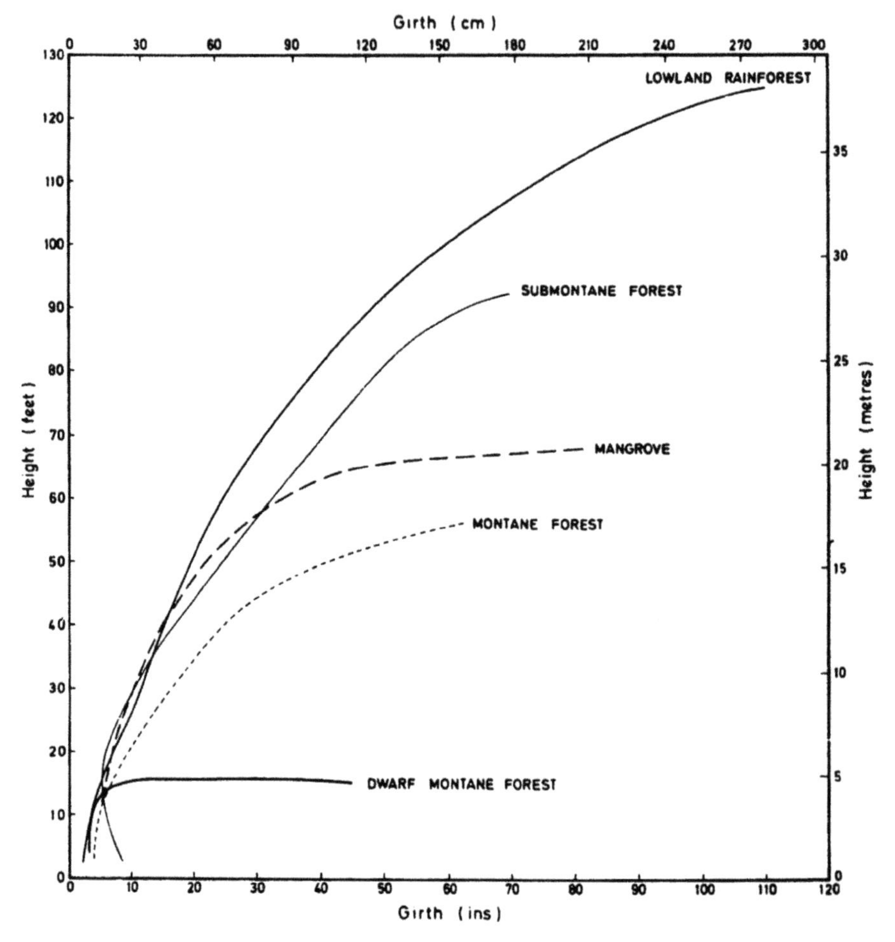

FIGURE 5. The relationship of tree girth to tree height (specimens with a girth of 10 cm or more) for vegetation types in Singapore and Malaysia.

zone, where much of the savana is actually man-induced, values are lower on black tropical soils than on the red-yellow ferralitic soils, though part of this difference is probably to be accounted for by somewhat drier conditions where such black soils are formed.

The presence of salts also tends to depress phytomass and production, perhaps in part because of the quite limited numbers of species adapted to this 'slightly-poisonous' environment. In the mangrove forests of the humid tropics, for example, no more than a dozen or so tree species exist, usually growing in relatively pure stands, in con-

TABLE II
Estimates of Annual Dry-Matter and 'Consumable' Production of Some Tropical Crops (t/ha/yr)

	Dry-matter	'Consumable'	Form
Rice (one crop)	2.5–12.0	0.5–4.5	grain
Maize (hybrid)	8.0–15.0	2.5–5.0	grain
Maize (standard)	6.0–12.0	0.5–2.5	grain
Millet (*Pennisetum*)	9.0–18.0	0.6–1.2	grain
Sorghum (*Sorgho*)	2.0–10.0	0.5–3.0	grain
Rubber (mature)	6.0–16.0	0.3–0.9	latex
Oil palm	30.0–45.0	3.5–5.0	oil
Sugar	40–120	4.0–18.0	raw sugar
Sown pasture	25	13–20	herbage
Natural grasslands	1.8–3.2	0.6–1.6	herbage

Compiled from various sources

trast to the heterogenous lowland forests nearby. In both the semi-arid and arid zones where both sodium and magnesium salts accumulate to form solonets and solonchak soils, phytomass and production are substantially below the means for these zones. Moreover, such soils can be reclaimed for agriculture only with considerable difficulty.

3. BIOLOGICAL PRODUCTIVITY AND AGRICULTURE

Though not all of the vegetation types enumerated in Table I have remained unaffected by man's activities, most reflect long periods of adaptation to the physical environment. Certainly montane vegetations have been affected by cooling during the Pleistocene glaciations as have the vegetations along the coast by sea-level changes consequent upon those glaciations. The vegetations of arid and semi-arid areas have also been influenced by changes in the global circulation both during and after the Pleistocene era. However the effects of these environmental changes are felt but slowly and it may be argued that within a time-span of hundreds of years these major types, though changing slowly, are essentially in balance with their environment.

If this is so, it becomes possible to use the values for biomass and production as a guide to the biological upper limits for which agricultural man might aim. Obviously agricultural man may be able to ameliorate water deficits, for example, but it seems unlikely that having done so, it will be possible to push production significantly beyond that of a similarly-located 'natural' vegetation, without employing energy subsidies, such as fertilizers—and perhaps not even then.

In practice, most agricultural systems are far below natural systems in terms of production, except where irrigation has been introduced. For example, Whitney (1979) has suggested that even the most highly-productive agricultural system in China, continuously-cropped rice, produces little more than half of the biological maximum represented by monsoonal forest, despite the use of irrigation and energy subsidies. Just how far short the production of agroecosystems falls short of that of natural ecosystems may be judged by comparing Tables I and II.

Table II shows that while a few crops, notably sugar, oil palm and sown pasture (all customarily grown with substantial energy subsidies) approach the annual production levels of 'natural' vegetations, most do not. But it should be recalled these are values for dry-matter, not actual substances consumable by man. The ratio of 'consumable' to 'non-consumable' dry-matter varies greatly from crop to crop. Natural rubber (*Hevea brasiliensis*) has only a tiny percentage of its biological production actually usable as latex. Some of the higher ratios are for modern varieties of long-used grasses such a sugar, 18% raw sugar, or *Sorgho,* 35% milled grain. So far as sown pasture is concerned the 'consumable' fraction is, of course, consumable by animals, not man, and the addition of another link in the food-chain depresses the final out-turn substantially. Cattle grazed upon well-managed tropical sown pastures give a final consumable production in the range 1.3-2.7 t/ha/yr.

Though in Table I the ranges of production levels within each vegetation type is not stated (for lack of data) it is unlikely that they are quite as large as those for the various crops where differences of a whole order of magnitude between the lower and upper parts of the yield range are not uncommon. Rice, for example, grown in the forest clearings of shifting cultivators, may yield around 0.6 t/ha of unhusked grain, yet on the best lowland farms with good irrigation and appropriate fertilizer applications, ten times that level is achievable even in the tropics where yields are still much below those of temperate rice-growing lands such as Japan or Australia. Thus while there are unquestionably environmentally-determined upper limits to production, for most crops these lie far above current levels. Future work will be not only in the direction of increasing total dry-matter but also in increasing the proportion of consumable material.

4. SOLAR RADIATION

In explaining areal differences of phytomass and production first place must be given to net radiation at the surface, otherwise known as the radiation balance, for it is to radiation, not temperature, that plants respond. Moreover, radiation greatly influences evapotranspiration, a basic factor in the environment of plants, as well as being basic to an understanding of atmospheric circulation.

Figure 6 shows the general pattern. Large values for net radiation are found over the tropical oceans, especially the Indian Ocean. Though cloudiness in such areas is

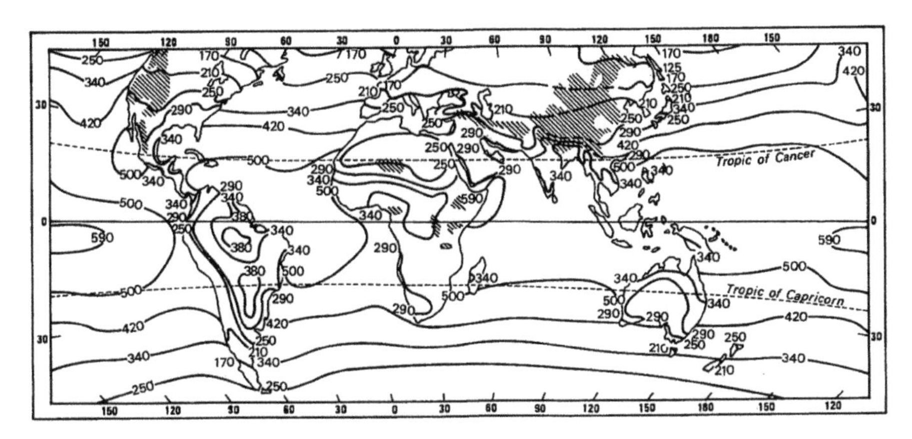

FIGURE 6. Mean annual radiation balance (KJ/cm^2) in tropical and near-tropical areas. (mainly after Budyko)

greater than near the Tropics, and this reduces to some degree the amount of sunlight reaching the surface, this reduction is less than the blanketing effect of the clouds which produces more long-wave thermal radiation. As a consequence, tropical seas are perennially warm, in equatorial regions being at about 26°C with mean annual sea surface temperatures falling to around 22°C at the Tropics where the annual sea-temperature range may reach 10°C. The only major exceptions to this pattern are areas of abnormally low sea temperatures off the west coast of southern Africa and off the west coast of South America extending from the Tropic of Capricorn to the Equator where the upwelling of cold water depresses temperatures and plays a major role in maintaining desertic conditions on the adjoining lands. Abnormally warm areas are less marked being found mainly in partly-enclosed shallow seas such as those of Southeast Asia and also the Caribbean where oceanic circulation is restricted. This great mass of warm tropical water has major effects upon the adjacent lands. Not only is it the source of moisture but it also acts as a great buffering system tending to even out seasonal differences, in temperature particularly. However its effects die out with increasing distance from the sea, especially where the atmospheric circulation is such as to cause off-shore winds to predominate as for example the North-east Trades off the western coast of West Africa.

Between the sea and the land there are abrupt discontinuities in net radiation generally amounting to 80-170 KJ/cm^2/yr. The continents therefore have more sunshine because of clearer skys but they also have greater nocturnal cooling because of reduced cloudiness at night. Day-time/night-time contrasts may exist even in equatorial maritime locations such as Singapore but the effects become much more marked as the centre of the continents and the Tropics are approached. In the Sahara, for example, day-time shade temperatures may exceed 40°C but may fall close to freezing point in the early morning as a consequence of marked radiation loss through the dry transparent air.

The reflectivity of the earth's surface also influences the radiation balance. In the equatorial zone the dark green colour of the rainforest is highly absorbtive of incoming radiation whilst reflection from a milliard raindrops after the frequent showers characteristic of this zone is probably insignificant. By contrast, the light colours of many desert soils (solonets and solonchak) serve to reflect a higher proportion of incoming radiation than darker soils such as those formed on basaltic rocks.

Net radiation not only has significant spatial variation but also seasonal variations basically reflecting changes in the angle of incidence of the sun's radiation arriving from space. Over the tropical oceans the contours showing net monthly radiation mostly lie parallel to the lines of latitude as might be expected, but over land the pattern is more complex as 'hot-spots' appear. In June for example, 'hot-spots' with net monthly radiation values ranging between about 35 and 60 KJ/cm^2 appear along the southern tropic, values at the higher end of this range being found in the western Atlantic and Caribbean (where they help to supply energy for tropical revolving storms), in the Arabian Sea and in mid-Pacific.

A significant reason for higher net radiation values in the summer is enhanced day-length which has important effects upon plants and animals. At the Equator days are of equal length but at the Tropics the difference between the longest and shortest days is just under three hours, not a great deal in comparison with higher latitudes to be sure, but sufficient to trigger off reproductive cycles, for instance, for many tropical organisms seem to respond to rather small differences in day-length. This scientific field is relatively little-researched in the tropics except in respect of some major crops. Traditional varieties of rice, for example, are virtually all photoperiod-sensitive so that the spread of the crop from its original home around the Bay of Bengal towards the Equator where days are shorter and its cultivation in winter when days are also shorter has necessarily involved selection for reduced sensitivity to day-length.

In tropical regions, the various components of the energy equation show rather greater seasonal variability than temperature—and it is to radiation and rainfall that biological activity responds. This is shown in Figure 7 which compares energy regimes with temperature and rainfall for a humid equatorial station, Mañaos and a monsoonal station, Ho Chi Minh City. Here S and R represent solar and terrestrial radiation fluxes respectively. LE is the term for evaporation (latent heat and rate) while C is the sensible heat flux by convection in the air. At the first station, where there is fairly constant moistening of the soil for eight months of the year, most of the net radiation is used in evaporation, but at the second station it rises rapidly during the winter dry season when cloud cover diminishes, despite reduced day-length. The pattern probably explains in part why when low precipitation is removed as a constraint to plant growth (by irrigation) crops grow particularly well at this season.

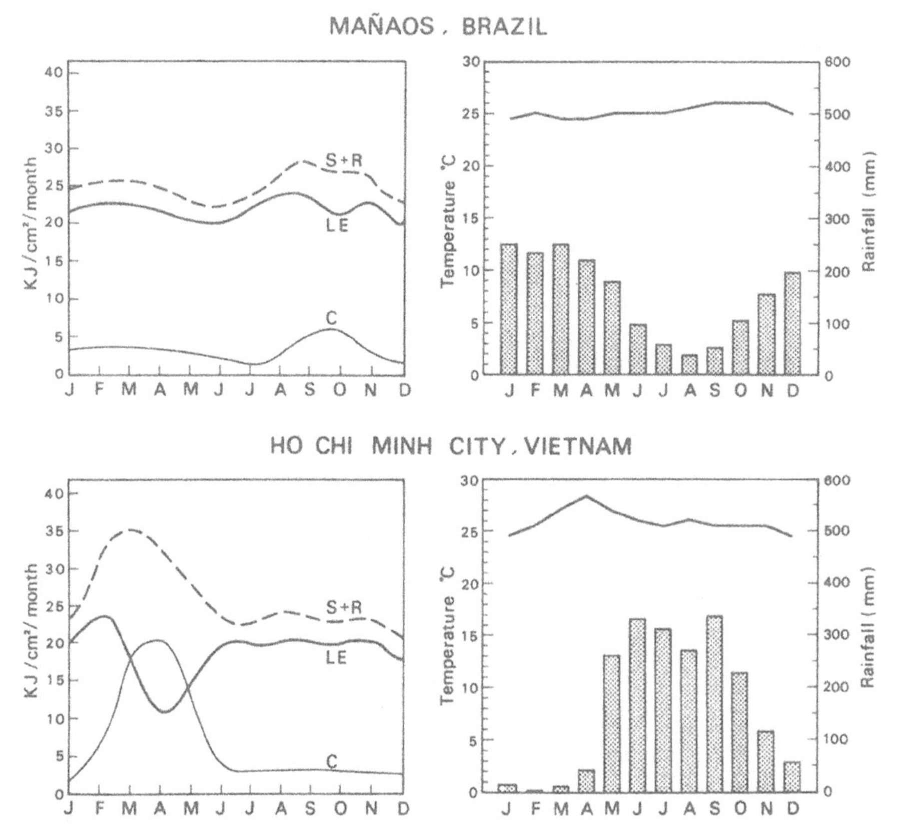

FIGURE 7. Comparison of mean monthly energy balance, temperature and rainfall for Mañaos and Ho Chi Minh City.

5. TEMPERATURE

Tropical lands were once said to comprise those parts of the earth characterized by year-long high temperatures. Temperatures, it was claimed, are always sufficiently high not to hinder plant growth. Neither assertion is strictly true. The former would clearly except mountain areas which in reality are tropical rather than temperate in every respect but one—temperature. This falls about one degree (C) for every 300m rise in elevation so it is perhaps not unreasonable to regard the temperature regime as 'temperate'. But in respect of virtually every other characteristic, biomass, production, radiation, seasonality, day-length, mountain environments are tropical rather than temperate.

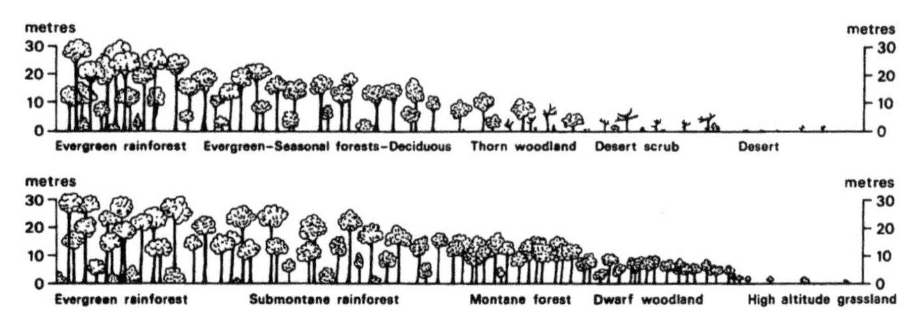

FIGURE 8. Vegetation zonation latitudinally and altitudinally (highly generalized).

If a principal criterion of a tropical climate be redefined as that in which the diurnal temperature range is greater than the annual temperature range then clearly tropical mountains have tropical, not temperate climates. That said, however, it must be admitted that the zonation of montane vegetation to some degree mimics that of a lowland zonation (see Figure 8). In the lowlands increasing aridity results in a reduction in stature, perhaps some increase in the horizontal space between trees especially, a reduction in the number of storeys present and an increase in the proportion of grasses. In tropical mountains these characteristics derive mainly from reduction in temperature with an assist from the common pattern of some diminution in rainfall above the condensation zone, which in the humid equatorial belt lies at about the 2000 m level.

Earlier in this section the assertion that in the tropics plant growth is not hindered by low temperatures was questioned. Scientific work in this area has been confined almost entirely to tropical crops. Despite limited evidence there is a strong suspicion that even in the lowlands some species respond to quite small variations in temperature, man being but one of them. The phenomenon of 'wintering' (leaf-fall) in *Hevea brasiliensis* may in part derive from temperature variation but the question here, as for other tropical plants, is extremely complex, with soil moisture and the radiation regime being but two of the many likely factors involved, amongst which it is extremely difficult to isolate any particular one.

Frost is a most obvious limiting factor in plant growth. The theoretical lower limit of frost lies at an altitude of around 2000 m at 15° north and south, rising to between 2800 m and 3000 m at the equator. In reality, freezing temperatures may occur spasmodically or regularly at levels substantially below these, though the observational network is far from satisfactory. In particular frosts are a fairly regular feature of seasonally-dry monsoonal areas such as northern India and northern Australia at levels below 300 m whilst even within ten degrees of the equator they have been reported on the higher uplands of central Java and central New Guinea. What the long-term effects upon natural vegetations may be is unclear and there is a strong possibility that deforestation in favour of such crops as tea or coffee may increase frost incidence.

But severe though the effects of frost may be, they are relatively localized. Much more general are the effects of coolness. For many tropical lowland species roughly 10°C seems to be a threshold below which phytosynthesis is severely curtailed and physical damage may ensue if such depressions of the thermometer are continued beyond a day or two. Cold has two major effects. First is the reduction in phytomass referred to earlier. While this cannot be documented in any detail, it is most evident on tropical mountains. Figure 9 shows a profile across Venezuela that illustrates this general pattern. The second major effect of increasing coolness is a reduction in the number of species present, particularly tree species. This is so even in the humid equatorial zone as is illustrated in Table III.

These analyses of vegetation can also be taken as an indication of biological diversity. In general, diversity decreases with increasing environmental 'difficulty' in which temperature is just one component, the reduction in diversity between lowland and highland being particularly striking. As will be discussed in more detail later, it is precisely those regions which are biologically diverse which are under heavy attack by human activities which, as the data for secondary forest hints, under the continued impact of man secondary vegetations tend towards simplification involving not only lower levels of phytomass but also the replacement of a good many species, each represented by relatively few individuals, by relatively few species, each represented by many individuals.

6. MOISTURE REGIMES

While temperature has some effects upon the growth and composition of tropical vegetation, it is the moisture regime that is reflected in considerable areal differences in that vegetation. It might be thought that those vegetations not significantly modified by man reflect patterns of rainfall. So they do but only in a very general manner. This is seen particularly in a continental mass such as Africa where, broadly speaking, isohyets lie latitudinally and are paralleled by vegetation zones (see Figure 10). The simple latitudinal pattern is affected by a whole series of other factors—the existence of atmospheric high pressure cells located more or less permanently along the maritime sections of tropics, the upwelling of cold currents off western African and South American shores to name but two factors leading to spectacularly-low rainfall levels. Similarly, convergence of tropical air-masses, together with strong radiation and the predominance of local circulation account for heavy rainfall in the humid equatorial zone. Topographical factors also come into play, especially where major ranges lie athwart the predominant direction of flow. This is particularly the case in Asia which

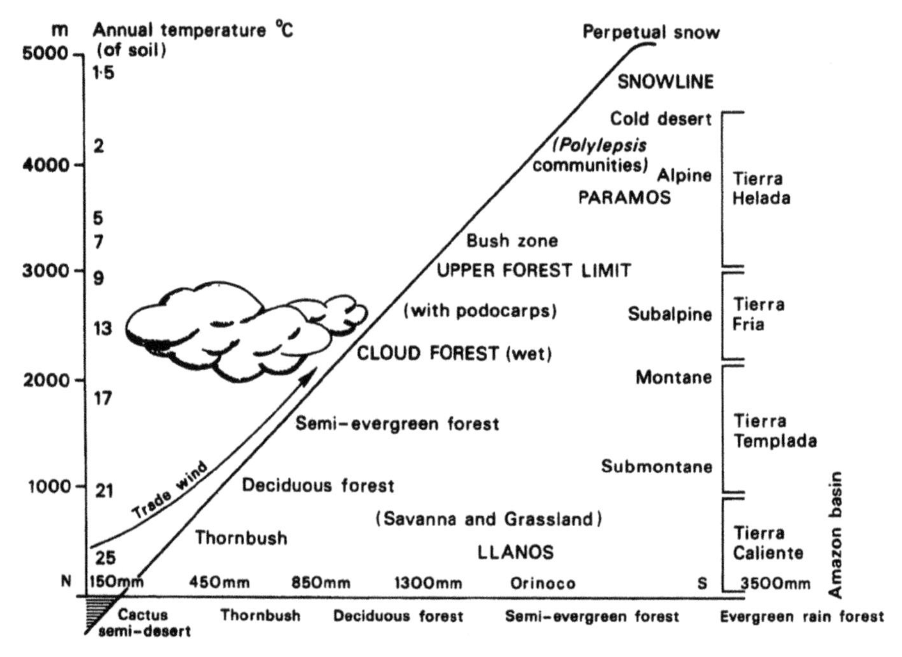

FIGURE 9. Vertical zonation of vegetation in Venezuela, a north-south transect showing relationship to temperature of the soil.

experiences those major annual reversals in flow termed monsoons. The windward slopes of India's Western Ghats, the topographical 'bag' formed by the Eastern Himalaya and the mountain masses of the India-Bangladesh-Burma border, the ranges of the Malay Peninsula and Sumatra all block the south-west monsoon to create strong windward to leeward rainfall gradients. Such gradients exist on quite a small scale as well. Several major plains of Southeast Asia are to some degree shut off from prevailing winds on both sides—the Irrawaddy and Chao Phrya valleys and the central Luzon plain—and are consequently to some degree dependent upon supplementary irrigation to ensure reliable crops of rice.

Rainfall also varies temporally so that two areas environmentally identical but for different patterns of annual rainfall distribution will have different vegetations and unlike agricultural cycles. This is so even in humid tropical areas. For example, in that portion of the Malay Peninsula north of the Isthmus of Kra, three months receive less than 10 cm of rain with only 30–40 rainy days in the four driest months. Further south all months have more than 10 cm of rain and more than 40 rainy days in the four driest months. These comparatively-localized differences are reflected particularly in a higher proportion of deciduous trees in the north and especially in species composition.

TABLE III
Analysis of Some Humid Equatorial Vegetations

Taxons	No. of Taxons	No. of Individuals Identified	Average No. Individuals per Taxon	Proportion of Individuals in Four Commonest Taxons (%)	Proportion of Taxons Represented by 1 Individual (%)
(1) Hill Dipterocarp Forest (Bukit Timah, Singapore)					
Families	35	220	6.3	33	7
Genera	69	220	3.2	19	22
Species	96	202	2.1	19	50
(2) Freshwater Swamp Forest (Nee Soon, Singapore)					
Families	43	220	5.1	48	20
Genera	63	220	3.5	30	29
Species	66	203	3.1	21*	31
(3) Old Secondary Forest (Botanic Gardens, Singapore)					
Families	31	235	7.6	51	7
Genera	62	235	3.8	37	30
Species	82	226	2.8	23*	48
(4) Submontane Forest (Gunung Bunga Buah, Peninsular Malaysia, 1060 m)					
Families	21	87	4.1	55	7
Genera	26	87	3.3	53	11
Species		Not determined			
(5) Montane Forest (Gunung Ulu Kali, Peninsular Malaysia, 1760 m)					
Families	13	134	10.3	74	3
Genera	16	134	8.4	74	6
Species	16	129	8.0	72	5

*Three commonest species. Fourth rank contains more than one species
Source: Author's data from 2 m belt transects. Identifications by courtesy Singapore Botanic Gardens staff.

Of the 139 species of Dipterocarp found in Peninsular Malaysia only 17 occur north of the Isthmus of Kra. In the same general area, rubber (*Hevea*) production is virtually continuous in Johor state at the southern tip of the Peninsula but under the more seasonal rainfall regimes of Cambodia and southern Vietnam, when these regions come under the influence of dry continental polar air sweeping down from the north-east, production entirely ceases during the three months of relative drought this change in circulation brings.

Vegetation also shows considerable variation in type as a result of factors that are not closely related to the hydrological regime but rather to the nature of the substrate and to its topographical position, whether on an interfluve, on a slope or in a valley. Even in the well-watered equatorial zone differences in vegetation type may be partly at-

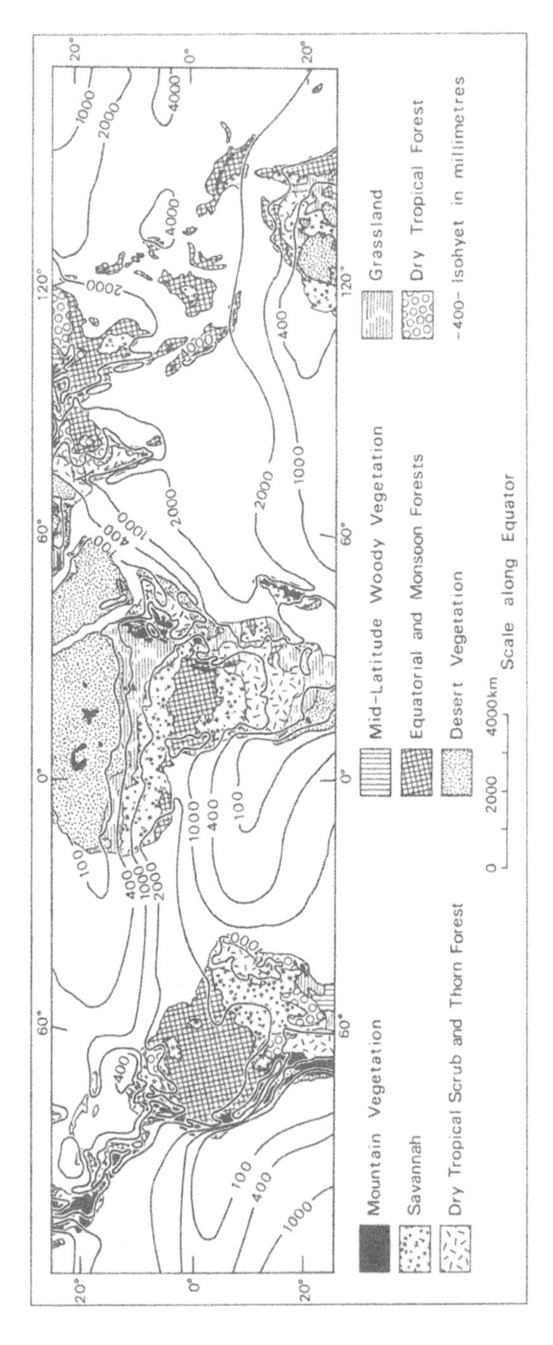

FIGURE 10. Vegetation types in relation to rainfall (Vegetation from Preston E. James. Compare with Figure 2).

tributed to such factors as is hinted at by comparing the swamp and hill forest data for Singapore where despite similarities such as the large proportion of Euphorbiaceae (15 and 9 per cent respectively) and Burseraceae (6 and 7 per cent respectively), there are also striking differences, in the proportion of Palms (20 as against 1 per cent) for example. The vegetation of freely-draining coastal sands is quite unlike that of hill forests, being much lower and containing fewer species while the vegetation of peat swamps in the region, the so-called *kerenggas,* is even more impoverished in species and phytomass.

However, it is on the semi-arid and arid margins of the tropics that edaphic (soil) factors become particularly prominent as is illustrated by Beard's study of northwest Australia. Though small bands of aborigines have regularly set fires thus modifying the woody species occurring in the grasslands in particular, Beard (1967) found that soil factors, rather than rainfall explained much of the variation in vegetation. The Canning Basin, consisting mainly of sandstone which has weathered to form dunes, well-covered with a grass steppe despite lack of surface water. The Pilbara block adjoining the Basin is little less arid but the dominant rocks produce little sand on weathering so that dunes are absent and much of the region is dominated by an *Acacia*-dominant shrub steppe or low savana (see Figure 11).

Finally, something must be said concerning sources of moisture other than rainfall and factors influencing local moisture regimes. Snow, of course, is of minimal significance being confined to the tops of major peaks such as Jaya (West Irian, Indonesia), Kilimanjaro and Kenya. Dew, however, may be of some significance as an environmental factor though measurements of it are quite scanty. In many parts of the tropics the dew-point is reached each night though the actual formation of dew depends upon the occurrence of low wind speeds at ground level. Just how much of this condensed moisture finds its way into the rhizosphere is problematic but dew unquestionably to some degree retards early morning heating of the ground layer as the incoming radiation is initially used for evaporation rather than heating. On tropical mountains occult precipitation, which results from the collision of clouds with solid objects, may be quite a significant source of moisture, estimates ranging between 10 and 20 per cent of the total. Again such moisture has a major effect in retarding heating and thus delaying the onset of evaporation, but for lack of research in tropical areas it is difficult to estimate its importance.

While it is true that in general the more rain falls the more water is to be found in streams and rivers, the relationship is a complex one involving rainfall, soil and vegetation characteristics. In the humid equatorial zone, the dominance of local circulation leads to high rainfall intensities, 25 cm per hour not being uncommon, and high frequency of rainfall. These lead to a run-off pattern and river regimes dominated by constant or slowly-changing base-flows upon which are superposed rapidly changing flows reflecting local weather conditions. As with any drainage basins, the 'peakedness' of discharge curves is also directly related to the size of the catchments with small catchments having greater peakedness than larger ones. High rainfall intensities are also associated with regional disturbances such as tropical cyclones and air mass convergence aloft. By contrast, low rainfall intensities are less common and generally derive from the penetration equatorwards of weakened mid-latitude depressions or on tropical mountains, from location in the actual condensation zone. Other factors being equal, low rainfall intensities are more effective in moistening the soil so that the tropics generally are disadvantaged in this respect by comparison with temperate lands. However, this is partly compensated for by quite high frequencies of rain—at least in areas with a fair total.

A further factor influencing rainfall effectiveness is the time of day when it falls. Obviously night-time fall is more effective since evaporation is reduced. Little research has been done on this aspect of raininess though in humid equatorial areas of Southeast Asia, heavy mid-afternoon showers are common in the intermonsoon season as a result of the dominance of local circulation.

Evaporation is a further factor affecting the supply of moisture and rates vary according to a number of interrelated factors. First is the degree of cloudiness which in turn influences the amount of solar radiation arriving at the surface. There is a broad and obvious correlation between cloudiness and rainfall. Long cloudy periods without rain are rare though in the condensation zone of tropical mountains it may be densely cloudy without rain necessarily falling. Second is the humidity since moist air can take up less moisture than dry air. Here matters are complex for the warm air characteristic of the tropics may be exceedingly dry as in deserts or exceedingly wet and everything in between. But generally the regions of high rainfall are also those of high humidity. There are exceptions, however, for warm moist air can be quite stable over considerable distances as where regional rather than local circulation predominates. For example, in southern China and Vietnam the summer monsoon, here southeasterly rather than southwesterly as it is further south, brings high temperatures (28-34°) and relative humidities usually above 60 per cent but without a great deal of rain other than that arising from disturbances in the circulation whether these be orographical (uplift of air masses as a result of hills or mountains) or arising from tropical cyclones.

Evaporation is also influenced by the vegetation though like many other environmental aspects the precise details are little known. A rainforest may trap up to a third of a light shower, possibly more, and this subsequently evaporates. Where rainfall intensities are low and much rain falls as showers interception may be considerable and removal of the vegetation can be expected to increase the water yield as it probably does in cities. On the other hand, where intensities are high, the entire or more usual partial removal of vegetation does not make much difference to total water yields though they certainly lead to greater peakedness in yield curves.

Transpiration also influences the yield of water though again little research on this question has been done in tropical lands. Reduced phytomass leads to reduced transpiration and in theory should lead to greater water yield in catchments. Since such a reduction also leads to increased sediment yields increased water yields are by no means always desirable.

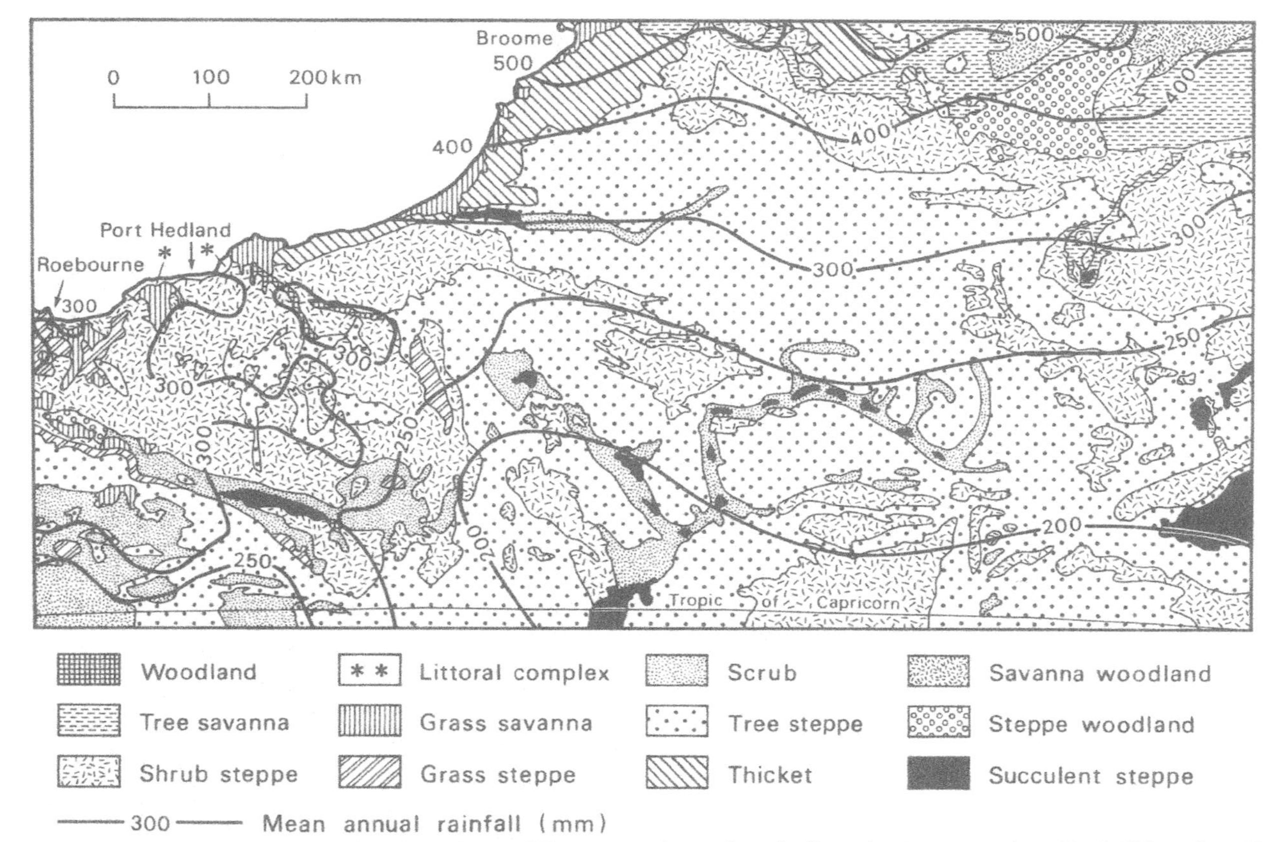

FIGURE 11. Vegetation types in the Great Sandy Desert—Pilbara area, north-west Australia. Vegetation does not correlate with rainfall but rather with soil parent material.

Infiltration rates are influenced by the rate at which water is delivered to the surface of the soil and this is in turn influenced by a number of factors. Up to a third of the rain arriving at the top of the vegetation canopy flows down stems. The rest drops slowly from leaf to leaf and layer to layer so that the vertical extent of the vegetation, its thickness, plays an important role in determining the rate of throughfall. (Obviously the slower such a rate the greater will be the evaporation from the cover). Any reduction in stature can be expected to increase the amount arriving at the soil surface. Here the litter intercepts it but since most tropical litters are thin as a result of high temperatures and rapid decay (taking about nine months in rain-forest) run-off is rapid.

The infiltration of water is influenced mainly by three further factors, the steepness of the slope, how much water the soil holds already and the size of the soil particles and soil aggregates. So far as is known the range of slopes in tropical areas is the same as that elsewhere so no significant differences can be expected on that score. So far as the grain size is concerned, however, matters are quite otherwise especially in the moist tropics. Here the combination of high temperatures and considerable moisture coupled with quite low rates of erosion in many locations lead to very deep weathering, 30 m in some cases, with the breakdown of most minerals, even the relatively-resistant quartz, into clays. Except where renewed by erosion, or the recent deposition of volcanic or wind-borne materials, most sedentary soils contain considerable clay. This dries out and wets with considerable difficulty. Clay also holds much more water than silt- or sand-sized materials delivering this water steadily to streams by groundwater flow. Clay soils, though they can hold much water, require severe suctions to withdraw it. Horizontal movement is very slow and consequently they are of little value as a source for irrigation while remaining excellent sources of water for plant nutrition.

7. CLIMATIC 'ACCIDENTS'

To base an overview of the tropical environment solely upon norms would be misleading. Though there must always be some doubt as to what constitutes an exceptional case, there can be no doubt that periods of exceptionally high or low temperatures, of severe winds, drought or flood may have effects that persist in the environment for many decades. So far as exceptional temperatures are concerned, there is little to note for extremes occur so rarely and moreover are so rarely sustained that lasting effects are minimal. In desert climates, air temperatures at over 40°C, day and night, have been sustained for periods well in excess of 100 days, at Marble Bar in northwest Australia for example, but probably without long-term damage from this cause alone. Low temperatures, as was noted earlier, are quite transitory in most tropical areas, again excluding winters in continental deserts at the tropical margins. Their effects upon crops, though rarely upon natural vegetation, can be quite significant as witness the periodical damage to tea planted at high altitude in Sri Lanka or to coffee in Brazil.

Severe winds occur with considerably regularity in the Caribbean, the Bay of Bengal and the South China Sea. Though particularly severe tropical cyclones may cause loss of life and property damage, it is far from clear that their effects are entirely negative. In many cases (the coastal areas of southern China and Vietnam is but one) tropical cyclones also bring substantial rainfall, a fact that ought to give serious pause to those who advocate their dispersal. The effects of cyclones on tropical agriculture is less clear. Certainly the physical damage to crops such as rice, whether directly or by the accompanying flooding may be of the order of tens of millions of dollars in some years in cyclone-prone areas such as Vietnam. Again, there can be little doubt that decisions to plant *Hevea brasiliensis* on a wide scale in some of Southeast Asia's 'lands below the wind' (south of the tropical cyclones) and not in cyclone-prone areas such as the central Philippines owe something to considerations of limiting wind damage. Such damage may persist for long periods where tropical cyclones stray from their accustomed paths. The forests of Kelantan (northeast Peninsular Malaysia) and the nearby Thai provinces to this day bear evidence of a disastrous storm in the 1880s.

But of all 'climate accidents' it is drought that must have the most serious effects though with the exception of the northwestern section of the Indian subcontinent these are confined largely to the dry tropical margins. There is not space to go into the details nor to argue the case that drought is not 'exceptional'. (Unquestionably drought is 'normal' but severe and prolonged drought is not). To oversimplify a complex situation it would appear that dry tropical margins are probably subject to long-term cycles of drought in response to changes in global circulation which in turn, some would argue, appear to be related to changes in radiation. Superimposed upon such natural fluctuations is a series of human activities leading to reduced moisture effectiveness, deforestation (mainly for domestic fuel) and over-grazing by livestock. In addition, civil disturbances, lack of basic agricultural infrastructure, sometimes improved infrastructure (such as tubewells for stock-water, improved access to urban markets), have tended to intensify the effects of drought.

Irrigation is often seen as an answer to such limitations though it too brings consequential problems such as the spread of water-borne disease like schistosomiasis (bilharzia) and salinization of the soil. It has been suggested, for example, that in the Punjab (India/Pakistan) as much land is annually lost to salinization, mainly as a consequence of inadequate subsoil drainage, as is added by new irrigation schemes. There can be little doubt that groundwater offers a great hope as a resource in semi-arid areas, though it is costly to exploit. Agroecosystems based upon it are necessarily fragile for excessive draw-down often results in diminution of total supply, deterioration of water quality and the need to drill deeper wells at greater cost. Such considerations will doubtless influence the utilization of a newly-discovered deep aquifer 1500 to 2100 m below the surface of parts of the pre-Himalayan foreland. Irrigation also has synergistic effects upon the local climatic environment. The irrigation of sufficiently large areas results in near-ground air mass transformation as soil and near-ground temperatures fall and humidities rise. These effects are a function of increasing size of the irrigated area and distance from its boundaries as well as such factors as depth of water applied and the timing of application, windspeed and albedo (reflectivity) of the wetted soils and crops.

8. RESOURCES

The degree to which irrigation modifies tropical environments is a reminder that resources are by no means fixed and immutable. Many are both space-bound and time-bound. They occur in some places and not in others partly in direct proportion to the degree to which they are searched for and appropriated. Groundwater is an example. They are time-bound not only in the sense that they may be exhausted—groundwater again—but also in the sense that both the technology and economics of resource use are in a constant state of change. To illustrate these considerations a return will be made to the *fons et origio* of all life, solar radiation, to biological production representing the transformation and storage of that energy flux and in turn to a tiny segment of that biological production, namely tropical plants of use to man.

At the global scale various measures can be used as indicators of climatic conditions. One common one is potential evapatranspiration which is that amount of water which would be evaporated and transpired from the soil and the plants growing in it were the water supply not a limiting factor. Penman (1948) and Thornthwaite (1954) have devised indices calculated from conventionally measured climatic parameters. Unfortunately these indices have not been mapped. Another approach is that of Budyko (1974) using radiation balance and mean annual rainfall data to construct a radiative index of dryness which fairly closely corresponds to geobotanical patterns as may be seen by comparing Figures 12 and 2. The formula used is $I = R/Lr$ where I is the index, R is the annual radiation balance, L is the latent heat of evaporation and r is the mean annual rainfall. The most humid conditions, where potential evaporation is very small, lie mostly outside the tropics but probably also occur on tropical mountains for which lack of data prevented calculated of the index. Humid tropical regions show intermediate values. While rainfall is high, so too is evaporation, point data indicating that at least one-third of rainfall is evaporated. Towards the tropics the value of the index increases with values of between 1.0 and 2.0 broadly corresponding to natural savanna grassland, 2.0 to 3.0 to semidesert and over 3.0 to desert.

In a broad way this index may be taken as a measure of environmental difficulty with those areas having a low index being less difficult, being biologically highly productive and diverse even though in detail other factors, such as the seasonal distribution of rainfall and generally impoverished sedentary soils, must modify the pattern. Thus it can confidently be asserted that the greatest physical resources of the tropics are its high levels of radiation and its potential for the conversion of that energy flux into biological production. The latter, however, is seriously limited by dryness, so that from a purely biological point of view it is the humid tropics which offer the greatest resource potential.

The realization of that potential is a major task not only technologically but economically. Few agricultural systems come remotely near the biological production levels of natural systems, except with substantial energy subsidies and many not even then. But these natural systems, especially the open and closed broadleaf forests are under heavy attack from several directions. First is by shifting cultivators. Though precise data are scarce, it seems likely that in many tropical lands, forest fallows are being reduced to the point at which soil nutrients are not fully restored following clearance, cultivation and abandonment. The few studies completed suggest that restoration of nutrients to pre-clearance levels takes in the region of 12–15 years under woody vegetation but that under grass it takes much longer. Health care and consequential human population increase play on important role in the shortening of fallow periods along with the partial commercialization of production, involving rubber in Borneo or beef cattle in Brazil, for example, as the area available for food crops is reduced.

Second is clearance for permanent agriculture and settlement. Given continued annual population growth levels in the rural areas of many tropical countries of 2.0–2.5% up to 4.0% in some parts of Africa, such clearance is inevitable and unstoppable given the political and social realities of the countries concerned. On the analogy of forest lands elsewhere, it is unlikely that clearance will be total because of remoteness and general unsuitability for agriculture. Moreover, there is every reason to believe that population growth will slow, though it may take half a century to do so in some cases, and that governments will increasingly be able to control clearance and to reserve forest and savanna lands for other purposes.

Similar considerations apply to lands from which timber is removed by commercial logging and for fuelwood, the latter accounting for about four-fifths of phytomass removals in the 76 tropical countries studied by Lanly (1982, 62). Tropical lands supply about 90 per cent of the World's commercial hardwoods, mainly for furniture, plywood and veneers and there is good evidence that annual cuts generally exceed annual increments. Moreover, there is increasing pessimism that forests logged by modern methods employing heavy tractors will regenerate in any reasonable time-span. Meijer's study of regeneration in Sabah (North Borneo) indicated that 20–40 per cent of the logged area was 'disturbed' and followed by the appearance of dense creepers and 'nomad' species (*Anthocephalus chinensis, Octomeles sumatrana, Duabanga moluccana*). Though these mostly disappeared after about 30 years, and the evidence points to the reconstitution of the species composition of mature forest after 40 years, the phytomass (stand volume) was substantially lower. Ironically, too, the

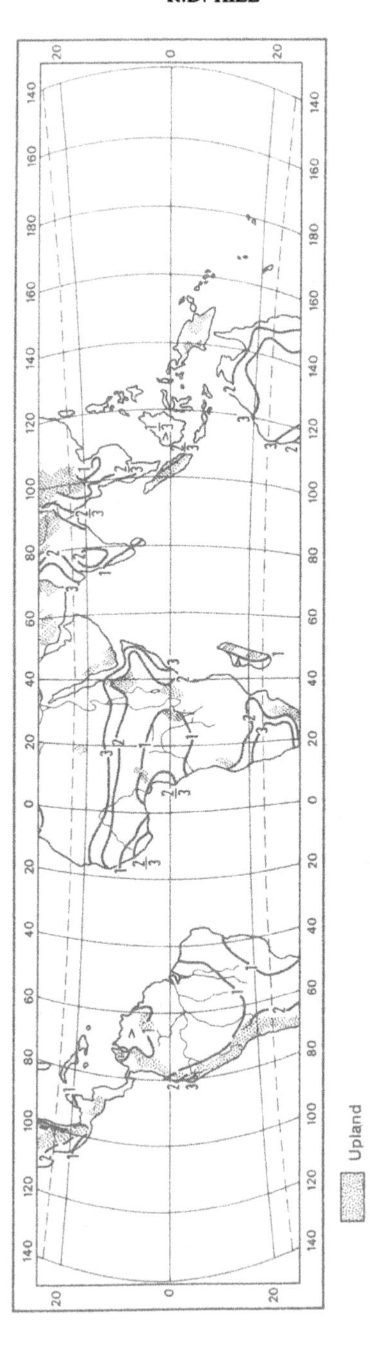

FIGURE 12. Radiative index of dryness for tropics and near-tropics (after Budyko).

TABLE IV
Estimated Areas of Broadleaf Forests in 76 Tropical Countries, 1980 and 1965 (000 Km²)

	Closed Forests					
	Undisturbed	Unmanaged Logged	Managed	Total Production	Unproductive	Forest Fallow
America						
1980	4530	535	*	5065	1474	993
1985	4371	547		4918	1420	1064
% change	−3.5	+2.2		−2.9	−3.7	+7.2
Africa						
1980	1182	419	17	1618	526	616
1985	1139	399	17	1555	523	667
% change	−3.6	−4.8	0.0	−3.9	−0.6	+8.3
Asia						
1980	973	584	362	1919	1001	672
1985	851	590	365	1806	1023	716
% change	−12.5	+1.0	+0.8	−5.9	+2.2	+6.5
Total						
1980	6685	1538	379	8602	3001	2281
1985	6361	1536	382	8279	2966	2447
% change	−4.8	−0.1	+0.8	−3.8	−1.2	+7.3
	Open Forests					
America						
1980				1429	741	617
1985				1368	739	630
% change				−4.3	−0.3	+2.1
Africa						
1980				1692	3173	1043
1985				1596	3152	1112
% change				−5.7	−0.7	+6.6
Asia						
1980				85	278	40
1985				81	219	41
% change				+4.7	+21.2	+2.5
Total						
1980				3206	4192	1700
1985				3045	4110	1783
% change				−5.0	−2.0	+4.9

Source: Lanly (1982)

silvicultural practice of felling defective Dipterocarps (the major timber tree family) delayed the succession (Meijer, 1970).

The spatial dimensions of vegetation change cannot be established for all parts of the tropics but Lanly (1982) has presented data for forests for 76 tropical countries. These show the current situation on a continent-by-continent basis (Table IV).

Lanly's data suggest an annual area deforested in the 76 tropical countries he examined of about 157 000 Km^2, a very large area to be sure and one only marginally compensated for by planting of quick-growing species, mainly conifers or by bamboo forests, the latter covering roughly 42 000 km of Asia in 1980. (Bamboo can substitute for timber in a considerable range of uses). Though it is foolish to extrapolate trends to their end point, 2057, as Guppy (1984) does, this judgement does not lessen the need for forest conservation. Nevertheless, amidst the clamour of conservationist interest it needs to be recalled that forests are very inadequate direct providers of foodstuffs, that secondary forests are by no means useless either as sources of fuelwood or as a protective cover and that there is as yet little evidence that crops and covers which to some degree mimic the structure of forests result in seriously increased rates of erosion and sediment yield.

Tropical forests are probably less well-known, because less-studied, than any other natural environment other than ocean floors. Tropical environments generally are the home of countless plants and animals that have been or are useful to man. For instance Burkill's monumental study of the economic products of the Malay Peninsula alone represents the compression for publication of some 17 000 listings (Burkill, 1966). Only a tiny fraction of tropical plants has been screened for possible pharmacological use, let alone any other. The question of potential use is an exceedingly large one and here only the briefest reference can be made. Table V, however, lists a selection of useful plants.

Given the great biological diversity of the tropics it would be astonishing if there were not many more useful living organisms awaiting discovery. Yeasts, for instance have scarcely been examined on any systematic basis. Though substances derived from tropical plants such as quinine (*Cinchona*), picrotoxin (from *Papaver somniferum*), cocaine (*Erythroxylon*) have long had a place in the pharmacopoeia, current rates of screening for new substances with medical uses are low, largely because drug companies and research institutes prefer to place their trust in biochemical recombinant techniques in the laboratory rather than in field investigation. This is perhaps understandable given the essentially long-term nature of biological screening whether for pharmacological purposes or any other. But it should be recalled that it would take the discovery of just a single drug as effective in the control of carcinomas as Penicillin was and is in the control of infections to pay for generations of search amongst tropical organisms. The tropics both physically and biologically remain one of humankind's greatest potential resources. The task is to make real those potentials and it is to that task that this paper makes a small contribution.

TABLE V
Some Cultivars of Tropical Origin

Key to abbreviations: A Foodstuff
B Beverage
C Flavoring, perfumery
D Medicinal (incl. narcotics)
E Construction
F Fibre
G Other industrial uses
H Other uses
The most important use is given first

Botanical name	Common Name	Uses
Acorus calamus, L.	Sweet flag	C
Adansonia digitata, L.	Baobab	A, F
Aechmea magdalenae, Andre	Pita	F
Aegle marmelos, Correa	Bel fruit	A, D
Aframomum, K. Schum.	African cardamom	C
Agave, L.	Sisal	F
Aleurites moluccana, Willd.	Candlenut	A, G
Allium cepa, L. var. *cepa*	Onion	A
A. cepa, var. *aggregatum* G. Don.	Shallot	A
A. sativum, L.	Garlic	C, D
Alocasia indica, Schott.		A
A. macrorrhiza, Schott.		A
Aloe vera, L.	Aloes	D
Alpinia galanga, Sw.	Galangal	C
Amaranthus, L.	Amaranth	A, D
Amomum, L.	Cardamom	C
Amorphophallus campanulatus, Bl.	Elephant yam	A
Anacardium occidentale, L.	Cashew	A, D
Anacolosa luzonensis, Merr.		A
Ananas comosus, Merr.	Pineapple	A
Annona, L.	Soursop, custard-apple, sweet sop	A
Antidesma, L.		A, C
Arachis hypogaea, L.	Groundnut, peanut	A, G
Areca catechu, L.	Betel-nut	H
Arenga pinnata, Merr.	Sugar-palm	A, B
Arracacia xanthorrhiza, Bancroft	Peruvian carrot	A
Artemesia, L.	Wormwood, absinthe	D, C
Artocarpus altilis, (Park.) Fos.	Breadfruit	A
A. heterophyllus, Lam.	Jack fruit	A
Averrhoa, L.	Belimbing, carambola	A
Azolla, Lam.	Water fern	H
Baccaurea, Lour.		A, B
Bambusa, Schreb.	Bamboo	G, A
Barleria prionitis, L.		D
Basella rubra, L.	Ceylon nightshade	A, G

(continued)

TABLE V (continued)
Some Cultivars of Tropical Origin

Botanical name	Common Name	Uses
Benincasa cerifera, Savi.	Wax gourd	A
Bertholettia excelsa, Humb. & Bonpl.	Brazil nut	A
Bixa orellana, L.	Anatto (turmeric)	G, C
Blighia sapinda, Koen.	Akee	A
Boehmeria nivea, Gaud.	Ramie grass	F
Bombax malabathricum, DC.	Silk-cotton tree, kapok	F
Borassus aethiopicum, Mart.	African fan palm	A, F
B. flabellifera, L.	Palmyra palm	A,B,F
Boswellia, Roxb.	Frankincense	H
Bouea, Meissn.	Plum mango	A
Broussonetia papyrifera, Vent.	Paper-tree	G
Butyrospermum paradoxum, Hepper	Shea-butter tree	A
Caesalpinia, L.		D, C
Cajanus cajan, Millsp.	Pigeon pea	A
Calamus caesius, Bl.	Rattan	G
Calocarpum sapota, Merr.	Mammey sapote	A
Calopogonium mucunoides, Desv.		H
Camellia sinensis, Kuntze	Tea	B
Canangium odoratum, Baill.	Ylang-ylang	C
Canarium, L.	Kanari nut	A, D
Canavalia ensiformis, DC.	Jack-bean	A
C. gladiata, DC.	Sword-bean	A
Canna edulis, Ker.	Queensland arrowroot	A
Cannabis sativa, L.	Hemp	F, D
Capsicum annuum, L.	Chili, capsicum	C, A
C. frutescens Roxb.	Bird chili	C
Carica papaya, L.	Papaya, paw paw	A, D
Carludovica palmata, R.&P.	Panama hat plant	G
Carthamus tinctorius, L.	Safflower	A, G
Caryocar, L.	Butter nut	A
Caryota urens, L.	Toddy-palm	F, B
Cassia, L.	Senna, laburnum	D
Ceiba pentrandra, Gaertn.	Kapok	F
Centrosema, DC.		H
Cephaelis ipecacuanha, Rich.	Ipecacuanha	D
Chenopodium guinoa, Willd.	Quinoa	A
Chrysophyllum cainito, L.	Star apple	A
Cicca acida, Merr.		A, D
Cicer arietinum, L.	Chick pea	A
Cinchona, L.	Quinine	D, B
Cinnamomum, Bl.	Cinnamon, 'cassia'	C,G,D
Citrullus lanatus, Mansf.	Water-melon	A
Citrus, L.	Orange, lime, pomelo	A,C,D
Clerodendron, L.		D
Clitoria ternatea, L.		G, D

TABLE V (continued)
Some Cultivars of Tropical Origin

Botanical name	Common Name	Uses
Cnidoscolus, McVaugh	Chaya, tree spinach	A
Cocos nucifera, L.	Coconut	A,G,F,B
Codiaeum variegatum, Bl.	Croton	D, H
Coffea, L.	Coffee	B
Coix lachryma-jobi, L.	Job's tears	A
Cola, Schott & Endl.	Kola nut	C, D
Coleus, Lour.	Borage	D, C
Colocasia esculentum, Schott	Taro	A
Commiphora, Jacq.	Myrrh	C
Copernicia cerifera, Mart.	Wax palm	G
Corchorus capsularis, L.	Jute	F
Coriandrum sativum, L.	Coriander	C, D
Corypha, L.	Talipot palm	D,A,B
Crotolaria juncea, L.	Sunn-hemp	F
Cucumis anguria, L.	West Indian gherkin	A
C. melo, L.	Melon	A
C. sativus, L.	Cucumber	A
Cucurbita maxima, Duch.	Squash	A
C. moschata, Duch.	Pumpkin	A
C. pepo, DC.	Pumpkin	A
C. foetidissima, HBK	Buffalo gourd	A
Cuminum cyminum, L.	Cumin	C
Curcuma domestica, Val.	Turmeric	C,G,D
C. zedoaria, Rosc.	Zedoary	C
Cyamopsis tetragonoloba, Taub.	Cluster bean	A, G
Cycas circinalis, L.	Cycad	A
Cymbopogon citratus, Stapf	Lemon-grass	C
C. martini, Stapf	Palmarosa	C
C. nardus, Rendle	Citronella	D
Cyphomandra betacea, Sendt.	Tree tomato	A
Datura, L.	Datura	D
Dendrocalamus, Nees	Bamboo	G, A
Derris elliptica, Benth.	Derris, rotenone	G, H
Dioscorea, L.	Yam	A
Dipteryx odorata, Willd.	Tonka bean	C
Durio, L.	Durian	A
Eichhornia crassipes, Solms	Water hyacinth	H
Elaeis guineensis, Jacq.	Oil palm	A, G
Elettaria cardamomum, Maton	True cardamom	C
Eleusine corocana, Gaertn.	Millet	A, B
Entada phaseoloides, Merr.	St. Thomas's bean	D
Eragrostis tef, Trotter	Teff	A
Ervatamia, Stapf	Rosebay	D, C
Erythrina, L.		H
Erythroxylon, L.	Coca	D

(continued)

TABLE V (continued)
Some Cultivars of Tropical Origin

Botanical name	Common Name	Uses
Eugenia aquea, Burm.	Watery rose-apple	A
E. caryophyllus B. & H.	Clove	C, D
E. jambos, L.	Rose-apple	A
Euphorbia, L.		D
Ficus religiosa, L.	Bodh-tree, pipal	H
Garcinia mangostana, L.	Mangosteen	A
Gigantochloa, Kurz.	Bamboo	H, A
Gossypium, L.	Cotton	F,A,G
Guilielma gasipaes, Bailey	Peach palm	A
Guizotia abyssinica Cass.	Niger seed	A, G
Hevea brasiliensis, Muell.-Arg.	Para rubber	G
Hibiscus abelmoschus, L.	Musk-seed	C
H. cannabinus, L.	Kenaf	F
H. esculentus, L.	Lady's finger	A
H. rosa-sinensis, L.	Shoe flower	G
H. sabdariffa, L.	Roselle	F, G
H. tiliaceus, L.	Dryland jute, kenaf	F
Hydnocarpus anthelmintica, Pierre	Chaulmoogra oil	D
Hyphaene thebaica, Mart.	Doum palm	A
Ilicium verum, Hook. f.	Star anise	C
Impatiens balsamina, L.	Balsam	G
Indigofera, L.	Indigo	G
Inocarpus edulis, Forst.	Tahitian chestnut	A
Ipomoea batatas, Lam.	Sweet potato	A
I. reptans, Poir.	Water-spinach	A
Jasminum, L.	Jasmine	C, D
Jatropha curcas, L.	Purging-nut	D
Lablab niger, Medik.	Hyacinth	A
Lagenaria siceraria, Standl.	Bottle gourd	H, A
Lansium domesticum, Jack	Langsat, duku	A
Lawsonia inermis, L.	Henna	G
Lepironia articulata, Domin	Sedge	F
Lucuma bifera, Molina	Egg fruit	A
Luffa cyclindrica, Roem.	Loofah	H, A
Lycopersicon esculentum, Mill.	Tomato	A
Macadamia ternifolia, F. Muell.	Macadamia nut	A
Madhuca longifolia, Macb.	Mahua	A
Mangifera, L.	Mango	A
Manihot esculenta, Crantz	Manioc, cassava, tapioca	A
Manilkara achras, Fosberg	Sapodilla	A
Maranta arundinacea, L.	Arrowroot	A
Melaleuca leucodendron, L.	Cajeput	G, D
Melia azedarach, L.	Persian lilac	C

TABLE V (continued)
Some Cultivars of Tropical Origin

Botanical name	Common Name	Uses
M. indica, Brandis	Nim-tree	D
Mesua ferrea, L.	Indian rose chestnut	H
Metroxylon, Rottb.	Sago palm	A
Michelia champaca, L.	Chempaka	H, C
Momordica charantia, L.	Bitter gourd	A
Morinda, L.		G, D
Moringa oleifera, Lam.	Horse-radish tree	C
Musa, L.	Banana, plantain	A
M. textilis, Nee	Abaca, manila hemp	F
Myristica fragrans, L.	Nutmeg	C
Nelumbium nelumbo, Druce	Lotus	A, H
Nephelium lappaceum, L.	Rambutan	A
N. litchi, Camb.	Lichee	A
N. longana, Camb.	Longan	A
Nicotiana rustica, L.	Nicotine tobacco	G
N. tabacum, L.	Tobacco	H, G
Nigella sativa, L.	Black cumin	C, D
Nymphaea lotus, L.	Water-lily	H, A
Nypa fruticans, Wurmb.	Nipa	G, A
Ocimum, L.	Sweet basil	C, D
Opuntia, Mill.	Prickly pear	A
Oryza sativa, L.	Rice	A, B
Oxalis tuberosa, Mol.	Oca	A
Pachyrrhizus erosus, Urban	Yam-bean	A
P. tuberosus, Spreng.	Yam-bean	A
Palaquium gutta, Blanco.	Gutta-percha tree	G, D
Pandanus, L.	Pandan, screw-pine	F,A,C
Panicum miliaceum, L.	Common millet	A
Papaver somniferum, L.	Opium poppy	D
Parthenium argentatum, Grey	Guayule	G
Paspalum scrobiculatum, L.	Kado millet	A
Passiflora edulis, Sims	Passion fruit	A
P. quadrangulis, L.	Grenadilla	B
Paulinia cupana, HBK	Guarana	B
Pennisetum typhoides, S.&H.	Bulrush millet	A
Persea americana, Mill.	Avocado	A
Phaseolus, L.	Pulse	A
Phoenix dactylifera, L.	Date palm	A, G
Pimenta dioica, Merr.	Pimento	C
Piper betle, L.	Betel-vine	H
P. cubeba, L.	Cubebs pepper	C
P. longum, L.	Long pepper	C
P. methysticum, Forst. f.	Kava	B
P. nigrum, L.	Black pepper	C
Pistia stratiotes, L.	Water lettuce	H

(continued)

TABLE V (continued)
Some Cultivars of Tropical Origin

Botanical name	Common Name	Uses
Plectranthus esculentus, B.E.Br.	Hausa potato	A
Pogostemon cablin, Benth.	Patchouli	C
Psidium guajava, L.	Guava	A
Psophocarpus tetragonolobus, DC.	Winged bean	A
Pueraria phaseoloides, Benth.		H
Punica granatum, L.	Pomegranate	A
Raphia, Beauv.	Raffia	F
Ricinus communis, L.	Castor	G, D
Saccharum officinarum, L.	Sugar cane	A, G
S. barberi, Jeswiet	Sugar cane	A
Schizostachyum, Nees	Bamboo	G
Sechium edule, Swartz	Choco	A
Semecarpus anacardium, L.	Marking-nut tree	G
Sesamum indicum, L.	Sesame	A
Setaria italica, Beauv.	Foxtail millet	A
Simmondsia chinensis, Schneid.	Jojoba	G
Solanum melongena, L.	Egg-plant	A
S. quitoense, Lam	Naranjilla	A, C
S. tuberosum, L.	Potato	A
Sorghum bicolor, Moench	Sorghum	A
Spirulina, Turpin		A
Spondias, L.	Tahitian apple, Spanish plum	A
Strychnos nux-vomica, L.	Strychnine	D
Styrax benzoin, Dryander	Benzoin	D
Tamarindus indica, L.	Tamarind	A, D
Theobroma cacao, L.	Cacao, cocoa	A
Ullucus tuberosus, Caldas	Ullucu	A
Uncaria gambir, Roxb.	Gambier	H, G
Urena lobata, L.	Congo jute	F
Vanilla fragrans, Ames	Vanilla	C
Vetiveria zizanioides, Nash	Vetiver	C
Vigna, Savi	Bean	A
Voandzeia subterranea, Thouars	Bambara groundnut	A
Xanthosoma sagittifolium, Schott.	Tannia	A
Zea mays, L.	Maize	A
Zingiber officinale, Rosc.	Ginger	C, D
Zizyphus mauritiana, Lam.	Indian jujube	D

Compiled from various sources

ACKNOWLEDGEMENTS

The author is most grateful for comments on early drafts of this work to his colleagues Professor Alan Griffiths and Dr. W.J. Kyle.

REFERENCES

1. Barry, R G and Chorley R J, 1982. Tropical weather and climate pp. 242–284 in their *Atmosphere, weather and climate,* 4th edn. London: Methuen.
2. Bazilevich, N J, Rodin L Ye and Rozov, N N, 1971. Geographical aspects of biological productivity. *Soviet geography,* 12, 293–317.
3. Beard, J S, 1967. Some vegetation types of tropical Australia in relation to those of Africa and America. *Journal of ecology* 55(2) 271–290.
4. Brunig, Eberhard F, 1977. The tropical rain forest—a wasted asset or an essential biospheric resource? *Ambio* 6(4) 187–191.
5. Budyko, M I, 1974. *Climate and life.* New York: Academic Press.
6. Burkill, I H, 1966. *A dictionary of the economic products of the Malay Peninsula,* Kuala Lumpur: Ministry of Agriculture and Co-operatives.
7. Cloudsley-Thompson, I L and Chadwick, M J, 1964. The desert environment pp. 1–25 in their *Life in deserts.* London: G.T. Foulis.
8. Flenly, J, 1979. *The equatorial rain forest: a geological history,* London: Butterworths.
9. Furley, J A and Newey, W W, 1983. The terrestrial biomes of the tropics, pp. 272–320 in their *Geography of the biosphere.* London: Butterworth.
10. Guppy, Nicholas, 1983–4. Tropical deforestation: a global view. *Foreign affairs,* 62, 928–965.
11. Harris, D R, 1974. Tropical vegetation: an outline and some misconceptions. *Geography,* 59(3), 240–250.
12. Hsuan Keng, 1974, How many vascular plants are there in W. Malaysia and Singapore? *Malayan Nature Journal,* 28, 26–30.
13. Lanly, J-P, 1982. *Tropical forest resources.* Rome: Food and Agricultural Organization.
14. MacArthur, R H, 1972. *Geographical ecology: patterns in the distribution of species,* New York: Harper and Row.
15. Macnae, W, 1968. A general account of the flora and fanna of mangrove swamps and forests in the Indo-West-Pacific region. *Advances in marine biology,* 6, 73–270.
16. Meijer, W, 1970. Regeneration of tropical lowland forest in Sabah, Malaysia, forty years after logging. *Malaysian forester,* 33, 204–229.
17. Myers, N, 1980. The conversion of tropical forests. *Environment* 22(6) 6–13.
18. Penman, H L, 1948. Natural evaporation from open water, bare soil and grass. *Proc. Roy. Soc. Land.,* ser. A, 193, 120–145.
19. Poore, Duncan, 1976. The values of tropical moist forest ecosystems and the environmental consequences of their removal. *Unasylva,* 28, 127–146.
20. Richards, P W, 1952. *The tropical rain forest,* Cambridge: Cambridge University Press.
21. Sommer, Adrian, 1976. Attempt at an assessment of the world's tropical forests. *Unasylva,* 28, 5–24.
22. Thornthwaite, C W, 1954. A re-examination of the concept and measurement of potential evapotranspiration. *Publications in climatology,* 7(1), 200–209.
23. Troll, Carl, 1958. Tropical mountain vegetation, *Proceedings Ninth Pacific Science Congress,* 20, 37–46. Bangkok, Department of Science, Thailand.
24. United Nations Conference on Desertification, 1977. *Desertification: its causes and consequences.* Oxford: Pergamon Press.
25. van Steenis, C G G J, 1958. Tropical lowland vegetation: the characteristics of its types and their relation to climate, *Proceedings Ninth Pacific Science Congress,* 20, 25–37. Bangkok: Department of Science, Thailand.
26. Whitney, J B R, 1979, Temporal and spatial changes in the productivity of Chinese farming ecosystems, pp. 183–213 in *China: development and challenge,* Lee N and Leung C K (eds), 2, Hong Kong: Centre of Asian Studies, University of Hong Kong.

Resource Management and Optimization
1990, Volume 7(1–4), pp. 39–52
Reprints available directly from the publisher.
Photocopying permitted by license only.
© 1990 Harwood Academic Publishers GmbH
Printed in the United States of America

TROPICAL SOILS: DISTRIBUTION, PROPERTIES AND MANAGEMENT

R. LAL
Department of Agronomy, The Ohio State University, Columbus, Ohio 43210

CONTENTS

Predominant soils of the humid tropics are Oxisols, Ultisols and Alfisols. These highly weathered soils have low cation exchange capacity, low plant nutrient reserves, low available water holding capacity, and are prone to compaction and accelerated erosion. Some soils have toxic levels of Al and Mn. Alfisols, Aridisols, Vertisols and Inceptisols are predominant soils of the semi-arid tropics. These soils are susceptible to wind and water erosion and suffer from frequent drought. Major soils of the arid region are Alfisols, Aridisols, Entisols and Vertisols. Drought, wind erosion, salinization, and desertification are among major constraints. There is also a strong interaction between soil on the one hand and vegetation and fauna on the other e.g. earthworms and termites.

Improved soil management technologies vary depending on soil type and ecology. The low input technology for soils of the humid tropics involves the removal of vegetation either manually or by shearblade; the frequent use of cover crops and planted fallows; no-till and mulch farming; alley cropping and the use of acid tolerant crops. In the semi arid tropics improved soil management technology includes components such as early sowing, planting on the flat after deep plowing, water harvesting and supplemental irrigation.

A logical approach to the development of technology for soils management in the tropics is that of eco-development. This approach evaluates both the potential and constraints of tropical soils and focuses on the development of suitable technology to realize this potential and minimize or alleviate ecological constraints.

TABLE I
Land Area of Predominant Soil Orders in the Tropics (Buringh, 1979)

	World Land Area		Land Area in the Tropics	
Soil Order	10^6 ha	(%)	10^6 ha	(%)
Alfisols	1730	13.1	800	16.2
Aridisols	2480	18.8	900	18.4
Entisols	1090	8.2	400	8.2
Histosols	120	0.9	—	—
Inceptisols	1170	8.9	400	8.3
Mollisols	1130	8.6	50	1.0
Oxisols	1120	8.5	1100	22.5
Spodosols	560	4.3	—	—
Ultisols	730	5.6	550	11.2
Vertisols	230	1.8	100	2.0
Highlands	2810	21.3	600	12.2
Total	13170	100.0	4900	100.0

1. INTRODUCTION

Before man's intervention, soil was a renewable resource. To meet the ever-rising demands for food, feed, fiber and fuel, vast areas of fertile and marginal lands have been brought under cultivation. The natural processes of soil renewal and restoration have been overwhelmed by the degradative processes e.g. accelerated soil erosion, salinization, desertification, etc. Large areas of once biologically productive lands have been rendered unproductive, sterile and barren. Many agricultural systems that were ecologically stable at low population density are now breaking down. It is estimated that about 2,000 million hectares of once productive land has been degraded and rendered unproductive irreversibly. This phenomenon of degradation is more prevalent in the tropics than elsewhere. Yet the demand for food and other basic necessities is estimated to be three times the present level by the time population stability is reached in the year 2110 (Dudal et al., 1982), bringing further stresses to this fragile environment. Man's intervention has, therefore, rendered soil a non-renewable resource.

2. MAJOR SOIL GROUPS AND THEIR DISTRIBUTION

Soils of the tropics are as diverse and varied as those of temperate regions. High spatial (macro) variability makes it difficult to generalise in relation to tropical soils and their distribution. Comparative distribution of major soil orders in the world and the tropics is shown in Table I. Predominant soils of the tropics are Oxisols, Aridisols, Alfisols,

Ultisols, Inceptisols, and Entisols. The relative distribution of these soils vary among ecologies.

Oxisols or Ferralsols are deeply weathered, old, acidic and well-drained soils characterized by an Oxic horizon. These soils occur extensively in the humid regions. The Oxic horizon is at least 30 cm thick and contains a fine earth fraction that retains 10 meq or less of NH_4-ions/100g clay; has a cation exchange capacity of less than 16 meq/100g clay; contains more than 15% clay and only traces of weatherable primary minerals (USDA, 1975). The clay minerals are predominantly kaolinitic with varying amounts of iron and aluminium oxides. The base saturation is low and CEC is mostly saturated with exchangeable Al^{3+}. The term "laterite" was originally used to describe them. Some Oxisols also contain "plinthite," which on exposure, is irreversibly hardened. Plinthite occurs on about 7% of soils in the tropics (Sanchez and Buol, 1975). In tropical Africa about 243 million hectares of land contain a plinthite layer at varying depths (Obeng, 1978). Major agricultural constraints of Oxisols include low level of plant nutrients, weak retention of bases and fertilizer amendments. Fine textured Oxisols have strong fixation and deficiency of phosphorus. Some Oxisols also contain toxic levels of aluminium. Degrees of constraint vary among different types of Oxisols. Coarse-textured Oxisols are susceptible to accelerated erosion.

Ultisols or Acrisols and Nitosols are relatively less weathered but are also acidic and underlain by a fine-textured acidic B horizon. These soils also have a low base saturation and occur in humid and subhumid regions, with a seasonal water deficit. Traditionally these soils were classified as red-yellow podzolic and reddish-brown lateritic soils. The predominant clay mineral is kaolinite but with somewhat higher CEC. Ultisols may also have a layer of plinthite. Major soil-related constraints to plant growth are similar to those of Oxisols e.g. low nutrient levels, aluminium toxicity, and susceptibility to soil erosion Also, under motorized farm operations, these soils are prone to crusting and compaction. Some Ultisols (Pale and Rhodic groups or Nitosols) have favorable soil physical properties, better moisture retention characteristics, and a low susceptibility to soil erosion.

Alfisols or Luvisols are predominant soils of the subhumid and semiarid tropics. These soils are less leached, have neutral pH, a high base saturation, and are characterized by an argillic subsurface horizon in which water is held at less than -15 bar tension during at least 3 months each year. The clay minerals are predominantly kaolinitic with some amounts of illite and smectite. Shallow and gravelly Alfisols are very common in sub-humid and semi-arid regions. Alfisols have moderate to high soil fertility and have less nutritional constraints. Soil moisture stress, severe susceptibility to erosion, crusting and compaction, and supra-optimal soil temperature regimes are major constraints to crop production.

Aridisols are soils of dry areas with possible accumulation of calcium carbonate, gypsum and other salts. In these soils water is held at less than -15 bar tension for more than 9 months a year. Soil moisture stress is the major limitation to crop production in these soils. Some soils also have toxic levels of salt accumulations, and are susceptible to wind and water erosion.

TABLE II
Distribution of Major Soils in the Humid Tropical Regions (After Moormann and van Wambeke, 1978)

General grouping	Soil Taxonomy Classification	Area (million ha)	Percent
1. Highly weathered soils with "low activity clays"	Oxisols Ultisols Alfisols	3479	71
2. Moderately weathered soils	Inceptisols Alfisols Mollisols	441	9
3. Hydromorphic or alluvial soils	Aquaepts Aquents Aquults	490	10
4. Others		490	10
Total		4900	100

Vertisols are dark, heavy textured, calcareous soils of the semiarid tropics. These soils have high shrink/swell properties, and develop cracks to considerable depth during the dry season. Most Vertisols contain predominantly montmorillonitic clay. Vertisols are severely constrained by their physical properties and soil moisture regime. The range of workable moisture content is very narrow. Poor trafficability and severe susceptibility to water erosion are major problems of soil management.

Inceptisols and *Entisols* are young soils of recent origin with no clear cut genetic horizon and with little profile development. Soils of alluvial origin along flood plains, and Andosols derived from the parent material of volcanic origin come under this category. These soils have a high fertility status. Hydromorphic soils are constrained by water management problems, and can be highly acidic. Soils with extremely low pH have toxic levels of aluminium.

2.1 Soils of the Humid Tropics

Predominant soils of the humid tropics are those that are highly weathered with low activity clays e.g. Oxisols, Ultisols and Alfisols. These three soils comprise about 71 percent of the land surface in the humid tropics (Table II). Moderately weathered Inceptisols, Alfisols and Mollisols comprise about 9 percent of the land area, and hydromorphic soils of alluvial origin another about 10 percent. Highly weathered soils have low cation exchange capacity, low plant nutrient reserves, low available water holding capacity, and are prone to crusting and compaction and accelerated soil erosion. These soils have traditionally been used for shifting agriculture and related bush fallow systems.

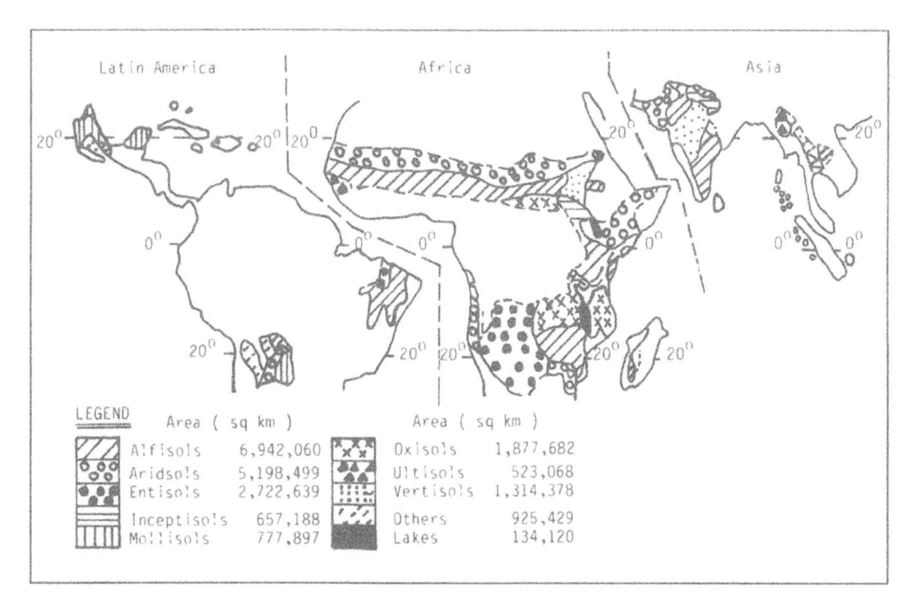

FIGURE 1.

2.2 Soils of the Semi-Arid Tropics

Alfisols, Aridisols, Vertisols and Entisols are the predominant so soils of the semi-arid tropics (Fig. 1). Alfisols in the semi-arid region are distinguished from those in the humid region by ustic rather than udic moisture regime. The ustic moisture regime implies dryness during part of the year, and is typical of the monsoon climate that has at least one rainy season lasting 3 months or more. In the Sudanian regions of West Africa, shallow Alfisols occur on the plateau uplands and middle portion of the toposequence. These soils are more fragile than equivalents Alfisols in southern India. The latter have mixed mineralogy in comparison with generally lower overall clay activity of Alfisols in dry regions of West Africa. Aridisols occur in the Sahel region of West Africa, Horn of Africa and in the Rajsthan of Western India. These soils have severe drought stress. Vertisols occur in Central India and parts of Africa (Fig. 1).

2.3 Arid Tropics

Soils of arid regions fall within five soil orders i.e. Alfisols, Aridisols, Entisols, Mollisols and Vertisols (Dregne, 1976) (Fig. 2). In comparison with Alfisols of the semi-arid tropics, these of the arid regions belong to Ustalfs and Xeralfs suborders. Predominant suborders of the Aridisols are Argid and Orthid. Both Alfisols and Arid-

FIGURE 2.

isols have low organic matter content, are very dry and may have toxic levels of salt accumulation. Drought, wind erosion, salinization and desertification are among major constraints to intensive arable landuse in arid regions. Some of the arid zone soils can be very productive, if irrigation were made available and salt accumulation in the root zone minimized.

3. INTER-RELATIONSHIP WITH FLORA AND FAUNA

3.1 Vegetation

There exists a strong inter-relationship between soils and vegetation. Soils of the humid region, with an average annual rainfall exceeding 2000 mm, support tropical rainforest. Tropical rainforest is a very diverse and complex ecosystem. The diversity of tree species may be as high as 180 species/hectare. These regions have little or no water deficit and most of the nutrient capital of this ecosystem is contained in the biomass rather than in the soil. A greater proportion of nutrient reserves are concentrated in the top few centimeters of the soil. Soils are deficient in cations, and the vegetation has developed special adaptive mechanisms to translocate nutrients before leaf fall and to directly absorb nutrients from decaying vegetation and leaf litter (Herrera et al., 1981; Moran, 1981). The most common soils supporting tropical rainforest are Oxisols, Ultisols and Alfisols containing predominantly low activity clays.

Semi-evergreen deciduous forests occur in regions with seasonal rainfall and a pronounced dry season. This vegetation is comprised of deciduous and semi-deciduous species and lack epiphytes. These forests are also termed monsoon forests. Soils supporting these vegetation are relatively less leached and comprise Ultisols, Alfisols and Inceptisols. These soils also contain predominantly low-activity clays.

Savanna and open woodlands are the vegetation of the semi-arid region. This is a region of prolonged dry season that supports grass-dominated vegetation with scattered trees and shrubs and an open canopy. This region is also characterised by repeated occurrence of man-induced fire. Soils that support this vegetation are Alfisol, Aridisols, Vertisols and Inceptisols.

3.2 Soil Fauna

Soil fauna plays a major role in influencing soil physical properties, organic matter decomposition, and nutrient dynamic. Soil fauna may be classified according to size as micro, meso, macro and mega-fauna. The macro-fauna comprise earthworms, termites and ants with a total population ranging from few to several thousand per m^2.

Fauna is mostly confined to the top few centimeters of tropical soils where organic matter reserves are also concentrated. The most important soil factor affecting animal population and its diversity are micro-climate, soil moisture, soil temperature, porosity and pore size distribution, and organic matter content. Cultural practices that influence these properties have also significant effects on soil fauna.

Earthworms are the most abundant invertebrate macrofauna in many tropical soils. Earthworms facilitate decomposition of organic matter and create channels and castings through their burrowing activity. The forest floor in some parts of West Africa is covered with a 2–3 cm layer of worm casts. Earthworm castings are rich in plant nutrients and have stable soil structure. Both physical and chemical properties of worm castings are different than that of the original soils (Lal and Akinremi, 1983; Lal and De Vleeschauwer, 1982; De Vleeschauwer and Lal, 1981).

As earth dwellers, termites influence soil properties through decomposition of plant residue, construction of mounds and forage galleries, and through soil turnover. Both physical and chemical properties of mound soil are drastically altered by termites. Termites influence soil properties in savanna whereas earthworms do so in the rainforest ecology.

4. SOIL MANAGEMENT TECHNOLOGIES

Appropriate soil management technologies for economic and sustained agricultural production are different for different soils and ecologies. Some of the improved technologies being piloted/practiced are described below.

4.1 Land Clearing and Development

Removal of forest/existing vegetation cover should be done in such a way that the delicate balance between soil-vegetation-climate is not disturbed. Improper methods of land clearing and development can rapidly degrade soil quality and initiate the processes leading to soil degradation. Deforestation causes significant alterations in many biophysical processes. The magnitude of alterations, however, is dependent on the method of land clearing used (Senbert et al., 1977; Lal and Cummings, 1979; Hulugalle et al., 1984; 1986). Recommended methods of land clearing include:

(i) Manual clearing, using improved tools e.g. chainsaw, is ecologically the most compatible method. The technique, being labor-intensive, is constrained by shortage of manual labor especially in tropical rainforest regions of Central Africa and in Amazonia. Labor-intensive methods, however, are commonly used in Asia e.g. Sumatra.

(ii) Shear-blade front-mounted on a crawler tractor is used to clear semi-deciduous rainforests and tree-savanna vegetation (Couper et al., 1981; Lal et al., 1986). Water runoff and soil erosion losses are generally less with shear blade clearing because this method of vegetation removal does not remove roots and stumps (Lal, 1981). Seeding a leguminous cover crop, however, is required to alleviate soil compaction caused by machine clearing and provide protective ground cover (Hulugalle et al., 1986).

(iii) Savanna vegetation is cleared with an anchor chain attached to two tractors (Allan, 1986). Compaction and accelerated erosion are the soil management problems following chaining that require ameliorative measures.

(iv) Chemical clearing has been used for development of pastures in savanna ecology. Ecological implications of chemical clearing, however, are not understood.

4.2 Management of Soil Fertility

Tropical regions are characterized by the non-availability of essential inputs (e.g. agro-chemicals, and farm power), by nutrient-deficient soils, by harsh and fragile ecologies, by lack of credit facilities and by isolation from commercial markets. Under these conditions, it is important to develop resource-efficient technologies. Therefore, soil management systems should minimize dependence on agro-chemicals and preserve natural resources. The basis of improved technologies to supplement soil fertility include the integration of livestock and woody perennials with food crop production, and the liberal use of mulch materials and organic wastes.

Acid soils of the humid tropics are characterized by low pH, low CEC and high levels of exchangeable Al and Mn. High levels of soluble and exchangeable aluminium in these soils often restrict root growth and development. Although liming at appropriate rates can raise the pH to neutral, high costs and low availability in many tropical regions are practical constraints. Productivity of acid soils may be increased by growing species and varieties of crops which are tolerant of low pH and high level of Al and Mn (Sanchez and Salinas, 1981; Kamprath, 1980). Because of the scarcity of lime sources in most areas of the humid tropics, the management of acid soils for more intensive food crop production should give high priority to selection of crop species and varieties that are tolerant to high soil acidity. Moreover, it is necessary to know the minimum rates of lime requirement for optimum crop growth as well as the residual value that can be expected.

Long term experiments conducted at IITA's high rainfall station at Onne in southeastern Nigeria and INIPA/NCSU station at Yurimaguas, Peru have shown that with crop rotation, relatively low rates of lime (e.g. 200 to 500 kg/ha) would be adequate to sustain crop yields under continuous cropping for a period of 5 years or more on the coarse-textured Ultisols (IITA, 1981; Friessen et al., 1982; Sanchez et al., 1982). Cropping systems recommended are cassava/maize intercropping, maize-cowpea rotation, and upland rice-groundnut rotation.

The fertility of acid soils can also be improved by adopting cultural practices that facilitate regular additions of large quantities of organic matter. Sanchez et al. (1982), Wade and Sanchez (1983) and other researchers from South America have shown that incorporation of plant matter results in improvement of soil properties and crop growth. Surface mulch provided by growing legume covers also improves fertility and soil physical properties e.g. soil temperature and moisture regimes. In Eastern Nigeria, Maduakor et al. (1984) and Hulugalle et al. (1986) have shown notable improvements in yields of yam, cocoyam and cassava by surface application of mulch. A practical technique of producing mulch is to grow it in-situ through planted fallows.

4.3 Soil Surface Management and Erosion Control

Objectives of soil surface management are to control soil erosion, favorably regulate soil temperature and moisture regimes, and stimulate biotic activity of useful soil fauna e.g. earthworms. The following practices are recommended for erosion control and prevention of soil degradation in different ecologies:

Seedbed Preparation By eliminating all pre-planting seedbed preparations and providing a mulch cover from the crop residue and chemically killed weeds, a no-till system improves soil and water conservation, soil organic matter content, and activity of soil fauna (Lal, 1983; Sidiras et al., 1983; Kemper and Derpsch, 1981; McCown et al., 1985; Jones and McCown, 1983). In addition to controlling soil erosion, the crop response to no-till systems is also favorable for humid and sub-humid regions, provided that weeds are adequately controlled and an adequate amount of crop residue mulch is made available. For acid Ultisols and Oxisols, however, no-till system may render it difficult to get lime incorporated within the rooting depth. Acid tolerant crops e.g. yam and cassava or cowpea that do not require liming also respond positively to a no-till system. Soil properties that favour the no-till system include; (i) coarse-textured surface horizon; (ii) low susceptibility to soil compaction and crusting; (iii) good internal drainage and (iv) friable consistency over a wide range of soil moisture content. These requirements are met for soils of the humid and sub-humid tropics where the length of the growing season exceeds 6 months per year.

For semi-arid and arid regions where the length of dry season exceeds 6 months a year, adaptation of no-till system is rendered difficult because of the (i) lack of adequate amount of crop residue mulch, (ii) soils susceptible to compaction and crusting, and (iii) poor seed/soil contact due to the lack of proper seeding equipment in an untilled soil. In semi-arid and arid regions soils with predominantly low-activity clays are particularly prone to hard-setting. These soils are structurally inert and require ameliorative measures prior to implementing no-till system. For these soils objectives of seedbed preparation are to improve total and macro-porosity to facilitate water infiltration and root penetration.

Research carried out in semi-arid and sahel region of West Africa has shown that deep and rough plowing at the end of the rainy season improves total and macro-porosity and water transmission characteristics of structurally-inert Alfisols and Arid-isols (Charreau, 1972; Nicou, 1974 a,b). Graded ridge-furrow and tied-ridging have also proven to be risk-avoidance systems in semi-arid ecologies (IITA, 1981; Prentice, 1946). Tied-ridging is effective on marginal lands of low-yield capacity, for structurally inert soils, and for regions prone to drought stress.

Lal (1985) developed a tillage guide to assess tillage needs on the basis of soil and environmental characteristics and crop requirements. While preparing this guide special attention was given to runoff and soil erosion control, crop stand establishment, and soil structure. No till techniques provide a suitable seedbed for friable, coarse-textured, self-mulching, structurally active soils. Ameliorative measures are needed for structurally inert soils of massive structure and hard consistency.

Cover crops Incorporation of planted fallows in the crop rotation is necessary to improve soil physical properties and the soil organic matter content. Cover crops have been demonstrated to increase macroporosity, infiltration rate and soil water retention and transmission for soils of East Africa (Pereira et al., 1954; 1967), West Africa (Wilkinson, 1975; Juo and Lal, 1977; Lal et al., 1979; Hulugalle et al., 1986), tropical America (Kemper and Derpsch, 1981; Sanchez, et al., 1982) and tropical Australia (McCown et al., 1985; Bridge et al., 1983). Deep rooted legumes are particularly beneficial in improving macroporosity. In Nigeria, Lal et al. (1979) observed beneficial effects of growing *Psophocarpus palustris* on soil structure. Hulugalle et al. (1986) have demonstrated the usefulness of *Mucuna utilis* towards improving physical properties of compacted Alfisols. In Northern Territory, Australia, Bridge et al. (1983), reported that growing *Stylosanthes hamata* increased macroporosity and infiltration rate. Deep-rooted pigeon pea is also effective in improving macroporosity (Hulugalle and Lal, 1986).

The choice of an appropriate cover crop depends on ambient soil conditions and subsequent landuse e.g. crops to be grown and soil and crop management techniques envisaged. Some desirable characteristics of appropriate cover crops are (i) ease of establishment (ii) vigorous growth and ability to provide a rapid ground cover, (iii) deep rooting capability (iv) determinate growth (v) some economic uses e.g. pasture, and (vi) easily suppressed. In fact, the successful adaptation of no-till systems partly depends on the use of an appropriate cover crop.

Alley cropping An alternative to growing a cover crop in a rotation for procuring mulch is to grow food crops in association with woody perennials. An appropriate woody perennial is planted on contour at 4 to 5m intervals and food crops are grown in the alleys between. Crops grown in this system of alley cropping (Kang et al., 1981) benefit from the mulch produced from the prunings of the woody perennial. The alley cropping system retains the basic features of bush fallow but allows continuous cultivation on the same land. The shrub or tree specie recycles soil nutrients, fixes nitrogen, and prevents soil erosion. Moreover, it provides fodder to small ruminants and firewood during the dry season.

Alley farming trials at IITA using *Leucaena leucocephala* and *Gliricidia sepium* over periods of 5 to 10 years have given sustained yields of the accompanying food crops such as maize, cowpea, yams and upland rice at various levels of fertilizer application.

Although the yield of food crops may be suppressed by 20 to 25 percent because of the shading and extra space taken to alleys, the system reinforces no-till farming in erosion control by providing additional mulch and by retarding the velocity of surface runoff.

Soil surface management technologies for semi-arid tropics and Vertisols
Research at the International Crops Research Institute for the Semi-Arid Tropics (ICRISAT) in Hyderabad, India, and by IRAT in West Africa have shown that for soils of the semi-arid regions, a particular emphasis is needed on early sowing, water harvesting and recycling, and in developing crop varieties that are more drought tolerant. It is reported that flat cultivation is on par with other land surface configurations with regard to crop yields but superior in controlling runoff and soil erosion from Alfisols (El-Swaify et al., 1983). Flat cultivation following an intensive primary tillage is essential for easily compacted Alfisols. However, supplementary irrigation is also necessary to stabilize crop production on these soils of low available water holding capacity. Studies at ICRISAT have shown that Alfisols have more potential for surface runoff collection and storage for supplemental irrigation than Vertisols (Pathak, 1980; Ryan and Krishnagopal, 1981).

The use of organic wastes and green manures has to be supplemented by chemical fertilizers to procure high yields. Because of low buffering capacity and high leaching losses, fertilizer banding and split applications are necessary for efficient use of the nitrogenous fertilizers (El-Swaify et al., 1983). Nevertheless, the liberal use of organic fertilizers improves soil structure in addition to supplying plant nutrients.

5. RESEARCH AND DEVELOPMENT NEEDS

With good soil management, resource-based subsistence agriculture can be transformed to a more productive and scientific agriculture in the tropics. Improved technology for intensive management and for economic and sustained production varies depending on soil type and ecology. The basic components of improved technology are:

(i) removal of vegetation for new land development either manually or by those mechanical devices that cause least soil disturbance e.g. shearblade,

(ii) liberal use of crop residue mulch, organic manures, and other biological means to improve soil physical and nutritional properties,
(iii) frequent use of legume covers and planted fallow in the crop rotation,
(iv) use of no-till or minimum tillage systems wherever feasible, and minimizing the number of passages of heavy equipment,
(v) effective nutrient recycling through growing food crops in association with woody perennials, and
(vi) water recycling for supplementary irrigation in the semi-arid and arid regions.

Packages of improved technology encompassing these principles need to be piloted and validated for a range of soils, rainfall regimes, and agro-ecologies of the tropics. The development of these packages requires research which is local and specific to given ecologies and requires coordinated approach between national and international organizations.

The evolutionary improvement of the traditional farming systems in the tropics requires a thorough study and understanding of soils, climate, economic constraints, human resources, logistic support and infrastructure available, and of the interaction between bio-physical and socioeconomic factors. This implies a sense of commitment to develop technology which specifically meets the needs of small farmers including:

(i) choice of crops and crop mixtures,
(ii) technology of land development and soil management that matches the potential and constraints of soils, climate and socioeconomic environments of the tropics, and
(iii) development of land and labor saving technology.

The most logical approach is that of "eco-development" that considers the potential and constraints of an ecosystem and relies on appropriate technology that maximizes energy flow, brings about structural improvement by energy input, optimizes input to output energy ratio, and utilizes resources according to their availability within the ecosystem. Furthermore, the improved system should be economically sustainable but also ecologically compatible.

REFERENCES

1. Allan, T.G., 1986. Land clearing in African Savannas. In *"Land Clearing and Development in the Tropics,"* ed. by R. Lal, P.A. Sanchez and R.W. Cummings, A.A. Balkema, Rotterdam, Holland.
2. Bridge, B.T., Mott, J.J., Winter, W.H. and Hartigan, R.J., 1983. Improvement in soil structure resulting from sown pastures on degraded areas in the dry savannah woodlands of Northern Australia. *Aust. J. Soil Res.* 21, 83–90.
3. Buringh, P., 1979. *Introduction to the study of soils in tropical and* subtropical regions, PUDOC, Wageningen, Holland.
4. Charreau, C., 1972. Problemes poses par l' utilisation agricole des sols tropicaux par des cultures annuelles. *L' Agron. Tropicale* 27, 905–929.
5. Couper, D.C., Lal, R. and Claassen, S., 1981. pp 119–130 in *"Tropical Agricultural Hydrology"* ed by R. Lal and E.W. Russell. J. Wiley & Sons, Chichester, U.K.

6. De Vleeschauer, D. and Lal, R., 1981. Properties of worm casts under secondary tropical forest regrowth *Soil Sci.* 132, 175–181.
7. Dregne, H.E. 1976., *Soils of Arid Regions*, 237 pp, Elsevier Scientific Publishing Co., Amsterdam.
8. Dudal, R., 1980. Soil-related constraints to agricultural development in the tropics. pp 23–40 in *"Soil-Related Cosntraints to Food Production in the Tropics."* IRRI, Los Banos, Philippines.
9. Dudal, R., Higgins, G.M. and Kassam, A.H., 1982. Land resources for the World food production. *Proc. 12th Int. Congress of the Int. Soc. Soil Sci.*, New Delhi, India.
10. El-Swaify, S.A., Walker, T.S. and Virmani, S.M., 1984. *Dryland Management Alternatives and Research Needs for Alfisols in the Semi-Arid Tropics*, 38 pp, ICRISAT, Hyderabad.
11. Friessen, D.K., Juo, A.S.R. and Miller, M.H., 1982. Residual value of lime and leaching of calcium in a kaolinitic Ultisol in the high rainfall tropics. *Soil Sci. Soc. Am. J.* 46, 1184–1189.
12. Herrera, R., Jordan, C.F., Medina, E. and Klinge, H., 1981. How human activities disturb the nutrient cycles of a tropical rainforest in Amazonia. *Ambio* 10, 109–114.
13. Hulugalle, N.R., Lal, R. and Ter Kuile, C.H.H., 1984. Soil physical changes and crop root growth following different methods of land clearing in western Nigeria. *Soil Sci.* 138.
14. Hulugalle, N.R., Lal, R. and Ter Kuile, C.H.H., 1986. Amelioration of soil physical properties by mucuna following mechanized land clearing of a tropical rainforest. *Soil Sci.*
15. I.I.T.A., 1981, *Research Highlights*, IITA, Ibadan, Nigeria.
16. Jones, R.K. and McCown, R.L., 1983. Research on a no-till, tropical legume-ley farming strategy. pp 103–121 in *"Eastern Africa-ACIAR Consultation on Agricultural Research."* ACIAR, Canberra, Australia.
17. Juo, A.S.R. and Lal, R., 1977. The effect of fallow and continuous cultivation on the chemical and physical properties of an Alfisol in western Nigeria. *Plant Soil* 47, 567–584.
18. Kamprath, E., 1980. Soil acidity in well-drained soils of the tropics as a constraint to food production. pp 171–187, *"Priorities for Alleviating Soil-Related Constraints to Food Production in the Tropics."* IRRI, Los Banos, Philippines.
19. Kang, B.T., Wilson, G.F. and Sipkens, L., 1981. Alley cropping maize and *Leucaena* in southern Nigeria. *Plant Soil* 63, 165–179.
20. Kemper, B. and Derpsch, R., 1981. Results of studies made in 1978 and 1979 to control erosion by cover crops and no-tillage techniques in Parana, Brazil. *Soil Tillage Res.* 1, 253–267.
21. Lal, R., 1981. Deforestation of tropical rainforest and hydrological problems. pp 113–140 in *"Tropical Agricultural Hydrology"* ed by R. Lal and E.W. Russell, J. Wiley & Sons, Chichester, U.K.
22. Lal, R., 1983. No-till Farming, 64 pp. *Tech. Bull.* 3., IITA, Ibadan, Nigeria.
23. Lal, R., 1984. Soil erosion from tropical arable lands and its control. *Adv. Agron.* 37, 183–248.
24. Lal, R., 1985. A soil suitability guide for different tillage systems in the tropics. *Soil & Tillage Res.* 5, 179–196.
25. Lal, R. 1989. Agroforestry systems and soil surface management of a tropical Alfisol. 8:I–V: 1–6, 7–29, 97–111, 113–132, 197–215, 217–238.
26. Lal, R. and Cummings, D.J., 1979. Clearing a tropical forest. I. Effects on soil and microclimate. *Field Crops Res.* 2, 91–107.
27. Lal, R. and Akinremi, 1983. Physical properties of earthworm casts and surface soil as influenced by management. *Soil Sci.* 135, 114–122.
28. Lal, R. and De Vleeschauwer, D., 1982. Influence of tillage methods and fertilizer application on chemical properties of worm castings in a tropical soil. *Soil & Tillage Res.* 2, 37–52.
29. Lal, R., Wilson, G.F. and Okigbo, B.N., 1979. Changes in properties of an Alfisol produced by various crop covers. *Soil Sci.* 127, 377–382.
30. Lal, R., Sanchez, P.A. and Cummings, R.W. Jr., 1986. (eds.) *"Land Clearing and Development in the Tropics"*. A.A. Balkema, Rotterdam, Holland.
31. Maduakor, H.O., Lal, R. and Opara-Nadi, O., 1984. Effects of methods of Seedbed preparation and mulching on the growth and yield of white yam on an Ultisol in southeast Nigeria. *Field Crops Res.* 9, 119–130.
32. McCown, R.L., Jones, R.K. and Peake, D.C.I., 1985. Evaluation of a no-till, tropical legume ley-farming strategy. pp 450–472 in *"Agro-Research for the Semi-Arid Tropics: North-West Australia."* ed by R.C. Muchow, Univ. Qld. Press.
33. Moormann, F.R. and Van Wambeke, A., 1978. The soils of lowland rainy tropical climate—their inherent limitations for food production. *Proc. 11th ISSS Congress* Vol. 2, p 272–291, Edmonton, Canada.

34. Moran, E., 1981. *Developing the Amazon.* Ind. Univ. Press, Bloomington, USA.
35. Mott, J., Bridge, B.J. & Arndt, W., 1979. Soil seals in tropical tall grass pastures of northern Australia. *Aust. J. Soil Res.* 17, 483–484.
36. Nicou, R., 1974a. Contribution to the study and improvement of the porosity of sandy and sandy-clay soils in the dry tropical zone: Agricultural consequences. *L'Agron. Tropicale* 29: 1100–1127.
37. Nicou, R., 1974b. The problem of caxing with the drying out of sandy and sandy clay soils in arid tropical zone. *L' Agron. Tropicale* 30, 325–343.
38. Obeng, H.B., 1978. Major soils of West Africa and their general suitability for crop and livestock production. *African J. Agricultural Sciences* 5(1), 71–83.
39. Pathak, P., 1980. *Runoff collection storage and ground water recovery.* Mimeo, ICRISAT, Hyderabad, India.
40. Pereira, H.C., Chenery, G.M. and Mills, W.R., 1954. The transient effects of grasses on the structure of tropical soils. *Emp. J. Exp. Agric.* 22, 140–160.
41. Pereira, H.C., Hosegood, P.H. and Dagg, M., 1967. Effects of tied ridges, terraces and grass leys on a lateritic soil in Kenya. *Exp. Agric.* 3, 89–98.
42. Prentice, A.N., 1946. Tie-ridging with special reference to semi-arid areas. *E. Afr. Agric. J.* 12, 101–108.
43. Ryan, J.G. and Krishnagopal, C., 1981. *Assessing the economics of water harvesting and supplementary irrigation: a simulation approach* Mimeo, ICRISAT, Hyderabad, India.
44. Sanchez, P.A. and Buol, S.W., 1975. Soils of the tropics and world food crisis. *Science* 188, 598–603.
45. Sanchez, P.A. and Salinas, J.G., 1981. Low-input technology for managing Oxisols and Ultisols in Tropical America. *Adv. Agron.* 34, 279–406.
46. Sanchez, P.A., Bandy, D.A., Villachica, J.H. and Nicholaides, J.J., 1982. Amazon Basin Soils: Management for continuous crop production. *Science* 216, 821–827.
47. Seubert, C.E., Sanchez, P.A. and Valverde, C., 1977. Effects of land clearing methods on soil properties of an Ultisol and crop performance in the Amazon Jungle of Peru. *Trop. Agric. 54,* 307–321.
48. Sidiras, N., Derpsch, R. and Mondardo, A., 1983. Effect of tillage systems on water capacity, available moisture, erosion and soybean yield in Parana, Brazil. pp 154–165. In ''*Notillage Crop Production in the Tropics*'' ed by I.O. Akobundu and A.E. Deutsch. IPPC Document 46-B-83, Oregan State Univ., Corvallis, USA.
49. USDA, 1975. *Soil taxonomy, a basic system of soil classification for making and interpreting soil surveys.* 754 pp, Agriculture Handbook No. 436, Washington, D.C., USA.
50. Wade, M.K. and Sanchez, P.A., 1983. Mulching and green manure applications for continuous crop production in the Amazon basin. *Agron. J.* 75, 39–45.
51. Wilkinson, G.E., 1975. Effect of grass fallow rotations on the infiltration of water into a savannah zone soil of northern Nigeria. *Trop. Agric.* 53, 97–103.

Resource Management and Optimization
1990, Volume 7(1–4), pp. 53–66
Reprints available directly from the publisher.
Photocopying permitted by license only.
© 1990 Harwood Academic Publishers GmbH
Printed in the United States of America

ECOLOGICAL PROCESSES IN TROPICAL FORESTS

FRANK B. GOLLEY
Institute of Ecology, University of Georgia, Athens, Georgia USA

CONTENTS

The systems approach is utilized for ecological modeling. Discussions of dynamic time and space variation are presented, along with the processes which govern ecology. Specific examples are used to illustrate points covered.

1. INTRODUCTION

This paper will be cast in terms of systems ecology which is a conceptual approach that allows us to make order of many different kinds of information. It is based on the concept of a system which is defined as an assemblage of components united by some form of interaction or interdependence in such a manner as to form an entity or a whole.[1] The paper was written initially as a beginning chapter for a book on tropical ecology. Since ecologists frequently are concerned with systems containing several thousands of species, as for example in a tropical rain forest patch of a few hectares, or individuals, as in a population of insects, we require a way to organize conceptually the data we collect through field observation. The systems concept, however, allows us to go well beyond the organizational role of, say, a library or an information network, since with the use of ecological modeling it permits us to represent the dynamic behavior of natural assemblages. And finally, from these system models we can draw general conclusions about the relationships and the processes in nature.

Ecological systems, or ecosystems, are usually defined as containing plants, animals, and microorganisms, all of which interact through the exchange of energy, materials, and information. A category such as plants, in turn, contains many taxonomic species, which may be arranged in groups with similar characteristics, such as trees, shrubs, herbs, fungi, algae, and so on. Thus, the ecological system can be subdivided or disaggregated into finer and finer detail.

Similarly, an ecosystem, such as a patch of forest, is part of larger systems such as a watershed or a landscape. And these are part of yet larger systems, until the whole earth, or the biosphere, is reached. Each of these levels of systems are interconnected and interdependent and we call this arrangement a hierarchy.[2] Our interest in this chapter will be to examine processes in tropical forest ecosystems, communities, and populations. But, the entire text can just as easily be considered an exploration of features of the hierarchy of natural and man-made tropical ecosystems.

As mentioned, the focus here will be on processes in tropical forests. What are processes? Ecologists distinguish process from structure in order to make clear the distinction between static elements and dynamic elements in systems. Static means that the time dimension is not considered. It is represented by the snapshot photograph. A forest is represented by the weight or mass of trees, the numbers of stems, or the kilograms of calcium and phosphorus in the soil. In contrast, process is represented by the motion picture, in that time is important. Process could include growth of the trees or change in weight over time, or the movement of calcium from the soil into the tree roots and incorporation as calcium pectate in cell walls in growing tissues. Movement of calcium might be measured in terms of kilograms of calcium taken up by the trees of the forest on one hectare in one year. Both structural and process data are required for full understanding of an ecosystem.

While process studies emphasize time, ecologists have difficulty measuring all the relevant temporal phenomena important to ecosystems. After all, an ecological study that lasts five or ten years is a long study. Yet, trees live for hundreds of years—so do many large animals. Humans with a life span of only 70 years finds it difficult to study, understand, and live with beings whose life span is much longer than theirs. We seem to destroy those species of the flora and fauna which are larger or live longer than ourselves, but that is another story. The problem here is that living systems have been operating on earth much longer than has humankind. Special paleontological studies are required to describe past events, while very long term contemporary projects are required to show the patterns of ecological processes today. Our environment is constantly varying in time and we need to understand the relationship between a fluctuating environment and highly adaptive and genetically changing species populations.

However, in addition to problems with time, we also must recognize that ecological processes vary over space. The surface of the earth is very heterogeneous. The parent rocks that create this surface texture differ in their physical properties especially those resisting erosion and their chemical properties. Soils which form on these surfaces differ across slopes and valleys. The species of organisms are distributed across these topographic gradients according to their tolerances and requirements. Together, they create new environments which characterizes a community of species. If one flies over a landscape, one can see the distribution of communities organized in a complex spatial mosaic.

Thus, when we study ecosystems we must be thinking in a dynamic time-space system. Ecosystems have no obvious beginning and end, and they are ever-changing as the constituent species adjust to the fluctuating environment and to each other. Ecosystems are also distributed over space in complex patterns, yet seldom are the boundaries of a system clear and distinct. Of course, everyone can tell if they are in a forest or grassland, but it may be difficult to tell where the forest ends and grassland begins because one finds small trees in the grassland and grass species in the forest. Thus, the spatial pattern is fuzzy and blending.

One can see that ecology is a different kind of science from engineering or chemistry. Ecologists study a natural world man did not make and of which he is but one small part. It is a fuzzy, changing, noisy kind of world without clear, simple patterns which can be stated as laws. While that makes ecology difficult and very frustrating, especially if you are trying to manage a river or a forest, it is real and exciting and very stimulating. It is through the study of process that this dynamic nature is most clearly shown.

2. PROCESS AT THE ECOSYSTEM LEVEL

2.1 Energy Flow and Production

Any system requires energy to carry out its work of production and maintenance of life processes. Thus, energy uptake and use become central to an understanding of ecology. Here we will consider the principles which underlie energy flow in communities and populations.

TABLE I
Mean Annual Balance Sheet of Organic Matter in Silver Springs, Florida[3]

	Dry Weight gms/m²/yr
Income	
Primary Production	6390
Import subsidy	120
Total	6510
Losses	
Community Respiration	6000
Downstream Export	766
Total	6766
Discrepancy	256

Utilization of energy follows principles which were developed in engineering and in physics. These are called the principles of thermodynamics. We will not develop a rigorous description of these principles since the interested reader can find them in physics and chemistry texts. Rather, we will develop models which can be applied to the understanding of ecological systems.

The first law of thermodynamics states that energy can be converted from one form to another but cannot be created or destroyed. This means that when we trace energy flow in an ecological system we should expect that all energy entering the system will equal all energy leaving the system, assuming no storage of energy within the system. In other words, energy input must balance output. This principle has given the ecologist a powerful tool in system analysis—if we have firm measure of the input, we can calculate output and vice versa (Table I). Further, we can measure output and input and compare our estimates. We know they should balance and if they don't, we suspect that our methods of measuring energy flow are faulty.

The second law of thermodynamics focuses on the transfer of energy between components. In such a transfer some energy which will be expended in the operation itself so that no transfer can take place with an efficiency of 100 percent. This means that we should pay close attention to the ratio of energy transfer between systems subcomponents, since we might be able to manipulate these transfer ratios. However, in most ecological systems the amount of energy degraded and lost for any further work is exceedingly high. For example, consider the input of energy through the process of photosynthesis in communities (Table II). In most cases the solar energy available to the community is used with an efficiency of less than five percent. We shall see later that the reason for this low efficiency is the quantum efficiency of photosynthesis and the light absorption and saturation characteristics of leaves.

All ecosystems process energy to accomplish work. In order to make the application of the thermodynamic law to these systems clear, let us consider the energy flow of an

TABLE II
Productivity in Selected Ecological Communities Data in KCal/m²/yr.

Community	Gross Primary Production	Net Primary Production	% Utilization of Solar Energy
Silver Springs (3)	20,180	8,833	5
Coral Reef (10)	35,040	20,440	5.8
Grass Field (11)	4,500	2,700	0.4
Rain Forest (12)	46,800	26,000	2.8

actual community ecosystem. The example will be the famous Silver Springs of Florida, studied by H. T. Odum in 1950s[3]. Odum divided the ecosystem into five subsystems: the producers, herbivores, carnivores, top carnivores and decomposers. Populations of plants, animals and microorganisms were placed into these subsystems. Producers refer to green plants, herbivores to those populations which feed on green plants, carnivores to those which feed on herbivores, top carnivors to those which feed on carnivores and decomposers to those which feed on dead material. Total energy available to the Silver Springs ecosystem was 1,700,000 kilogram calories per square meter per year. Of this amount of available energy the producers converted 1.2 percent in photosynthesis. The plants themselves used about 60 percent of the energy input in their own maintenance. There is a noticable diminution of energy as it is transferred from producers to herbivores, to carnivors and so on. Actually, this loss is so great that insufficient energy remains to allow further expansion of the chain of components.

Finally, we should note that almost all of the energy fixed in Silver Springs is utilized in maintenance of the ecosystem. Only about 10 percent is exported to the river which flows from the spring. There is essentially no community growth or storage of energy. This phenomenon of almost total utilization of fixed energy in system maintenance is typical of many forests, grasslands, tundra, and other mature communities. Crop communities channel a share of the energy converted in photosynthesis to yield, which is harvested by man. In this latter case we would add the energy stored in the system or harvested as yield to community production and maintenance energy flows to balance the accounts.

2.2 Productivity

Production can be treated in a variety of ways. First, we may consider the conversion of solar energy by photosynthesis and the synthesis of the basic organic material for the system. This is called Gross Primary Production (GPP). Second, after some of the photosynthesis has been used by the plant system to maintain itself, the remainder, which we see as plant growth, is termed Net Primary Production (NPP). Net primary production feeds all the animals and microorganisms in the system. Each of these consumer or decomposer units also may produce tissue and this latter production is

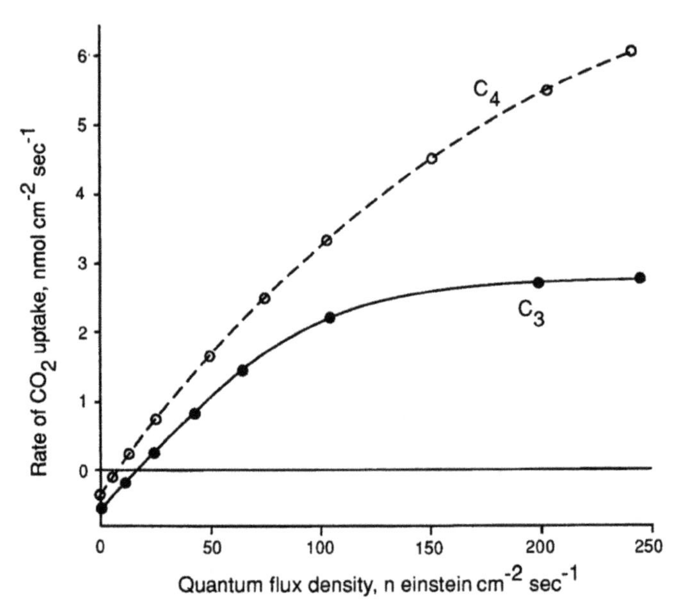

FIGURE 1. Rate of CO_2 uptake as a function of incident light (quantum flux density in n einsteins per cm^2 per sec.) of heat adapted C_3 and C_4 species (redrawn from 7).

termed Secondary Production. In crop ecology we are mainly concerned with net primary production, or the portion utilized by humans called Yield.

Gross primary production is a function of solar energy, water, temperature, growing season, nutrients and other factors. However, if we assume that all these limiting factors were available in optimum amounts what is the potential production we could expect on an area of land?

To answer this question, we must consider the efficiency of the photosynthesis process itself. In this process:

$$6\ CO_2 + 12\ H_2O + n\ \text{quanta} = C_6H_{12}O_6 + 6\ O_2 + 6\ H_2O$$

Bonner[4] has estimated that about 10 quanta are required to reduce one CO^2 molecule. Since 10 quanta supply about 520 kilocalories of energy mole of CO^2 and one molecule of CO^2 stores about 105 kilocalories, the quantum efficiency of photosynthesis is about (105/520 = 0.20) twenty percent.

These calculations assume that the photosynthetic material is arranged in an optimum configuration with the light. However, we know that the process of photosynthesis is not directly related to light intensity. Rate of photosynthesis increases with light intensity to a point (Figure 1), then becomes independent of light intensity, until the heating effect begins to inhibit photosynthesis. The former point occurs at about

1/10 to 2/10 of full sunlight in C-3 plants. C-4 plants, such as sugar cane, can utilize higher light intensities and are more efficient at taking up CO_2 and reducing loss of carbohydrates through respiration.

In contrast to photosynthesis, light absorption by the leaves is directly related to light intensity. Typically sun leaves absorb 80% of the incident light, while shade leaves absorb from 60–70%.

With this information let us recalculate the efficiency of the production process. If leaves absorb 80% of the light, and use 20% with an efficiency of 20%, then the overall efficiency is about [(0.20 × 0.20) (0.80) = 0.032] three percent. Assuming that the solar input to the vegetation averages about 500 kcal/cm^2/day, then the production process would yield (500 × .032 = 16) 16 kcal/cm^2/day. Assuming that plant material has a caloric value of 4000 kcal/g dry weight, then (16/4000 × 10,000 cm^2/m^2) 40 g dry weight is produced. This maximum estimate of production will vary depending upon the choice of efficiencies and caloric values used in the calculations; other estimates range up to 71 g/m^2/day maximum production. We conclude that it is possible to achieve a gross production of about 40–70g/m^2/day. These estimates do not include the effect of C_4 or CAM pathways of photosynthesis which may boost the potential level further.

Beyond the limits of the chemistry and physiology of the production process, what limiting factors concerned with the plants or vegetation should be evaluated? We have seen that the efficiency of photosynthesis varies with the light intensity and we know that light intensity is reduced as it passes through the leaf mass. Light attenuation by the canopy follows a predictable relationship. The leaf mass is typically distributed with the major portion in the mid section of the canopy (Fig. 2).

Light interception also follows a typical relationship. Natural vegetation has evolved so that the incident light energy is used with optimum efficiency. Since the individual plants compete for light, there has developed a variety of complicated structural units which support the photosynthetic machinery. These structural components must be supported by primary production so there is a balance between the amount of structure and production in natural vegetation. Given optimum environmental conditions, forests can only grow to a certain height. One of our aims in genetic selection is to produce crop systems which similarly use light with high efficiency.

The influence of plant architecture on primary production has been considered in a variety of ways. Besides the weight of leaves, we also can consider the area of leaves or the chlorophyll. Crop ecologists have found that a leaf area index, which is the area of leaves over a specified area of the ground surface, of 4 or 5 is the upper limit for high productivity of crops—above a LAI or 4 or 5 shading results in less than optimum utilization of light energy. In forests the LAI is often as high as 12–15 m/m^2 and a maximum record is 22 m/m^2 for a bamboo forest in Thailand[(5)].

One further feature of the plant system is its utilization of the gross primary production in respiration to maintain the plant tissues. The respiration of plants varies with such factors as type of tissue, size of plant and age. For example, in forests respiration increases with tree diameter eventually reaching an asymptote in very large trees. Respiration of woody tissue is directly related to the volume of the sapwood. The

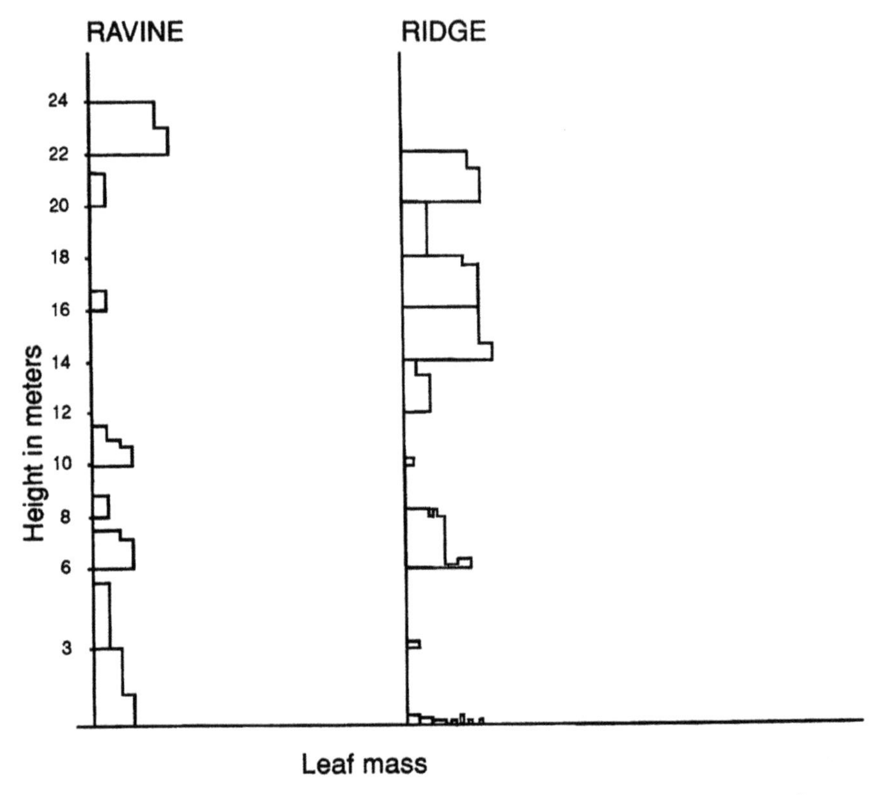

FIGURE 2. Distribution of Leaf mass with height in a montane tropical forest in Puerto Rico.[8] Size of square indicates mass of vegetation.

heartwood which increases in volume as the tree grows, contributes little to the overall respiration. Therefore, a forest made of many small trees may consume a greater proportion of the primary production than another stand of bigger trees having the same biomass. Thus, respiration of the stand is influenced by distribution of tree size, as well as age of trees.

The result of this type of an effect is that mature forests appear to use a larger proportion of the gross primary production in respiration than do forest plantations, grasslands, crops or tundra. The frequency distribution of actual field measures of these relationships indicate that forests use about 70 percent of gross primary production as respiration, while the other communities use about 40 percent.

With these data we can correct our potential gross production estimates and obtain potential net production estimates to compare with the abundant field data. Seventy percent of our estimate of optimum GPP of 40 g/m^2/day is 28 g/m^2/day, leaving about 12 g/m^2/day potential net primary production. Forty percent of 40 is 16 g/m^2/day,

TABLE III
Peak Daily Rates of Net Production

$g/m^2/day$	Xeric	Mesic	Hydric
arctic	1.2	11.1	3.7
temperate	0.2	9.5	28.0
tropical	—	17.0	16.0

leaving 24 $g/m^2/day$. We should have a range of from 12 to 24 $g/m^2/day$ potential net primary production.

Actual peak daily rates of net production are shown in Table III. It is clear that vegetation does operate at the expected peak level some of the time. However, a variety of environmental factors act to reduce this peak performance. Primary factors include solar energy and rainfall or water availability. Nutrients are of great importance in crop systems, while in natural vegetation the effects of nutrient limitation seem less significant.

The consequences of control of these environmental factors is a range of peak yearly net production values which differ greatly from those in Table III. The range of variation from arctic to tropical within a moisture column is about 30 ×. The range across moisture levels within the temperate region is about the same. Maximum net primary production is achieved by tropical reed swamp communities and such crops as sugar cane, which are C_4 plants and are especially productive because of their photosynthetic and respiration physiology.

Finally, it will be useful to consider the actual data on gross primary production (Table 2).

An estimate of 120 mt/ha/yr for rain forest is one of the highest values ever reported for a terrestrial plant community. Coral reefs and some estuarine ecosystems have gross production rates as high as rain forests. Crop systems usually do not produce at the maximum rate of the unmanaged community.

How do these data compare with our estimate of the maximum rate of production? Recall that we calculate that from 40 to 70 $g/m^2/day$ was a probable maximum level of gross production. Seventy times 365 is 25,550 $g/m^2/yr$ or 255 mt/ha/yr. Forty times 365 is 14,600 $g/m^2/yr$ or 146 mt/ha/yr. A Thailand rain forest with 12 mt/ha/yr comes very close to this lower estimate.

The efficiency of utilization of solar energy by these communities ranges from 3.1 percent for sugar cane to 0.37 percent for mature grassland. Tropical forest has a photosynthetic efficiency near 3 percent. The obvious conclusion is that relatively little of the available solar energy is utilized in the primary production.

2.3 Biogeochemical Cycling

Ecological systems are composed of a variety of chemical elements. There are more than 100 elements that might occur in the environment but those of interest to the ecologist are reactive under the normal conditions of temperature and pressure of the earth's surface. The basic six are carbon, hydrogen, oxygen, nitrogen, sulfur, and phosphorus. The elements differ in their abundance in the lithosphere, atmosphere, hydrosphere and biosphere.

There are two major transfers of elements between these spheres. The first, called the geological cycle, results from uplift of deposited sediments, volcanism and rock formation, the erosion of land forms and transport of eroded materials through water and the atmosphere to places of deposition, where ultimately uplift may again expose the materials to cycle again. In certain parts of the tropics, for example the Amazon basin, materials have been uplifted, eroded, transported, re-eroded and retransported, leached and deposited so that the present soils may be almost devoid of the mobile elements. These conditions produce sterile soils.

The other major transfer is through the biosphere. This cycle is frequently termed biogeochemical cycling or mineral cycling and is a major topic of ecology, forestry and agriculture. In a biogeochemical cycle elements are taken from the soil and atmosphere or hydrosphere, incorporated into the tissues of the biota and then ultimately returned through death or metabolism. Cycling rates may be rapid or slow depending upon many factors in the environment and the biota. The uptake of elements from soils is partly dependent upon the energy which causes evaporation and transpiration at the leaf surface. The pull of water through the plant by this physical force provides the mechanism for movement of elements from the soil water, through the roots to the growing tissues. Active uptake may also be occurring at the root-soil interface and in many plants fungi are associated with the roots which also mediate uptake. All the other organisms, the plant-eaters, predators and soil biota, obtain the chemicals they require for growth and reproduction from their food and the atmosphere and hydrosphere.

We know relatively little about the nutrient requirements of the biota. Examination of the chemical content of plants and animals shows that they differ one from another in many ways.

For example, members of a species appear to share a chemical constitution but those living in different chemical environments may differ in their composition. There are several themes of mineral cycling, two of which are of special interest to ecologists.

First, the strategies of nutrient cycling can be divided into several patterns. On one hand are those species which grow rapidly and have fast rates of cycling. These organisms are typical crop plants and they also typify succession. On the other hand are organisms with slow growth, slow cycles, and large storage of nutrients in the biomass. These typify mature systems. One easy way to see this contrast is to compare a crop field and a mature forest. The crop grows in one year, takes up nutrients quickly, losses them easily, the cycles are turning over rapidly throughout the year. Nutrients are stored at low levels since the biomass is small and the soil is subject to leaching and erosion. In contrast, the forest has a large biomass that changes very slowly, tying up large quantities of elements. The cycles move slowly as a consequence. Of course, there are rapid cycles in a forest—the leaves may develop, grow and die in one year, but the bulk of the biomass turns over relatively slowly. The forest conserves nutrients and alters the temperature and water regimes, reducing the impact of erosion. Of course, all plants don't fall easily into these two strategies but they do suggest a contrast in the ways elements are cycled.

The second important theme involves conservation of nutrients. In some circumstances the soils or waters contain inadequate quantities of nutrients for plant growth, especially those plants with the rapid cycling strategy. Nutrients not only enter the ecosystem from the soil, but also from the atmosphere and from rainfall so that if the system has a mechanism to capture these essential materials and hold them, they could live on these inputs. Thus, it is important to understand how mechanisms to conserve nutrients could evolve in the biota. If we visualize a diagram of mineral cycling, we can ask if there are any known ways to control the exit from one compartment to another or to the environment. An examination of this sort reveals controls throughout the system. For example, the leaf can lose nutrients by leaching of rainwater, by feeding through a herbivore, or by loss of the leaf as it falls to the ground. Leaves may evolve structural controls of water loss, such as a scherophyllous surface. They may accumulate toxic chemicals, such as tannins, which make the leaf unpalatable to herbivores. Nutrients may be taken up from the leaf before abscission. As another example, the herbivores may concentrate nutrients through their activities which can conceivably benefit the plants. In Panama, sloths crop trees intensively (one tree may lose 20% of its leaves to sloths), but they defecate on the ground beneath the tree. The sloth feces form a long-lasting source of nutrients for the tree roots, which invade the fecal mass. Similarly, it has been shown that leaf-cutting ants built refuse dumps of leaves, which, in turn, were invaded by plant roots which could recycle the nutrients in the leaves.

These types of conservation mechanisms can be observed in many different ecosystems. Where the sources of nutrients are poor there tends to be more mechanisms in evidence. The result may be closed cycles, highly efficient cycles, large storages in the living tissues, and a capacity to absorb and hold atmospheric inputs. The Amazonian forest on highly leached soils at San Carlos de Rio Negro, Venezuela is an example of a system that has many of the mechanisms[9].

3. PROCESS AT THE COMMUNITY LEVEL

We will turn now to a focus exclusively on interactions between members of the biota. Two kinds of interactions are of special interest to our examination of process in tropical systems. These concern the flow of energy and nutrients through feeding relationships and the control of reproductive and growth processes.

3.1 Feeding Relationships

Within a simple community there may exist thousands or even tens of thousands of individual species of plants, animals, and microorganisms. No ecologist or team of ecologists can study all the biota at one time. Therefore, we frequently try to group the species into catagories that have common properties and study their processes collectively. At the most abstract, we can group organisms using the same source of food. These groups are called trophic levels and they usually are listed as producers, consumers, predators, decomposers. Producers are green plants, consumers consume green plants directly, predators consume live animals, decomposers consume dead tissues. Another way to organize the species is to group those who fill a similar role. For example, birds feeding on insects in the canopy are distinguished from those feeding on the trunks and those feeding on the ground surface. These groups are called guilds.

What ever classification device we use, one can visualize a complex set of relationships in space and time in which arrows could indicate the flow of food or information or impact from one group to another. These complex diagrams look like spider webs (although they are less organized) and they are frequently called food webs. Where one group feeds on another and is, in turn, fed on by yet another we can visualize a chain of feeding relationships or a food chain.

We may also consider feeding relationships from the point of view of the food. How does feeding effect the survival, growth and reproduction of the organism being fed upon? The food and the feeder are coupled together in a feed-back system and they adapt to one another and over time coevolve.

Organisms feeding upon vegetative parts use a food that differs in a fundamental way from that of carnivores. Plant foods contain a large quantity of species–specific chemical defense materials and, therefore, are not easily exchanged from one feeder to another. The production of these protective compounds is a cost to the plant and for a full accounting of the feeding relationship must be considered.

The impact of feeding on fresh leaves and stems can be especially serious if the growing tip is removed or if a new leaf is removed just at the point where photosynthesis is beginning. Thus, while extreme feeding resulting in total defoliation is less often observed in the tropics—although it does occur—the impact of a single bite may also be important to the fitness of a plant. Chemical defenses may be mobilized at the site of feeding to reduce further feeding and it is possible that the chemicals produced by the feed (in the saliva, for example) may influence the plant. What we see here is an exceedingly complex chemical interaction system that can be very subtle, requiring small quantities of energy or materials yet having large impacts in the life history of the forest.

The plant-feeder specificity is especially pronounced in tropical insects and must have something to do with the large numbers of species found in canopies of tropical forests. Further, the distance and spacing of the plants can influence the rate of feeding. If plants are far apart they can escape feeders, although they may also have difficulty attracting pollinators. Plant specificity also influences feeding of vertebrate herbivores but to a lesser degree.

Organisms that feed on seeds also have a great impact on tropical vegetation. Seed predators are divided into those feeding on developing embryos, those eating mature seeds and those serving as post dispersal seed consumers. The cost to the plant increases across this classification because other sources of mortality act and reduce the probability of the plant reproducing successfully. The plant has invested least in the developing embryos and most in the seed that is dispersed and is germinating at a site where growth of a new individual may occur. The impact of seed predators depends upon the size of the seed crop, the proximity of other seed crops in space and time, and the ability of the habitat to support seed predators. Very large crops may exhaust the capacity of the predator to consume all of them, thereby allowing seeds to germinate and grow. In contrast, seeds produced in very small amounts or for very short periods, or in scattered locations may be effective strategies of escape since predators require time to find the seed, and require enough seeds to sustain them and allow them to carry out their life cycle. These relationships suggest that the animals in the forest communities are partly responsible for the time of production and the location of the plants in the forest. What must be the impact of the large-scale extinction of the grazing fauna of South America during the Pleistocene, or the removal of large mammals by hunters today, on the tropical plant communities?

3.2 Pollination Biology—Providing Biological Services

In the tropics wind pollination tends to be less important than it is in other biomes. Rather, animals provide important services to plants in pollination, as well as in seed dispersal. The central issue involves the degree of transfer of genetic information to the next generation. The role of inbreeding or outbreeding on speciation can be exceedingly important, and therefore, ecologists have studied if tropical trees (and other life forms) are self or cross fertilized. Outcrossing may be promoted by mechanical or structural developments of the flower parts or by changing the timing of anther or stigma activity. Self incompatability may prevent inbreeding, as will production of staminate or pistillate flowers on separate plants.

The actual operation of pollination involves an animal encountering pollen which is then carried to the stigma of another flower. Anthesis usually occurs at a particular time of day or night and lasts for 24 hours. The large number of nocturnal flowering plants means that nocturnally active animals such as moths and bats may be of special significance in tropical forests. Of course, the length of time pollen is available varies greatly. Orchids are famous for the length of life of their blossoms.

Insects that provide these pollination services to tropical plants include beetles, bees, wasps, moths, butterflies, and flies. Large solitary bees are especially important pollinators of woody tropical plants and some are capable of making long flights of over 20 km in a day, allowing for widespread dispersal of pollen among widely distributed and rare tropical plants. Trap-lining is a term to describe the visiting of a large number of widely dispersed plants by insects.

A variety of birds also serve as pollinating agents in tropical forests. Hummingbirds are well known for this trait, but a variety of passerine birds may also be important as well. Among the mammals, bats are important pollinators. Bat pollinated plants are frequently trees or epiphytes and often have hanging inflorescences which permit a flying bat to approach and feed.

All of these organisms are attracted to flowers to obtain rewards, which may be nectar or pollen. Nectar contains sugar, proteins, amino acids, lipids, and a variety of other organic compounds. The concentrations and amounts of these organic materials can very widely between species of plants and at different periods.

Acknowledgement: I acknowledge the assistance of Carl Jordan, who provided data and information.

REFERENCES

1. Patten, B. C. A primer for ecological modeling and simulation with analog and digital computers, pp. 3–121. In, B. C. Patten, ed. Systems Analysis and Simulation in Ecology, Volume 1. Academic Press. NY. (1971).
2. O'Neill, R.V., DeAngelis, D.L., Waide, J.B. and Allen, T.F.H. 1986. A Hierarchy Concept of Ecosystems. Princeton University Press, Princeton, N.J. 253 pp.
3. Odum, H.T. 1955. Trophic structure and productivity of Silver Springs, Florida. Ecological Monographs 27:55–112.
4. Bonner, J. 1962. The upper limit of crop yield. Science 137(3523):11–15.
5. Ogawa, H., Yoda, K., and Kira, T. 1961. A preliminary survey of the vegetation of Thailand. Nature and Life in Southeast Asia. 1:21–157.
6. Odum, EP. 1971. Fundamentals of Ecology. W.B. Saunders Co. Philadelphia.
7. Osmund, CB.; Bjorkman, O; and Anderson, D.J. 1980. Physiological Processes in Plant Ecology. Springer Verlag, Berlin.
8. Odum, H.T., Copeland, B.J., and Brown, R.Z. 1963. Direct and optical assay of leaf mass of the lower Montane Rain Forest of Puerto Rico. Proc. Natl. Acad. Sci. 49(4):429–434.
9. Jordan, C.F. 1982. The nutrient balance of an Amazonian rain forest. Ecology 63:647–654.
10. Odum, H.T. and E.P. Odum. 1955. Trophic Structure and productivity of a windward coral reef community on Eniwetok atoll. Ecological Monographs 25:291–320.
11. Golley, F.B. 1965. Structure and function of an old-field broomsedge community. Ecological Monographs 35:113–131.
12. Hozumi, K., K. Yoda and Kira, T. 1969. Production ecology of tropical rain forests in southwestern Cambodia, II. Photosynthetic production in an evergreen seasonal forest. Nature and Life in Southeast Asia. 6:57–81.

Resource Management and Optimization
1990, Volume 7(1–4), pp. 67–95
Reprints available directly from the publisher.
Photocopying permitted by license only.
© 1990 Harwood Academic Publishers GmbH
Printed in the United States of America

TROPICAL FOREST RESOURCES

E.F. BRUENIG
Department of World Forestry, Institute of Forestry, University of Hamburg, 91 Leuschnerstrasse, Hamburg 80, Federal Republic of Germany

CONTENTS

Environmental preservation and supply of useful products are the main utilities of rain forests to man. European foresters more than half a century ago warned that destruction of tropical forests would adversely affect the environment and the remaining forests. Today this is widely accepted as a very serious problem. Sustained silviculture of tropical forests must be based on the choice of option which has the least chance of being wrong.

1. PRODUCTS AND FUNCTIONS

The sustainable utility of rainforest to man lies mainly in two areas: environmental preservation and supply of useful products. The environmental utility mainly refers to climate and hydrology. Tropical moist forests influence the environment most intensively and foresters very early warned of the consequence of deforestation on the environment and the residual forests. European foresters suggested half a century ago that removal of tropical lowland rainforest and montane cloud forest would irreversibly change site climate and hydrology and reduce self-generated convective precipitation to such extend that restoration would be impossible by natural or artificial means.

The floristic, biochemical and architectural diversity of the tropical forests makes it a store and source of a vast number of divers products and functions of which many are actually or potentially beneficial to man. Fig. 1 illustrates the variety of useful products which come from tropical forest trees. Nationally and globally the most important forest products are and will remain wood products, especially timber for local use, domestic markets and export. In the predominantly evergreen moist forest biome, the supply of fuel in the form of wood or charcoal is ample and prices are usually low. In the seasonally dry tropics, wood fuel supply is usually insufficient, availability low and prices high. Non-wood products may equal or exceed the value and importance of timber and fuel in certain areas or in certain years, e.g. if there is a bumper crop of illipe nuts in Malaysia.

A presently much emphasized function of possible importance of tropical forests is as reservoir of pharmaceutically useful plants. However, on scrutinizing past histories of such plants and products, it appears doubtful that the longterm commercial values are as high as some experts believe. The discovery of marketable natural drugs is in any case a very rare event. If one has been found and proved useful, the pharmaceutically active compound usually has been rapidly synthesized and replaced by a artificial substitute in industrial production. The same applies in principle to cosmetics, dyes and condiments of high value. Powerful computer techniques for simulating structural molecule models are rapidly developing. These may not only speed the replacement of discovered compound but even replace the natural plant materials as blue-prints for pharmaceutical chemical engineering.

Another product of tropical forests is genetic stock for breeding. Examples are rubber, coffee, cocoa, tea and many other crops. It is presently fashionable to estimate highly the present and future value of tropical forest as stores and reservoirs of genetic information which could in time be commercially utilized. It is also fashionable to accuse even selective logging of impairing seriously the genetic value of the tropical forests. The trouble is, that vast reservoirs had to be established for very low chances of jack-pot hits, while the value of the other functions, especially of wood production, could not be realized. Also, judgment on the reality of assumed potentials and dangers are without the necessary substance as long as forest dynamics and age structure are hardly understood. Another uncertainty arises from our inadequate understanding of

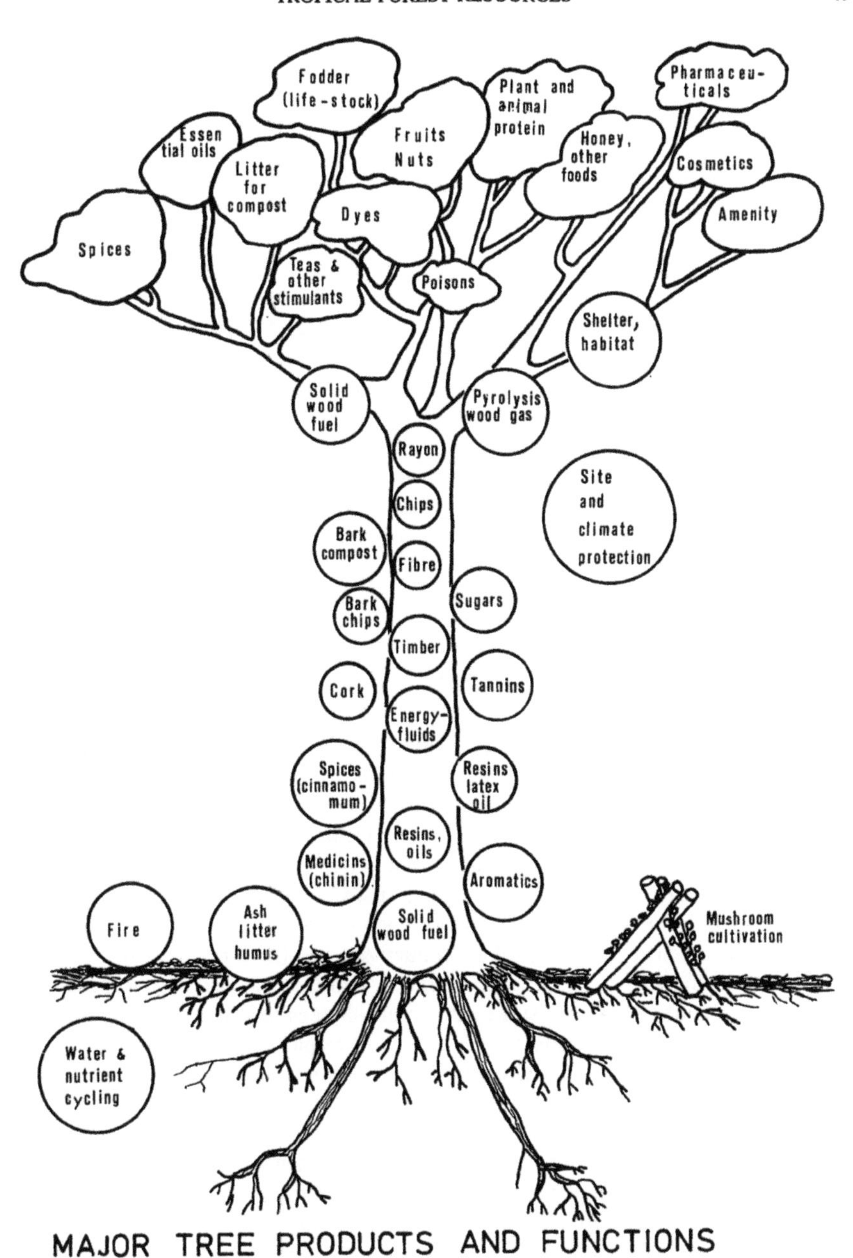

FIGURE 1. The major products which can be obtained from tropical trees and forests and some of their environmental and social functions.

the genetic patterns of tropical forests and the effects of logging and silviculture on them.

The value of tropical moist forests as a source of vegetable food, spices, condiments and fruits for the local population is generally high. Sago from a number of wild or semi-domesticated palm species is staple food of many tribes of nomadic gatherers and hunters and of settled agriculturists in wetland and flood-plain areas. Nutritionally important dietary supplements are many types of fruits, nuts and leaves. Animal protein, important as it is as a dietary supplement, is of such low level of supply, excepts in areas with very low human population densities and extensive river systems with plenty of fish, such as parts of Borneo, New Guinea and Amazonia, that it's actual contribution to human livelihood is small. The reason is, that the proportion of mammals and large birds in the terrestrial biomass is low in dense tropical evergreen and deciduous moist forests (Fig. 2).

One of the presently most important, but also most destructive and damaging uses of tropical forests is to produce ash, litter and soil humus which are used to fertilize agricultural crops on the spots or in nearby fields. Also, the use of fire to clear the site, kill pests and contain weeds by slash-and-burn pratices is a use of the phytomass of tropical forest which is equally important and wide-spread, but often overlooked when the products and services of the tropical forests to mankind are considered.

2. POTENTIALS AND LIMITATIONS

The average potential vegetable productivity of the tropical predominantly evergreen moist forest is several times greater than the productivity of the cold boreal evergreen and cool temperate mixed forest formations on edaphically comparable average productive forest sites. Within the area of the tropical forests productivity changes along ecological gradients. It is greater in the better energy-supplied but not excessively hot and oceanic Malesian region than in the Congo basin. It is still lower in the less energy and water supplied Amazon-Orinoco basin which is also geo-chemically poorer. Within each region, vegetable productivity and the proportion of potentially merchantable phytomass change with edaphic, hydrological and physiographic site conditions and altitude. As a result of this multi-dimensional pattern at regional and local scale, estimating the potential productivity of tropical forests is difficult. Table I gives a summary of the presently available best estimates of productivity and phytomass stock.

The Net Primary Productivity (NPP) is the theoretical upper limit of potentially utilizable yield. This maximum level can never be reached in forestry management because a substantial part of the production must be diverted to cover the soil and maintain soil conditions. In addition natural causes, mis-management, wasteful harvesting practices, market preferences and price/cost relationships reduce the actually utilizable and sustainable yield to half and less of the potential.

Assuming that the productive forest area declines as shown in Table II and that the

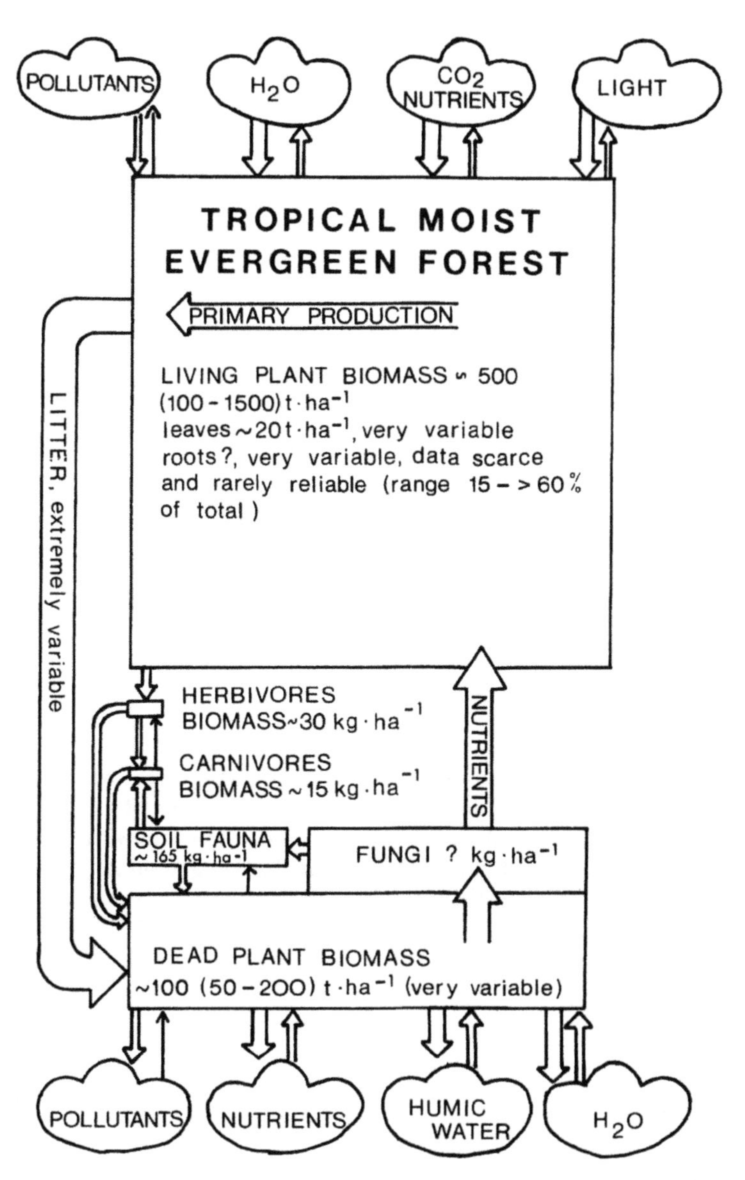

FIGURE 2. The proportional amounts of the very large stock of living and dead phytomass (plant biomass) and the tiny amounts of biomass of herbivores and carnivores in the terrestrial ecosystem of closed evergreen moist forest. The ratios indicate that vegetable biomass stock, turn-over rates and productivity predominantly of the dry-land forest ecosystem are very much larger than those of animals, particularly of those of potential economic value. The amounts and effects of pollutants entering the system are yet unknown.

TABLE I

Net Primary Productivity and Phytomass Stocking in Metric Tons Dry Matter per Hectare. The Figures Are Very Broad Approximations Based on Very Scanty Data from the Few Available Reliable Biomass Sample Plots and Apply to Natural Closed Forest, Excluding Secondary Regrowth After Deforestation.

	Net Primary Productivity		Phytomass Stocking		Total Phytomass
Forest Formation	**t/ha**	**Range***	**t/ha**	**Range***	**t .10⁹ (1980)**
Predominantly Evergreen Moist	25	(1) 10–35 (2) 3–40	450	(1) 300–800 (2) 50–1500	200
Predominantly Deciduous Moist	15	10–20	300	200–500	195
Predominantly Deciduous Dry	7	3–12	200	100–250	130

*Approximate range related to site factors in the "mature climatic climax" forest according to data in BRUENIG (1966, 1971, 1977), KURZ (1982)

(1) Climax forest formation.

(2) Edaphic climax formations included.

TABLE II

Development of the Area of Open (Approx. 1/3 of the Area, Mostly in Dry Forests) and Closed Natural Forest Under the Impact of Continued Agricultural Expansion, Assuming an Optimistic Scenario with Successful Agricultural Consolidation and Political Stabilization and an Alternative and Possibly More Realistic Scenario of Delayed Improvement.

	Forest Area in km^2 $.10^6$ (= ha $.10^8$) in the Years			
Forest Formation	**1965**	**1980**	**2000**	**2050**
		optimistic scenario		
Predominantly Evergreen Moist	5.5	4.9	4.0	2.7
Predominantly Deciduous Moist	7.5	7.0	6.2	4.8
Predominantly Deciduous Dry	7.5	7.0	5.9	4.5
Total, optimistic	20.5	18.9	16.1	12.0
Total, pessimistic	20.5	18.0	13.5	8.0
Man-Made Forests	0.06	0.12	0.25	0.50

proportion of very infertile and inaccessible sites and purely protective forests increases at the same time, the best we can hope for by 2050 would be between 1.3 and 2.0 · 10^9t d.m. from predominantly evergreen tropical forest, 1.2 and 1.9 · 10^9t d.m. from predominantly deciduous moist forest and 0.6–0.9 · 10^9t d.m. from deciduous dry forests. The total which plantations could contribute would be between 0.4 and 0.7 · 10^9t d.m.. But it is very doubtful that the 50 mill. ha productive plantations will be established by 2050 and adequately managed to achieve a sufficiently high productivity. How much of the secondary growth on abandoned agricultural land will be commercially and socially available is impossible to say. Secondary forests may make a substantial contribution to meet local fuel needs. Adding all up, the most likely level of sustainable yield from tropical forests will be around 4.10^9 d.m. to supply 8.4 · 10^9 people in the developing world (estimate for 2050 in the 7th Development Report of the World Bank). This would be just enough to meet forseable requirements, but leaves little margin for unexpected developments and no room for complacency.

3. USE AND MISUSE

The tropical predominately evergreen moist forest has most probably been the site on which human evolution initiated. The complex and divers environment offered the varied and rich diet, shelter and tranquility needed for acquiring capabilities by preadaption which were needed to conquer the more exacting environments of the deciduous and open vegetation in drier and colder climates. Man affected the evergreen moist forest very little. Locally, certain plant species may have been exterminated, other plant species might have been enriched accidentally or purposely planted. On a whole, man remained an integrated, locally disturbing but not degrading and destroying component of the ecosystem. The characteristic land use system which evolved during that period is the family or village forest garden of many species and life-forms. These forest gardens on good soils of easy access required little maintenance and hardly any hard work. They could be run by the women, while males remained more addicted to roaming the landscape for food and adventure. They became the prototype of the present day forest gardens, which as individual home garden or communal village garden are wide-spread in the humid tropics and especially well developed and sophisticated in Southeast Asia.

Destruction of forests and degradation of sites began with the development of slash-and-burn techniques in the deciduous tropical forests where conditions were more favourable for man to acquire the technical skills and experiences needed for field cropping. The new capability to grow annual field crops and the technical skills and physical strength required to clear the land initiated profound social and cultural changes in the societies which lived in the emerging man-made savanna biotop.

Slash-and-burn shifting agricultural could for technical reasons not have originated in the predominantly evergreen moist forest. But once man had acquired the technical capability to clear land of tall forest he could cope with the problem of felling the large evergreen top-canopy trees without which the slash would not be able to dry sufficiently for a certain burn in the equatorial climate. The scene was now set for the return in force to the evergreen moist forests biome. Population increase, warfare and political pressures, land degradation and climatic events in the dry tropics set a massive migration into motion which still continues to this day. This advancing agricultural front is the biggest threat to the rainforest biome to which the shifting slash-and-burn agriculture is ecologically not adapted. The damage remains less obvious and is repairable as long as the fallow rotation system is combined with eventual and timely shifting of village location to new forested land. This is only possible in areas where population density is still low and plenty of virgin land is available. Under such conditions, traditional native agricultural systems (e.g. the various forms of rotation-cum-shifting cultivation) still work reasonably well on a subsistence farming system level. High labour inputs, the unavoidable decline of land and labour productivity in successive rotations and the disruptibe effects of shifting make this system economically and socially unattractive as soon as alternative options become available and aspirations rise.

Over-use of land and forest resources with grave and lasting effects is the inescapable consequences of:

—transition from rotation-cum-shifting cultivation to settled rotational fallow systems;
—settlement schemes in which too many locally inexperienced people are squeezed into too little space on fragile humid tropical soils;
—conversion to pasture at large scale for beef raising at industrial scale in the predominantly evergreen moist tropical forest biome;
—inefficient range management and over-grazing in the deciduous moist and dry forest biomes, aggravated by destructive forest utilization practices (fuel wood collecting, charcoal burning, mechanized clear-felling), especially in hill and catchment areas.

The result is generally increase of instability of the landscape, progressing deforestation and degradation of site toward greater micro-climatic aridity.

An example for the potential hazards of ill-considered land use is the catastrophic fire which raged through 4.2 ± 0.6 mill. hectares of Mixed Dipterocarp, peatswamp and kerangas forests, secondary forest and agricultural land in Kalimantan/Indonesia, when the Pacific El Nino weather anomaly in 1982/83 made the ill-used landscape particularly susceptible to fire.

The present rate of tropical deforestation is estimated to be at least 10, but possibly nearer to 15 mill. ha per year. About 2/3 of this occurs in closed, mostly moist, forests and 1/3 in open, mostly dry, forests (FAO, 1982 a and b). At the same time, about 20 mill. ha at least of agricultural land are lost by desertification. Both processes are interacting and accelerated by unadapted land development schemes in all parts of the tropics.

Exploitation of forests for timber, fuelwood and minor produce is by comparison much less destructive, in whatever form and intensity it is conducted, than shifting agriculture. Extraction from tropical forests of timber, wax, resin and luxury items for domestic and export market has a very long history. Tropical timbers found their way to Egypt and Europe many thousands of years ago (GONGGRYP, 1942). Bornean Ironwood (*Eusideroxylon zwageri* T. & B.) has been logged in Bornean alluvial forest and exported across the South China Sea at least for the last 1000 years. *Haematoxylon brasiletto* Kat. (Caesalpinia echinata), the famous Brazil wood, was exploited and exported to Europe since 1503 where a brick-red dye was extracted and traded until synthetics replaced it. Another famous early export from the Amazonian region is *Aniba rosaeodora* Ducke which produces the precious rosewood oil from its wood and leaves.

Selective exploitation of special cabinet or heavy duty timbers at greater scale began with the European conquest and the establishment of colonial empires. But even then, the general functionality and structure of the forest ecosystems were hardly affected. Timber exploitation and trade were of minor scale and mostly local economic importance. Only locally over-use caused decline of growing stock density and productivity of preferred species. In some cases, this tendency was reversed by silvicultural means since the mid-18th century (e.g. teak in Burma). Also, early attempt were made to conserve forests for protective and productive reasons and to introduce the principle of sustained yield to tropical forestry (SCHLICH, 1922). The real changes came with independence in the fifties and sixties of this century. But even now, at a global trade scale, tropical timber trade and export makes only a tiny contribution to world timber consumption and trade. The annual volume of timber reported as extracted from the natural tropical productive broadleaved forests for export in recent years fluctuated around 110 mill. m^3 hardwood logs, to which 15 mill. m^3 softwood logs are added from coniferous forests (Table III). About two-thirds of the hardwood logs come from closed primary forest, the rest mainly from relogged forest and much less from open forests. Only 15 % of the softwood harvest come from pristine forest (FAO, 1982b, p. 18). These statistics are extremely unreliable and actual fellings are much higher. But even then the annual removals are a tiny fraction of 0.1 % of the gross timber volume of growing stock (Table II). The 4.3 mill. ha area from which they are estimated to have been harvested are less than 0.3 % of the still existing closed forest area (Table II).

There is considerable uncertainty about the amounts of timber cut, removed and sold on the domestic market. There is even less known for certain about the amounts removed for immediate local consumption and used in buildings and for making farm implements and boats. Even more difficult to assess are the amounts removed from primary and secondary forests to supply local fuelwood demands.

With these difficulties and uncertainties in mind, the best estimate of global consumption of wood from tropical forests is between 1.3 and 1.5 bill. m^3. Of this about 85 % are used as fuel and charcoal and 10 % as timber within the countries of origin. Between 3 and 5 %, or about 0.05 to 0.06 bill. m^3 roundwood equivalents, are exported, more than half of this from Southeast Asia to Japan.

The by far greatest wood consumer is agriculture. Trees are cut and burned to exterminate weeds, pests and diseases and to provide fertilizing ash to permanently and intermittendly cultivated fields. How much wood and litter is consumed this way is difficult to say. It is certainly not less than 1 bill. tons dry matter or at least 1.5 bill.

TABLE III

Annual Removal of Saw and Veneer Logs from Tropical Forests Under Legal License and Recorded in the Official Timber Removal Statistics. The Second Column for the Period 1976-79 Refers to the Proportion Which is Estimated to Have Originated from Natural (More or Less Primary, Pristine Virgin) Forests and Shows the Negligible Amount of Timber Produced in Tropical Plantations. The Additional Amount of Timber Removed for Direct Use or Local Markets and Thereby By-Passing Statistics is Substantial but Difficult to Assess (Abstracted from FAO, 1982 a and b).

Region within Tropics	1961–65	1966–70	1971–75	1976–79	1976–79 (%)
America (23 countries)	22.4	26.4	32.3	42.4	33.4 (78.8)
Africa (37 countries)	10.3	13.7	15.8	16.7	15.5 (92.8)
Asia (15 countries)	32.1	47.0	65.3	80.0	76.3 (95.4)
Sum (75 countries)					
- coniferous	11.5	14.6	17.4	23.7	14.9 (62.8)
- broadleaved	53.3	72.5	95.9	115.4	110.3 (95.6)
Total recorded tropical timber removals	64.7	87.1	113.3	139.1	125.2 (90.0)

m^3 per year, but possibly even as much as 2 bill. tons d.m. or approx. 3 bill. m^3 per year in the moist forests alone. Of the total consumption of wood in the tropics, therefore, at least 50 % are for agriculture, and at most 43 % for fuel, 5 % for domestic timber supply, and only about 2 % for export.

4. ECOLOGICAL AND ENVIRONMENTAL IMPACTS

Ecological impacts of very light single-tree creaming or cautious selective logging are small and structural and functional features of forests are very little affected. Species diversity and architectural complexity may even increase and the often feared genetic erosion is most unlikely to happen. Such light interferences equal natural ''catastrophies'' at small to medium gap size scale. They are ecologically indifferent, if not beneficial. They add to structural and floristic diversity and accord with the natural dynamics of the tropical forests. Even the commonly very heavy logging damage to the residual growing stock may impair the production of merchantable timber volume but ecological implications are ephemeral and possibly within the margins of the ecosystem elasticity to which natural catastrophes have trained it. Ecologically and environmentally much more serious consequences are caused to soils and hydrology by careless loading and skidding. Also consequential from a management point of view is the interaction between selective logging and the natural patchiness of the phasic and age-class distribution pattern of the forests. Exploitation favours patches with good stocking of well sized medium-weight hardwood trees. Without silvicultural treatment, selective logging therefore may increase the patchiness of the forest by selecting medium stages and avoiding late and final stages of forest development.

An alternative and practicable approach to selective logging, which is still within the limits of natural events, is carefully planned and executed complete logging of commercial timber trees which removes between 20 and 50 percent of the above-ground phytomass. Inspite of being a drastic interference with forest structure and functions, it would not necessarily be harmful, if soil damage is avoided. The profound changes of stocks and turnover rates initiate successional developments which in some respects resemble natural sucession after large-scale destructive disturbances. Unless natural sporadic or man-caused fire interfers, adequate protective ground cover will be fairly rapidly restored. More time is needed to replenish nutrient stocks which may require a few to many decades depending on the soil, the kind of vegetation and the elements concerned.

The ecological situation is different in single-species dominated stands. Complete harvesting of all timber trees in such stands amounts is almost to clear-felling and removes up to 50 percent, in chipping projects even more, of the phytomass. In stands on rich alluvial flood plain soils recovery will be rapid, but at the other extreme of the site gradient where single-species dominated stands are common on extremely oligotrophic dry and wet sites the story is very different. Complete removal of merchantable growing stock in the Shorea albida peatswamp forests and the Agathis dammara kerangas forests of Sarawak or the Eperua spp. forests on white sands in Surinam leaves residual vegetation of very weak regenerative capacity. Loss of organic matter and nutrients and unfavourable micro-climatic and soil conditions make recovery a very slow process. The regrowth is on these sites particularly susceptible to fire. One severe wildfire may already cause final collapse and the formation of a degraded, arrested disclimax. Even without such catastrophy, according to GRAAF (1982), expensive silvicultural treatment would become necessary, including ploughing, planting and perhaps fertilizing, to speed recovery, but this would on these soils be impracticable on economic terms. The regeneration of clear-cut shorea albida forests in Borneo is yet an unsolved problem.

Logging and silvicultural manipulation of any type more drastic than the selective logging and improvement system (e.g. such as the "Malayan Uniform System", MUS, in comparison to the Philippine "Selection and Improvement System", SIS), will lower the floristic diversity, but not necessarily the species richness, and inevitably result in a simplification of the architectural forest structure. The aerodynamic roughness of the canopy will be reduced and, as a consequence, the albedo will increase, the radiation balance and evapotranspiration decline and the surface water run-off, soil leaching and erosion increase at least temporarily. The consequent changes in the chemical, physical and biological soil conditions will be more lasting and amplified by the changes in the biochemical nature and amounts of litter fall from the more simple crop. Also, changes in the rooting depth and intensity may be expected, but little is yet known about this. But whatever the initial changes are, eventually the system will restore its former stature, structure and ecological and environmental functionality, provided natural regeneration and regrowth, aided or not by silviculture, can take its course. However, irreversible degradation and irreparable damage may be caused if wildfires disrupt the course of natural succession. The final arrested fire dis-climax is fundamentally different in structure and functions from primary or secondary forest ecosystems.

Very similar, if less disruptive, will be the consequences of conversion to uniform short-rotation monoculture. Such man-made crops eventually will develop instabilities which to control will require very high and increasing inputs of chemicals, manual labour, managerial skills and research. This applies particularly to monocultures in the predominantly evergreen moist tropical forest biome. The very climatic factors, high rainfall and temperature and weak seasonality, which cause the very high biological activity and production potential, also make the soils chemically, structurally and biologically fragile. The greatest asset of the average, loamy-clayey ultisol/oxisol soils on reasonably favourable parent material is their structure, which permits deep rooting and extends the space in which the water and nutrient cycling systems can operate. The maintenance of this structure depends largely on biological activities (exudates from roots and micro-organisms, physical and digestive activities of larger animals such as earthworms, termites and mites, root growth and death e.t.c.) and supply of ample and divers litter. Changes in the microclimate, hydrology, rooting habits and litter properties as they happen, if plantation are established and managed as monoculture, carry the risk of initiating soil degradation which through a latent phase may eventually develop a phase of chronic damage to acute collaps. Particularly dangerous are plantations of species which possess features of leafage and crown and canopy architecture, such as a sclerophyllous monolayered uniform canopy of great height, and ground cover (no understorey, little soil cover) which may not reduce but increase the impacts and stresses on the soil after canopy closure. Examples are among others some pine and eucalypt species. The use of such species and stand structures must be avoided if soil and site protection are desired targets.

Environmentally, the major effects of change of forest structure and deforestation are on water quality, water discharge and flow patterns from catchments to populated plains, on sediment yield especially in regions of high relief energy, and on the atmospheric conditions. A major concern in the latter respect is the effect of exploitation, silviculture and forest land conversion on the carbon dioxide and moisture contents of the atmosphere. While the former is more of global relevance, the latter possibly is of great significance in cases where large tracts of forests are modified, converted or deforested in the path of moisture carrying winds. The lower evapotranspiration is suspected to reduce the recycling of moisture and consequently the rainfall in the more distant, possibly already drier, areas. Examples of most likely major effects in this direction are China and West Africa, and potentially Amazonia.

5. RATIONALE FOR CONSERVATION AND MANAGEMENT

The removals of logs for industrial use from tropical forests has since the end of the second world war continually risen (Table III). This trend is forecasted to continue far into the next century. The same applies to removals of timber and fuelwood for domestic use. World-wide wood demands and supplies are expected to rise. In many areas gaps between supplies and demands will develop and existing gaps widen as a result of local deforestation, distance from source and poverty. This development will increase existing problems of supply to critical proportions in many parts of the tropics, especially in the more densely populated parts of the less densely wooded dry tropics, but also in the moist deciduous and locally in the moist evergreen forest biomes. It is therefore very likely that demands for tropical wood production and wood removals from tropical forests will remain high and may even steeply increase. It seems, therefore, good policy to aim at the highest level of wood production in the eventual permanent productive moist forest area (Table II) which is compatible with ecology, economy and the environment. This means preserving and improving the productivity of tropical moist forest vegetation, of their soils and sites at the lowest possible costs in a micro-economic, national economic and social context. The first measure in this direction is to stop the presently ongoing, unnecessary destruction of the forests and their sites caused by expanding agricultural activities, and the forest degradation by careless and wasteful timber exploitation. Better adapted and more efficient agricultural and logging technologies are available and should be adopted. Secondly, the remaining permanent forests and degraded land allocated the forestry must be put to full use under efficient silvicultural management in order to produce the large quantities of wood which are needed in addition to the essential environmental benefits.

Tropical forest conservation and management are desastrously far behind needs. At present, 25% of the biome's land surface are covered by 11.5 mill. km^2 closed evergreen and deciduous forest. Of this only 0.4 mill. km^2, or 3.6%, are under formal forest management plans (Table IV), almost exclusively in India and Malaysia. But only a small portion of the management plan areas are probably effectively silviculturally managed in practice. The question is how far exploitation without formal management procedures of utilization and silviculture is a threat to sustained productivity and environmental functionality of the forest resource. The answer to the question is with all likelihood that well-planned and careful harvesting will leave soil and vegetation capable of quick recovery of protective functions. Securing full production in terms of economic goals, however, requires silvicultural interference to release and guide regrowth. The answer will be the reverse, if the exploitation is done in the usual haphazard, inadequately planned fashion, characterized by inordinately heavy machin-

TABLE IV
Areas of Closed Natural Broadleaved, Coniferous and Bamboo Forests and Their Management Status at the End of 1985, in mill. km² (Extracted from FAO 1982 a and b).

		Productive forest				
Category	**Total forested**	**Total**	**Virgin**	**Logged**	**Managed**	**Unproductive**
Mill.km²	11.5	8.5	6.4	1.7	0.4	3.1
Percent	100	74	56	15	3.6	27

ery, careless non-directional felling, unsystematic placement of skidtrails and unnecessarily destructive hauling. The recovery of soil and vegetation is a slow process and long lasting erosion and site degradation affects a large portion of the area. The seriousness of the situation is aggravated if lack of management and protection invites intrusion of wild-fires, shifting cultivators and semi-settled squatters. In this event erosion, deforestation and land degradation could become a serious challenge not only to the value of the forest resource, but also generally to environmental and social stability, if large tracts of land are affected, as it commonly happens.

In a review of the role of forestry in the economies of newly independent nations in the tropics, WECK (1955) recognized 4 major types of silviculturally managed tropical moist forests:

1. intensive industrial timber plantation
2. montane protective and productive forests
3. lowland mixed forest (e.g. Malesian Mixed Dipterocarp forests)
4. afforestations on degraded land near cities.

The objectives and priorities are different in each case. Intensive industrial timber plantations must be restricted to good soils and almost flat sites and are linked with definite projects of forestry industry. They are created to supply defined products for specified purposes at short (5—10 yrs) to medium (10—20 yrs) time scales of economic planning. Therefore, they are usually narrowly optimized monocultures with short time horizons of planning and expectations. The consequences are described in sect. 6.

However, for industrial development projects at the usual time scales of technical and economic planning in the order of 10 to 30 years such monocultures usually meet the requirements as reliable source of raw-material. In addition, opportunities to take advantage of the manifold benefits of pioneer industry status and to obtain tax reliefs both in the developed country, where capital usually comes from, and in the developing country, where the plantations are established, make such afforestation schemes financially attractive even if silviculturally they would fail. Another aspect which rationalizes reforestation by planting is the more recent phenomenon that planting is made a condition of timber licenses. In such cases, the planting and its costs are by the licencee regarded as part of the conditions of the timber exploitation agreement and as such considered a cost of exploitation rather than an investment to earn returns. In both cases, longterm considerations of sustained yield and functionality do not apply as criteria of success.

Montane productive and protective forests serve primarily hydrological functions. The canopy extracts moisture from the air by combing drifting clouds, effectively intercepts and stores precipitation. The phytomass and soil store water, regulating surface and groundwater discharge to the lowlands. This requires maintenance of an aerodynamically rough canopy and dense vegetation of high combing, filtering and storing efficiency. This excludes any but cautiously selective silvicultural systems which are economically viable if the relatively low yield is obtained at correspondingly low cost, restricting logging to well accessible sites and easy terrain, and particularly excluding steep slopes. The functionality of montane protective forests much depends on the success of maintaining the canopy physiognomy and structure and the potentially deep rooting of the natural forest.

The "climatic climax" lowland mixed forests on average soils (well structured ultisols and oxisols, good rootability, rooting depth between 2 and deeper than 4 m) on well drained sites in undulating to hilly country are capable of high rates of production during dynamic building-phases which can be guided by silvicultural treatment to produce economically potentially valuable crops at very low cost. Condition is, that the prior exploitation was well planned and cautiously executed. The productive function does not antagonize the protective and environmental functions. Especially the maintenance of edequate hydrological and edaphic conditions is possible with little or no yield sacrifice and no additional cost. All that is required is to protect and guide the natural development of regrowth into a divers, complex crop of potentially useful trees. More is said about this in sect. 7.

The last of WECK's four types is primarily aimed at rehabilitating degraded sites. Adequately protective soil cover, adequate litter supply and restoration of organic matter and biological activity in the soil is in most cases of degraded sites most effectively achieved by natural succession. Succession may be aided by planting pioneer shrub and tree species in recolonization nuclei at strategy points. The essential condition for success of such approach is the removal of the external disturbing factor or factors which have originally been responsible for the degradation and prevent recovery from the arrested disclimax stage. Most commonly these are fires laid for hunting purposes or pleasure, or escaping from burning agricultural fields. In the dry tropics particularly, fodder, fuelwood and litter collecting and grazing are major degradation factors which also prevent natural rehabilitation. If the objective of rehabilitation is primarily to contain erosion, the establishment of a dense grass cover may be sufficient and more adequate than blanket afforestation with pioneer tree species. Planting of pine, eucalypts or other pioneer species on degraded land is justified if a major immediate objective is fuelwood and timber production. But their protective value is often low, sometimes even negative, and their hydrological effects often undesirable, particularly on deep soils with great water storage capacity or access to groundwater. For multiple purposes and long-term environmental and economic functionality, the silvicultural stand type must be more complex, species-rich and divers. The silvicultural approach in this case again should utilize the natural dynamics of succession to achieve the silvicultural target efficiently and effectively. Fig. 3 illustrates the general features of such approach.

In summarizing we may conclude that the expected future demands for forest products, especially timber, fuelwood and locally special minor products, such as fruits, will reach levels which require the orderly management of the remaining permanent productive forest area. Together with the unproductive forest area on mountains, hills and infertile soils this permanent forest area will possibly be sufficient to fulfill the protective and environmental functions adequately, without which climatic and hydrological conditions on the agricultural lands are in danger.

The area of mostly already deforested fertile land (alluvial flood-plains, upland sites on rich parent material) and land of medium quality, but suited for adapted forms of agriculture, is adequate to meet land requirements of agriculture. However, technological and social improvements and financial assistance will be needed to reclaim the degraded and under-used lands. Therefore, the present apparent conflict for land between forestry and agriculture can and must be solved to mutual advantage if the permanent forest lands are to be successfully protected and adequately managed and the agricultural lands rehabilitated and better used than at present.

6. OPTIONS IN PLANTATION FORESTRY

Intensively managed tree plantations in agriculture (rubber, oil-palm, fruit trees etc.) and forestry (short-rotation pulp-wood plantations, medium-rotation combined pulpwood and saw-timber plantations) have been since about 200 years traditionally designed world-wide to maximize narrowly defined yield functions. This has the following consequences:

1. Narrow optimizing of crop design targetted at unrealistically high production levels and ignoring risks;
2. "unexpected, unpredictable" events occur (disturbances caused by biological, climatic, technological, economic social changes or extreme episodic events);
3. the narrowly optimized ecosystem cannot buffer and adapt because it is not sufficiently elastic and flexible;
4. damage and degradation of crop and soil cumulate;
5. internal swinging tendencies feed-back with exogenous impacts and amplify the system's reactions under stress;
6. the system's condition becomes instable and performance erratic;
7. ad-hoc "problem-oriented" sectoral research cannot find solutions, but produces only further sectoral, isolated repair strategies which fail;
8. as a result, silvicultural management stumbles from one emergency to the other, management is determined by catastraphies rather than the manager;

FIGURE 3. Two fundamentally different approaches to plantation design. On the left the traditionally narrowly optimized monoculture of pine for pulpwood and sawtimber production. On the right biocybernetically designed crop which uses the natural dynamics of the trees and vegetation communities to achieve objectives fast and safe. The design applies the basic laws of interactions between structure (tree and leaf shapes, canopy architecture, rooting depth and intensity) and functions (evapo-transpiration, productivity, protection) to achieve better performance at lower risk of damage to the crop and of degradation of the soil. The high degree of biological automation and self-regulation keeps silvicultural costs low. The example is a succession series of pine, broad-leaved species, conifers such as Araucaria spp., which may be followed again by clear-felling and pine planting, or be naturally regenerated.

9. finally, the ecosystems at forest stand and at management unit level loose their viability and eventually reach a state of ecological and economic collapse and a general state of chaotic confusion.

The risks and hazards of monocultures are known in Amazonia since the early colonial period. The cultivation of sugar cane in northeastern Brazil in the 17th century proved the nonsustainability of such type of monoculture. OLIVEIRA COSTA (1983) reports that "rarely would two generations pass without a plantation changing location or owner". Similarly discouraging has been the experiences with rubber, ginelina, Acacia and pine plantations which failed as a result of inherent instability, ecological neglect and poor management and policies.

Underlying these and similar failures and their continued repetition is the ignorance of the properties of complex dynamic ecosystems and their behaviour in a variable environment. System dynamics are not understood, sensitive interactions are not realized, the conditions of dynamic stability not known and the interrelationships between natural, economical and social ecosystem levels not seen and understood. Alternative designs of plantations are available which are more likely to perform ecologically and economically more satisfactorily by applying basic bio-cybernetic principles (BRUENIG, 1979, 1984, 1985):

1. The functionality of the crop is improved by complexity and aerodynamic roughness adjusted to site conditions and independend of volume growth;
2. broadly defined production targets to achieve versatility in preference over single-purpose products (e.g. one large tree of 1 m^3 volume is better than 2 trees of 0.5 m^3);
3. versatility and adaptability is further improved by combining suitable and compatible species in such manner that each finds a niche in which to perform adequately in relation to the management objectives (e.g. mixture in storeys or in succession);
4. growing space of each species is regulated in such manner that the target is achieved at low cost and in the shortest possible time;
5. mixture and complexity aims at the highest possible rates of biological activity and turn-over of energy, water and phytomass, vigorous recycling and effective self-regulation, combined with ecological and economic elasticity and adaptability;
6. as much as possible of the water cycle should be allocated to effective transpiration to support productive and protective functions by means of complex structure, deep rooting and appropriate choice of species and a design of canopy architecture which optimizes the atmospheric exchange processes.

In the example in Fig. 3, the narrowly optimized crop produces theoretically the highest volume yield of pine, but suffers losses from damage and eventually degradation of the soil and reduced functionality of the ecosystem as a whole. On the right, a combination of mixture and succession of commercial tree species (e.g. pine, broadleaved species, Auraucaria) as main crop, grown as fast as possible to a desired size, opens many options of utilization. Soil cover and auxillary crops (e.g. nitrogen fixing ground-cover species, shrubs and trees producing ancillary products) further increase versatility, resilience, elasticity and high level of performance which reduces the risk to fail ecologically, silviculturally and economically.

A special case are afforestations of bare, usually degraded land. The option of natural succession as alternative to planting if site protection is the objective has been discussed in sect. 55. The need to consider the external social and political conditions is particularly evident not only with respect to identifying linkages which have been responsible for the deforestation and site degradation in the past and may still be active, but also with respect to unexpected side-effects in future. Large-scale fuelwood plantations in the area of the deciduous dry tropical forest may not only put stress on the hydrological regime and exhaust water reserves, but may also, by removing contraints in the energy and cattle feed sectors, cause increases in population and lifestock densities which eventually will lead again to instability, strain and collapse, only at higher levels of suffering.

7. FEASIBILITY OF NATURAL SILVICULTURE

One problem has caused and still causes concern with disagreement among tropical silviculturists: "what kind of crop are we to grow, which species should we consider acceptable, what are the criteria of sustained yield". On the answer depends the future of tropical moist evergreen forest as an ecological, economic and environmental resource, more specifically the verdict on the feasibility of natural regeneration and natural silviculture. The narrower the production goal is defined, the fewer species are accepted, the less likely is natural regeneration to be considered feasible. In turn we may ask ourselves how far it is feasible and justified to:

—insist that the future crop contains the same volumes and assortments of the same species as the original virgin forest;
—restrict the list of acceptable species to those which happen to be current by considered to be commercial and merchantable?

In searching for an answer we must realize the essentially indeterministic and dynamic nature of the forest ecosystems on one side and equally of the economic and social-political systems on the other. There are plenty of examples of fundamental changes of crop performance and in the technological and market sectors which happened suddenly, unexpectedly and to some extend unpredictably. The very heavy, naturally durable hardwoods declined in desirability except for some rare cases of very special prime timbers such as *Eusideroxylon spp.* in Borneo and *Ocotea spp.* in America. The medium light hardwood Ramin (*Gonystylus bancanus* (Miq.) Kurz.), was poison-girdled in Sarawak/Borneo as a useless weed species in favour of the chicle-producing jelutong (*Dyera lowii* Hook.f.) before the second world war. After the war, Ramin became the number one export timber while Jelutong lost its market to substitutes from tropical America and to synthetics. But then, another problem arose. After logging and contrary to expectations, intermediate residual trees and seedling regeneration of Ramin, which is a gregarious species of the most peripheral and dynamic phasic mixed peatswamp association and therefore was expected to be aggressive, did not respond and perform to expectations. The dilemma of silvicultural planning in this as in other very dynamic, complex forests is to decide whether we should trouble to enforce regeneration, perhaps by artificial means, or should we conform with the natural trend of the forest regrowth and leave development to nature and chance. How far should we reduce chance effects and influence this development toward certain mixtures of species and sizes to reach the broadly defined production goal while still keeping costs and ecological and economic risk low. Why insist on certain species and assortments at higher costs if there is great risk of silvicultural failure and an unpredictable future market prospect?

In a situation of such uncertainty the most rational approach is to choose the alternative which is the least likely to go wrong, that means which requires less investment, produces an adaptable crop of technologically versatile timbers of good size at low risk. This is a mixed crop which grows rapidly after cautious, little damaging first exploitation to an ecologically well functioning and self-regulating forest in which commercial production is, by silvicultural treatment, concentrated on timber trees of medium properties but good dimensions, from which overmature relics, defective trees and technologically disadvantaged species are eliminated as far, but not farther, than the release of major crop trees requires. This broadly defined objective, and to do just enough to achieve it, but not less nor more, is the rationale for natural silviculture in natural tropical lowland forests.

As we have to make decisions in face of an unpredictable future, sustained silviculture of tropical forests must be based on the principle to choose the option which has the highest likelihood to prove the least wrong. This means as we have already argued in the previous sections:

—broad optimization to assure flexibility and adaptability;
—diversity of structure and functions;
—function orientation instead of narrow product orientation;
—low input requirements for silviculture and management.

These conditions are met by crops which develop after cautious logging (uniform or selective) of virgin natural tropical closed forest. An important advantage of broadly optimized natural silviculture for long-term development planning is that it requires low initial investments, spreads risk over many species and functional pathways and leaves many options open for later adjustments to changed conditions. It produces flexible, adaptable and resistent stands which effectively buffer the effects of ecological, economic and social perturbations, thus economically balancing the effects of their possibly but not necessarily lower volume increments of certain species. GRAAF (1982) considers natural silviculture with a cautious girth-limit selection system and low-cost silvicultural treatment economically viable even at the very low annual com-

TABLE V

Estimated Harvestable Mean Annual Increment in Field Practice. Natural Silviculture of Good Quality Saw-log and Veneer Timber of Presently Commercial Species in Two Types of Mixed Dipterocarp Forest (MDF) in Peninsular Malaysia and One Type in Mindanao, Philippines Under the Malayan Uniform System (MUS, 70 yrs Rotation) and Under Selection-Improvement System/Tebang Pileh Indonesia (SIS/TPI, 30 yrs Rotating Felling Cycle) and for Comparison High-yield Pine or Broadleaved Plantations on the Same Site. Plantations: the M. A. I. for Plantations Refers to Large Tracts on Suitable, Good Sites and Efficiently Managed Pulp-plus-saw Log Regimes Assuming Unavoidable Short-comings of at Least 20–30 % from Site and Species Potentials for Wood Production. The Maxima Refer to Prime Forest Sites and Excellent Management.

	Natural Silviculture		Plantation	
Forest Type	MUS R = 70 years	SIS/TPI R = 30 years	Pines	Broadleaves
Single species dominated				
MDF, Malaya	7	2.4	10–25	15–35
Meranti-	(max. > 15)	(max. > 10)	(max. > 40)	(max. > 50)
Keruing MDF; Malaya	7	3.5		
Mixed MDF, Mindanao	7	3.8		

mercial volume increment of about 1 m^3 which he regards realistic at 20 yrs cutting cycle on poor, acid sand in the Mapane forest in Surinam. Very much higher increments and yields can be achieved and sustained on better soils and sites with commercially advantaged forest growing stock (WDELL, 1989).

Potential timber yields from natural silviculture of mixed forests are usually underestimated while yield expectations from monocultural tree plantations are usually overly optimistic. Common mistakes are to confuse potential biological net primary above-ground wood matter productivity and realizable net merchantable wood production. The difference may be 2 : 1 due to climatic variation, pests, disease and abiotic damages. In plantations, choice of unadapted genetic stock and poor management may increase the ratio to 3 : 1 or more. The potentials and limitations of humid tropical trees and forests to produce timber have been critically discussed by DAWKINS (1963). BRUENIG (1967, 1971) assumed sustainable annual timber increments between 2.4 and 7 m^3 in naturally regenerated lowland Dipterocarp forest on average forest sites (Table V), while KIO (1983), based on field experience, considers 10 m^3 feasible. Taking the mean annual yield figures in Table V as reasonable, rather conservative questimate, very simple linear programming of resource allocation shows that natural regeneration is feasible, but that an optimum use of land and capital is achieved by a combination of natural regeneration and planting. After first logging the whole area, a larger portion of the forest land is naturally regenerated at low cost, while a very much smaller portion on the most suitable sites is planted at high cost with high-yield species. The ratios between the two alternatives is decided by the ratio between costs and revenues considering ecological and environmental constraints.

Still higher economic and environmental benefit are obtained if the whole area of logged forest is naturally regenerated and the plantation is established on already deforested waste land outside the forest area, using the capital which is left after completely regenerating the logged area naturally.

The choice between MUS (Malayan Uniform System) and SIS/TPI (Selective and Improvement Systems) depends on the original growing stock structure, the capability and success to reduce logging damage to the residual stand and the soil, and finally on the species preferences and the interpretation of the principle of sustained yield. A comparative analysis of the financial criteria "annual forest rent" = net cash flow, "opportunity costs" and "net discounted revenues" of MUS and SIS/TPI gave the following results (BRUENIG, 1967). Creaming under the SIS in Mindanao with no prescribed limitation on the annual felling area gave highest net cash-flow to the logging company but a low cut and low subsequent increment per hectare. On the other hand, the SIS/TPI systems require much more land for the same volume yield than the MUS. Also, they involve high social costs because potentially merchentable growing stock is retained for another 30 to 40 years which could have been harvested in the MUS. Under Malayan lowland forest conditions, the highest net cash-flow return per hectare and for the whole annual felling area is obtained in the MUS if the yield is controlled by area. It is therefore understandable if companies prefer the SIS in Mindanao and the MUS in Malaya. If the annual cut is only controlled by volume, limited by working plan prescription, company's capacity or market conditions, the highest annual net cash return to the company is obtained by application of the SIS/TPI system, even if the cut per hectare is fairly low, for example below 20 m^3 ha^{-1}. The example demonstrates the close interdependence between forest structure, economics, governmental policy and choice of logging and silvicultural system.

In summarizing we may conclude, that natural silviculture is a fully feasible approach provided logging is well planned, adequately supervised and cautiously executed and provided a realistic and rational view is taken of future prospects. It's production potential is not only determined by the variation of site and stand (ASHTON and BRUENIG, 1975), but also by the variable and changing climatic, economic and social environmental conditions. Future sustained production per unit forest area and of the whole productive tropical forest area (Tables I and II) are difficult to assess because:

1. the future states of timber markets and wood technologies are unpredictable and it cannot yet be said which species, timber sizes and qualities will be in demand;
2. the size of forest area available in the future is uncertain;
3. macro-climatic conditions may change;
4. air-pollution may cause yield increases or damage to trees and soils, which cannot yet be predicted;
5. changes in silvicultural and exploitation technologies may affect net volume and money yields.

8. INTEGRATED FOREST MANAGEMENT AND CONSERVATION

NEIL (1981), elaborating on a suggestion by DAWKINS, distinguishes 5 phases of development of tropical rainforest management:

1. Before 1850	Pre-Management
2. 1850—1900	Indo-Burman (Franco-Teutonic)
3. 1900—1950	Afro-Malaysian
4. 1950—1980	Pantropical-Exploitative
5. 1980—onward	New Phase (system-oriented)

Similar to shifting cultivation in pre-historic times, tropical forest management and silviculture in the 19th century (phase 2 of NEIL) has begun and developed in the dry deciduous and moist deciduous forests where conditions for exploitation and natural regeneration are easier than in the evergreen moist forest. From India and Burma, tropical silviculture and management spread to eventually all other countries which possess tropical forests in Asia, Africa and America. From the start in India and Burma, the environmental and social functions of forestry were realized and motivated the formulation and implementation of forest policies which aided at land-use integrated forest management and conservation.

As well considered and vigorous these 19th century forest policies were, their implementation met with increasing difficulties and resistance as the 20th century went by.

HESKE (1931), as many other foresters before and after, argued that forest management both for sustained forest yield and sustained environmental forest functions in tropical countries was essential not only to secure tropical timber supplies to the world, but also and particularly for the benefit of the tropical countries themselves. However, forest management continues to lag far behind the progress of logging and deforestation. The underlying causes for this obvious failure to develop adequate action from available scientific knowledge and practical experience are primarily political. Most former colonies possess, often very adequately formulated, forest policies, forest ordinances and forest rules. Their application, however, is obstructed by socio-political factors. Since the fifties, politicans in the tropical world have increasingly shifted priorities away from agricultural and forestry development to prestige and power oriented industrialisation, military armament, political power struggle and finally war-making. In addition in recent years politicians acquired active interests in forest exploitation and forest land titles. The consequences are decline of agricultural production and acceleration of deforestation. This is amplified by the side-effects of controlled low food prices, foreign food aid and ill-conceived governmental agricultural policies. The decline in most African countries exemplifies the effects of these interacting mechanisms. The consequence to the forest resource is desastrous. The rate of deforestation by slash-and-burn agriculture increases and is accelerated by ruthless and careless exploitation of forests for timber. While timber companies unwillingly render a social service to the landless poor by providing easy access to desparately needed fresh land, they also open the road to deforestation to their own eventual disadvantage in the long run. Agricultural plantation and rangeland developments add locally to forest destruction. Globally, however, shifting cultivation is the major destructive force and the least accessible to control.

The other political obstacle to implementation of a policy of sound forest management is insecurity of land tenure. Uncertainly of future opportunities makes timber companies refrain from investing care and money into forest protection and silviculture. In addition, timber companies are increasingly owned by self-interested nationals, often politicians, members of government or government agencies. The combined effect is that forests are not managed but mined, that careless logging continues to cause excessive loss and damage to timber, forest stands, climate and soils, and what is left falls to the landless poor shifting cultivator. The effective implementation of land-use integrated sustained forest management and forest conservation is therefore more a moral and intellectual issue at the political level than a problem of science and technology. We know now enough about the interactions between forests, forestry, natural, economic and social environment, of the societal long-term requirements and of ecological, economic and social side-effects of forest destruction and the benefits of sound, balanced forestry developments to be able to formulate long-term concepts of integrated development of forestry, agriculture and industry. The complex interactions, interdependies and critical feed-back loops in the network which forests and forestry are embedded (Fig. 4), are reasonably understood, sufficient at least for basic policy decisions. The major obstacle is the very unwillingness for very personal reasons of those in power to draw the conclusions, accept the consequences and take the necessary action.

At the micro-economic and technical level, the implementation of integrated approaches to land-use development also is primarily a question of successful education and training of those who participate and are immediately affected at company, village or farming unit levels. In comparison it is much easier to solve the mainly technical problems of how to analyse the potentials of silvicultural and agricultural, or combined production systems, and how to design stable, functional and efficient cropping and land-use systems. It is more difficult to find efficient ways of implementing plans effectively. The reason is that this involves basic problems of attitude, understanding and acceptance by the people affected and involved at all levels of the socio-political hierarchy. The general pattern by which forest conservation, forestry management and the various agricultural cropping systems can be harmonized at the tropical landscape level is illustrated in Fig. 5. However, in formulating problem-solving technical strategies and in making the appropriate planning provisions for forestry development, immediately a score of conflicts become apparent of economic, social and political nature. The network of major conflicting interests, expectations and essential needs as seen from the point-of-view of the forest land and of the private or public forest owner is demonstrated in Fig. 6. Both examples in the 2 illustrations primarily refer to predominantly evergreen moist tropical forest but also applies in principle to the moist and to the dry deciduous forest biome, where additionally desertifying effects of overgrazing and fuelwood collecting are major issues and mounting threats to human survival which require special solutions and undelayed action. Figure 7 illustrates rates

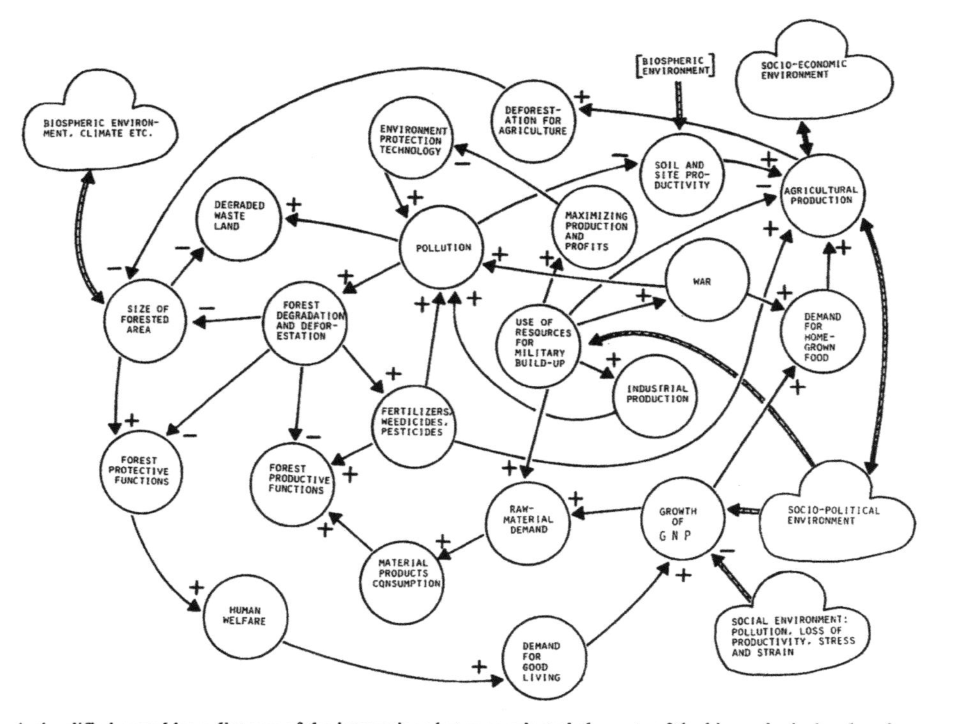

FIGURE 4. A simplified causal loop diagram of the interactions between selected elements of the bio-ecological and socio-economic ecosystems which are involved in the process of deforestation and forest land degradation. Mayor driving forces are human demand for wealth, political quest for power, both leading to preference for industrial prestige and profit-generating projects and military build-up at the expense of investing into measures to reduce pollution and to improve agricultural perfomance, which in turn increases deforestation and forest degradation, which adversely affects the national economy (less timber), the environment and the agricultural productivity.

+ increase of the issuing element increases the receiving element; —the elements are counteracting, if one increases, the other decreases; cloud: external system which influences the respective element.

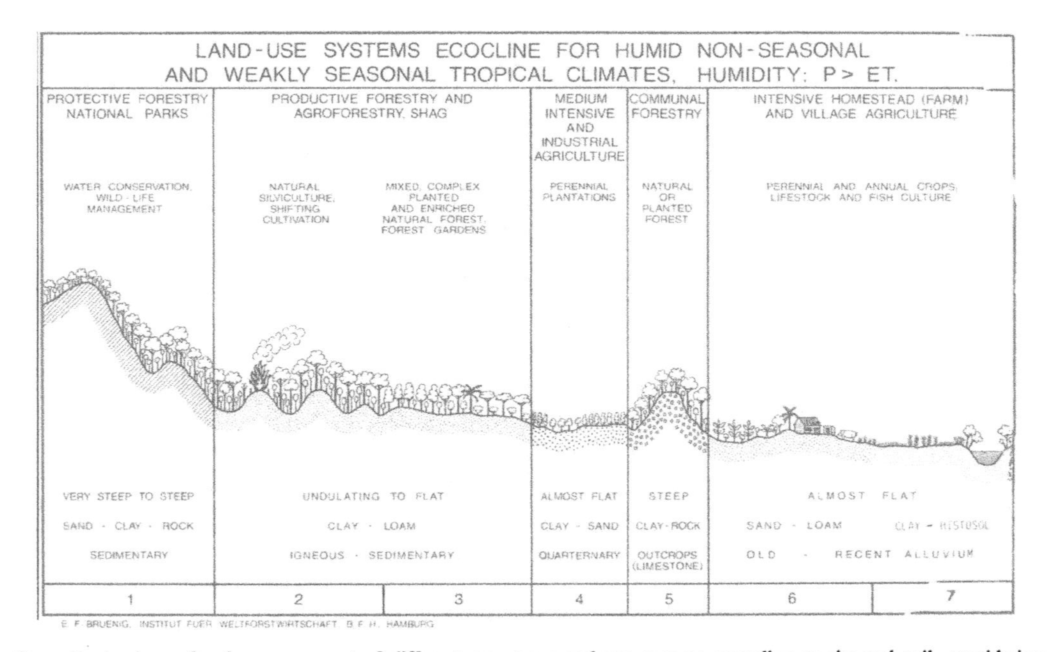

FIGURE 5. Generalized scheme for the arrangement of different crop types and use systems according to site and soil, considering economics of infrastructure and accessibility, preservation of environmental conditions and adequate cost/benefit ratios at landscape unit level.

1: absolute forest sites, only protective forestry is admissable on the ulti-oxisole soils on steep slopes and spodosols on steep scarps and gentle dip-slopes; 2: relative forest sites, system choice depends on landform, physiography, soil type; typical region in which currently shifting cultivation is practiced to the detriment of sites and environment. This land-use should be replaced by more adapted forms of agriculture, agroforestry and forestry; 3: relative forest sites on landforms and soils suitable for more intensive forestry and agroforestry, including more complex and adapted forms of tree plantations; 4: amost flat sites with good soil conditions, fully suited for intensive agroforestry and perennial crops in agriculture but usually too distant for very intensive cultivation of annual crops; 5: absolute forest site suitable for pretective forestry and adapted forms productive communal forestry by natural silviculture; 6: high-productivity sites, as in 4., but but the easier access makes intensive agriculture more feasible, including annual crops; 7: very high-production sites with potential for irrigated agriculture, trees planted around houses and in shelterbelts for fuel, minor produce, soil improvement and shelter; on the river banks broad protextive forest strips are maintained to protect the river, river bank and fields.

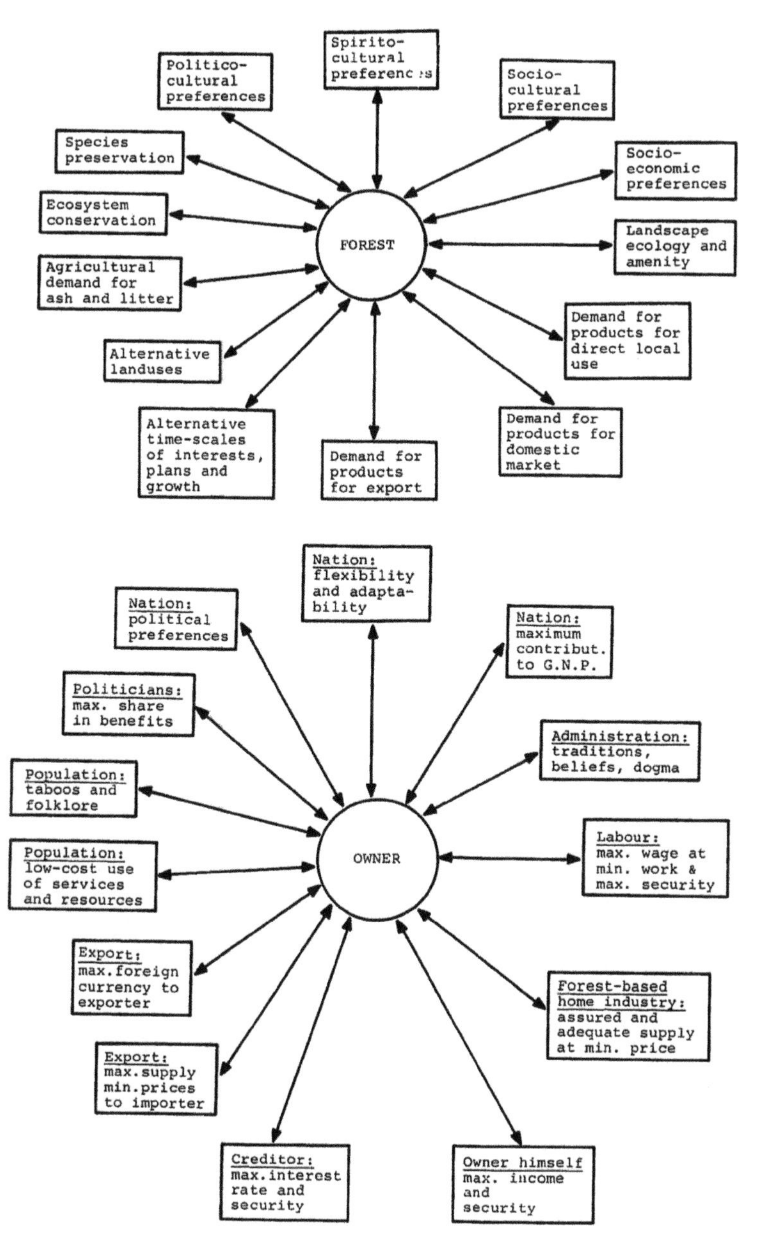

FIGURE 6. The conflicts between the forest (top) and the forest owner (bottom) and some of the divers and among themselves competing own and external interests and limiting constraints (adapted from BRUENIG, 1971).

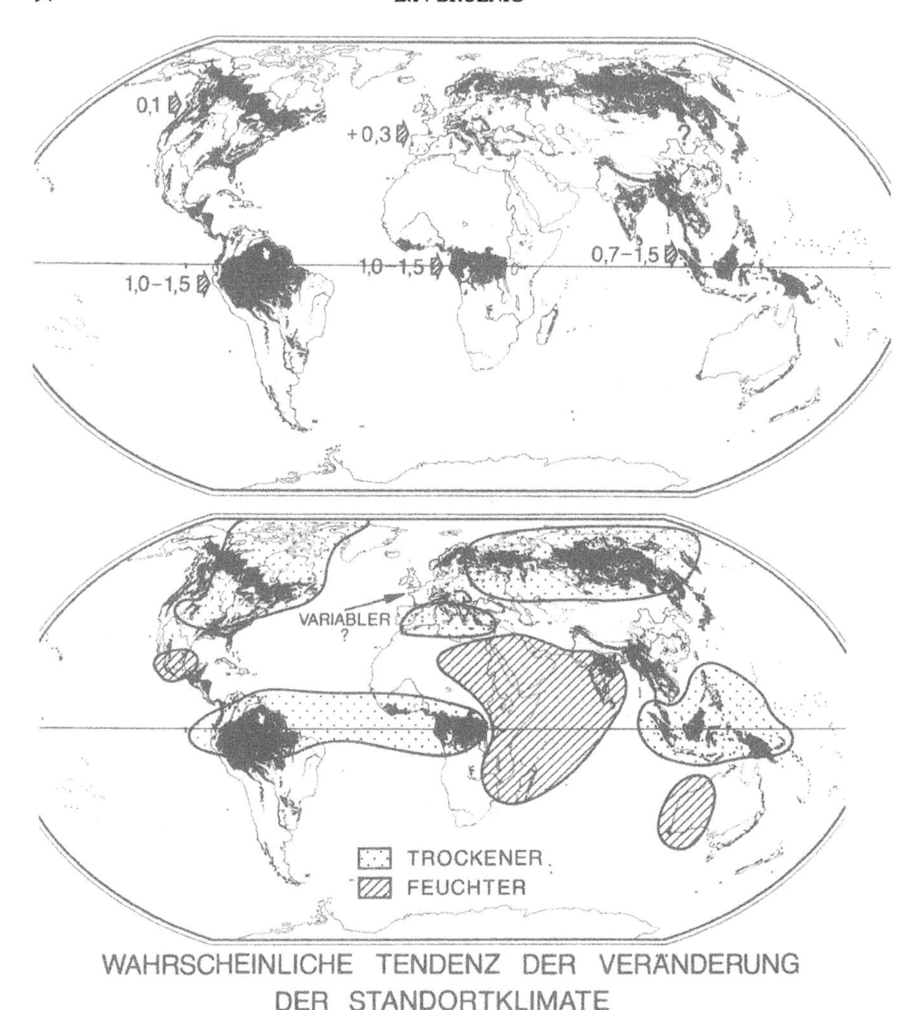

FIGURE 7. Actual forested area (black) and ranges of national rates of change of productive natural forest area in the tropics as a result of human activities (top) and possible trend of the impact of climatic changes due to deforestation and air-pollution on the moisture regime of site climates (bottom, adapted from KELLOG and SWARE, 1983).

of change of productive natural forest areas as a result of human activities and the probable impact on climate as a result of deforestation and air pollution on moisture regimes of site climates.

REFERENCES

1. Ashton, P. S. and Bruenig, E. F. (1975). The variation of tropical rainforest in relation to environmental factors and its relevance to land-use planning. Mitt. Bundesforsch.anast.Forst-Holzwirtsch. No. 109: 59–86.
2. Bruenig, E. F. (1966). Der Heidewald von Sarawak und Brunei (The Heathforest in Sarawak and Brunei—a study of its vegetation and ecology). *Habilitation Thesis*, University of Hamburg, pp. IV + 117.
3. Bruenig, E. F. (1967). Financial aspects for planning the conversion of natural Mixed Dipterocarp forest by natural regeneration or plantation establishment. In: DSE (ed.) *Report on a seminar "Planning of Forestry and Forest Industries in Tropical Regions"*, Manila, Cyclostyled, pp. 31, (available at Institute for World Forestry, Hamburg).
4. Bruenig, E. F. (1971). *Forstliche Produktionslehre*, H. Lang Verlag, Bern—Frankfurt, 318 S.
5. Bruenig, E. F. (1977). Leistungen und Leistungsgrenzen der Wälder der Erde, *Mitt. Bundesforsch. anst.Forst-Holzwirtsch.*, Hamburg, Nr. 118, 43—52, also transl.: Utilization of the world's forests: possibilities and limitations. *Applied Sciences and Development*, Tübingen, Vol. 14, 15–25.
6. Bruenig, E. F. (1979). The means to excellence through control of growing stock. In: *1st Weyerhaeuser Science Symposium*, Tacoma, April 30-May 3, 1979. Tacoma, Weyerhaeuser, 1980, pp. 201–224.
7. Bruenig, E. F. (1984). Designing ecologically stable plantations. In: Wiersum, K. F. (ed.). *Strategies and designs for afforestation, reforestation and tree planting*, Wageningen, Pudoc, 348–359.
8. Bruenig, E. F. (1985) Forestry and agroforestry system designs for stustained production in torpical landscapes. In: Embrapa/Sudam (ed.), *Proceedings of the First Symposium on Humid Tropics*, Belém.
9. Dawkins, H. C. (1963). The productivity of tropical high-forest trees and their reaction to controllable environment. *Thesis*, Commonwealth Forestry Insitute, Oxford.
10. FAO (1982 a). Tropical forest resources, *FAO Forestry Paper 30*, Rome, FAO, 106.
11. FAO (1982 b). Conservation and development of tropical forest resources, *FAO Forestry Paper 37*, Rome, FAO, 122.
12. Gonggryp, J. W. (1942). Die Holzzufuhr aus den Tropen nach Europa, *Intersylva 2*, 232–247.
13. Graaf, N. R. de (1982). Sustained timber production in the tropical rainforest of Surinam. In: Wienk, J. F. & Wit, H. A. de, *Management of low fertility acid soils of the American humid tropics*, Surinam, Dep. Forestry, 179–189.
14. Heske, F. (1931). Probleme der Forstwirtschaft in unterentwickelten Ländern als Lehr- und Forschungsgebiet, *Inangural Lecture*, Institute for Foreign and Colonial Forestry, Forestry College, Tharandt, *Thar. Forstl. Jahrb.*, 1.
15. Kellog, W. W. & Sware, R. (1983). Society, science and climatic change. *Dialogue*, 3, 62–69.
16. Kio, P. R. O. (1983). Management potentials of the tropical high forest with special reference to Nigeria. In: Sutton, S. L. et al. (eds.). *Tropical rainforest : ecology and management*. Oxford, Blackwell Scientific Publ., p. 445–463.
17. Kurz, W. A. (1982). Biomasse eines amazonischen Feuchtwaldes: Entwicklung von Biomasseregressionen, Hamburg, *Mitt. Bundesforsch.anst. Forst-Holzwirtsch.*, Nr. 139, pp. III + 98.
18. Neil, P. E. (1981). Problems and opportunities in tropical rainforest management. *C. F. I. Occasional Paper Nr. 16*, Oxford, Department of Forestry, Commonw. For. Inst., Univ. of Oxford, pp. 127 + 49.
19. Oliveira Costa, J. P. de (1983). History of the Brazilian Forests: an Inside View. In: *Ecological structures and problems of Amazonia*. IUCN, Commission of Ecology Papers, No. 5: 50–56 (reprinted from The Environmentalist, Vol. 3, Suppl. 5, 1983).
20. Schlich, Sir. W. M. (1922). *Forestry Policy in the British Empire*. London, Bradbury, Agnew & Co., XI + 342 pp.
21. Weck, J. (1955). Aufgaben und Möglichkeiten des Waldbaus im Gefüge der Volkswirtschaft autonomer Staaten in den Tropen. *Zeitschr. Weltforstwirtschaft* 18, 3:92–105.
22. Woell, H. J. (1988). Struktur und Wachstum von Kommerziell genutzten Dipterocarpaceenwäldern und die Auswirkungen von waldbaulicher Behandlung auf deren Entwicklung, dargestellt am Beispiel von Dauerversuchsflächen auf den Philippinen, Hamburg, Ph.D. Thesis, Univ. Hamburg; Mitt. Bundesforsch.anst.Forst-Holzwirtschaft, in press.

Resource Management and Optimization
1990, Volume 7(1–4), pp. 97–114
Reprints available directly from the publisher.
Photocopying permitted by license only.
© 1990 Harwood Academic Publishers GmbH
Printed in the United States of America

TROPICAL WILDLIFE RESOURCES

S.K. ELTRINGHAM
Department of Zoology, University of Cambridge, Cambridge, England

CONTENTS

Tropical wildlife, particularly monkeys, rodents and insects, is widely used for food by local people. Small antelopes and pigs are also trapped, especially in forests. Only the large mammals of the grasslands have a potential for commercial meat production, but attempts to crop wild ungulates in Africa have generally failed owing to problems in harvesting, processing and marketing. Game-viewing and sport-hunting are more profitable ways of exploiting wildlife, particularly in East Africa. The cropping of large mammals for their skins is potentially profitable, and there is a flourishing trade in leather made from the skins of Asian reptiles. There is also an international trade in live birds centred mainly in Asia and west Africa. The wildlife of tropical America is of little commercial value but in Australia there is a significant international trade in the meat and skins of kangaroos, many of which come from the tropical north. The principal threat to tropical wildlife in habitat loss caused by commerical forestry and agricultural development.

1. INTRODUCTION—THE HABITATS

The term "wildlife" has a number of meanings. Here it refers to all animals that are not domesticated. The tropics are particularly rich in animal life but only a small proportion of the species constitutes a resource in the sense that they are exploitable by man. However, uses may be found for animal products not now exploited. This potential of all wildlife is one reason for its conservation.

The habitats in which tropical wildlife occurs can be broadly, divided into forest and savanna. Forest generally refers to a vegetation type in which trees grow close enough together to form a closed canopy or to reduce the amount of light reaching the ground such that extensive grass cover is precluded. There are many definitions for savanna (Bourlière, 1983). Originally, it was considered to be tropical grassland devoid of trees. Although grasses are an essential component if a habitat is to be considered a savanna, most modern authors accept that trees and bushes may also be present. Bourlière & Hadley (1970) list three criteria in their definition of savanna: it is a tropical or sub-tropical vegetation type (1) in which grass cover is continuous and important although occasionally interrupted by trees and shrubs; (2) where grass fires occur occasionally; and (3) where the principal growing periods are closely associated with alternating wet and dry seasons. The extent of the cover in a savanna can vary widely and there is no sharp distinction between forest and savanna, which grade into one another. Alternatively, "savanna" may be rejected as a descriptive term and habitats named according to their vegetative composition (*cf.* Pratt *et al.* [1966]). Yet the term is too convenient to be abandoned completely.

The significance to animals of this broad distinction is that, in general, forests support browsers, which fed on broad-leaved vegetation, and savannas support grazers that eat grasses. The difference is not merely taxonomic but has relevance to the potential of wildlife for commercial exploitation.

As animal food the two plant types present different problems of palatability and digestibility. Plants have evolved numerous devices to deter animals from feeding on them. In general, broad-leaved plants rely primarily on mechanical and chemical defences whereas grasses depend more on rapid growth to replace tissues lost to grazers. Grasses can do so because of their continuously growing basal meristem that permits a leaf to regenerate itself. Foliage lost from a dicotyledonous plant can only be replaced by the formation of a new bud or the growth of a new shoot. Grasses can also lay down mechanical defences, such as silica, and secondary components to deter grazers, but the production of these substances is "expensive" and it is often cheaper for the grass to rely on rapid growth to cope with heavy grazing.

The structures of the two types of leaf also have a profound influence on the type of animals found in forest or savanna. The leaf of a dicotyledon is supported by the twig from which it has sprouted, and has little need for mechanical support. A grass leaf, on the other hand, must support itself. Consequently it is higher in fibre and other structural components. From the perspective of the consuming animal, browse is more easily digestible and is richer in protein compared with grasses, but because of chemical defences most leaves are somewhat poisonous and can be eaten only in small quantities, unless the animal has evolved enzymes to break down some of the poisons. Grasses, on the other hand, are abundant and generally non-poisonous, so that they can be eaten in quantity.

Such differences will obviously have profound effects on the ecology of the animals, but a distinction must be made between invertebrate and vertebrate herbivores, mainly because of their disparity in size. Whereas a large vertebrate can move easily from plant to plant and take a great variety of species so that it ingests only a small amount of any poison, an insect, such as a caterpillar, is usually restricted to a single plant and perforce must evolve an enzyme to deal with its poison. An insect could, however, store the poison during its short life span.

Among large mammals browsers must subsist on food that is in short supply but which is highly nutritious, whereas grazers must cope with an abundance of rather low quality food. Grazers, therefore, must be large to accommodate a large amount of poor quality food while it is broken down by micro-organisms in the gut, for no big animal can digest cellulose. Large animals cannot hide easily on grasslands so they tend to congregate into large herds for mutual protection. Thus the large, grasslands-dwelling mammals are most easily exploited by man, because they can be more easily harvested than the smaller, forest dwellers. There are some large browers, such as the black rhinoceros (*Diceros bicornis*), but most are small and solitary, living as they do in thick country. Consequently, they are not well suited to cropping. It follows that the savannas have a much greater potential than the forests for the human exploitation of their animals. The forests, of course, are of great value for timber, and they are by no means lacking an economic potential in their animal resources.

2. FORESTS

There are a variety of types of tropical forests. Tropical rain forests develop where the annual rainfall exceeds 2,000 mm and is evenly distributed throughout the year. Most trees are evergreen with leaves heavily protected from browsing by a variety of mechanical and chemical defences. The largest area of rain forest occurs in the Amazon Basin and there are other substantial areas in West and Central Africa, eastern Madagascar, Southeast Asia and smaller patches in western India, Ceylon and eastern Australia. Rain forests do not develop where rainfall is seasonal, even though the annual precipitation may reach or exceed 2,000 m. Instead, a deciduous forest develops. Where rainfall in less than that required to support forest the vegetation grades into open woodland or savanna. An altitudinal gradation from rain forest to open country occurs with increasing elevation in tropical mountains, although trees are limited by temperature rather than precipitation.

2.1 Forest Resources—Ungulates

The aboriginal human inhabitants of tropical forests formerly fulfilled their nutritional and economic need from the wildlife. Even today they depend heavily on wild animals for their regular food. Meat is obtained from a large number of small antelopes, rodents and monkeys. Of 183 carcases on sale along a roadside in Liberia, 70.5% were duikers, 12.6% primates, 10.4% antelopes, pigs and other ungulates, 2.7% small carnivores, 2.7% reptiles and 1.1% pangolins (Jeffrey, 1977). The only mammals of any size were a pygmy hippo *(Choeropsis liberiensis)* and two giant forest hogs *(Hylochoerus meinertzhageni)*.

Large mammals are hunted where available, thus, the pygmies of the Central African rain forests kill elephants *(Loxodonta africana)* and the okapi *(Okapia johnstoni)*, a large rain forest ungulate, which is restricted to eastern Zaïre.

Pigs are typical of the rain forest, with two species found in Africa and three in Asia. The African bush pig *(Potamochoerus porcus)* is not confined to the forest, but it is always found in thick country. The other African species, the giant forest hog is typically found at the forest edge. A third African pig, the warthog *(Phacochoerus aethiopicus)*, is more a creature of the savanna. The common wild boar *(Sus scrofa)* is one of the Asian species but it too occurs in other habitats. The other two, the bearded pig *(Sus barbatus)* and the babirusa *(Babirussa babyrussa)*, are mostly confined to forests. The pig family is represented in South America by the peccaries, one of which, the white-tailed peccary *(Tayassa pecari)*, is an inhabitant of the rain forests. All are hunted but none is important as an internationally exploitable wildlife resource. The pigs are of local economic significance, however, as most can be agricultural pests.

The hippopotamus is related to the pigs and although the common hippopotamus *(Hippopotamus amphibius)* is a grazer, the smaller species, the pygmy hippo, is a true forest animal. The latter is solitary and largely terrestrial, unlike its larger relative. It is, however, found near rivers, in whose banks it constructs short tunnels for rearing the young.

The chevrotains, a group of forest-living animals, show affinities with antelopes, deer and pigs, probably because they have changed little from the ancestral forms of those groups. The water chevrotain *(Hyemoschus aquaticus)* is the African representative, and as the name suggests is found near water. The three Asian species *(Tragulus* spp.) are smaller and are often called mouse deer.

As mentioned earlier, forest-dwelling mammals are usually small, partly because they are browsers and partly, no doubt, because small animals can move more easily through the thick vegetation of the forest floor. The African rain forests are characterised by a great variety of small antelopes, many of which are duikers of the genus *Cephalophus*. Other small African forest antelope include the 5.5 kg Bate's pygmy antelope *(Neotragus batesi)* and the smallest of all, the royal antelope *(N. pygmaeus)*, which weighs only 3–4 kg. A more normal-sized antelope is the bushbuck *(Tragelaphus scriptus)*, which is typical of the rain forest although it also occurs in wooded savannas. A much bigger antelope, unusually so for a forest species, is the bongo *(Boocercus euryceros)* an elusive striped relative of the bushbuck which weighs up to 225 kg. Like the okapi, these larger antelopes are striped or spotted, so that they are well camouflaged. Cryptic coloration is typical of forest animals and is in marked contrast to the uncamouflaged skins of the savanna-living herbivores. The small antelopes tend not to be patterned but they are camouflaged nonetheless, for their grey or rufous coats blend into the background. Rufous coloration is typical of forest animals and is seen in the bush pig and buffalo *(Syncerus caffer)* as well as in the antelopes.

Antelopes do not occur in Asian rain forests, their place being taken by deer and cattle. Other large mammals include tapirs and rhinoceroses. The tapirs form a curious group of primitive ungulates, distantly related to the rhinoceros. A curious feature is their remarkably broken distribution. The Malaysian tapir *(Tapirus indicus)* lives in Southern Burma, Thailand, Malaysia and Sumatra, whereas the only other tapirs are the three species of South America. Only two of the latter are forest dwellers. The third, the mountain or woolly tapir *(Tapirus pinchaque)*, inhabits the mountains of northwest Colombia. The tapirs illustrate the probable ancestral form of the perissodactyls and may have remained largely unchanged because of their stable forest habitat.

The three Asian rhinoceroses are predominantly forest animals, unlike the two African species which occur on grassland or savannas. All three are rare and only the Great Indian rhinoceros *(Rhinoceros unicornis)* is not threatened with imminent extinction. The Javan rhino *(R. sondaicus)* is most at peril, with only a few dozen remaining in one small reserve in Java. The situation of small Sumatran rhino *(Dicerorhinus sumatrensis)* is little better, but a few hundred remain in parts of South-east Asia.

The cattle family probably evolved in forests, and many species inhabit the rain forests. Only one species, the Cape buffalo *(Syncerus caffer)*, occurs in Africa. It lives both in forests and on grasslands and shows a wide range in size and colour. Rain forest cattle tend to be small and rufous whereas the plains buffalo are huge and black with massive horns. Two sub-species are often recognised with the forest buffalo, or bush cow, being assigned to the sub-species *nanus*. Grubb (1972) believes that there are three distinct groups of buffalo although they overlap in range and show a clinal gradation. These are the *nanus*-group of small forest buffaloes, the *brachyceros*-group of west African savanna buffaloes and the large buffaloes of the *caffer*-group on the savannas of East and southern Africa. It is possible that the latter two groups represent independent evolutionary attempts by forest buffaloes to colonise the grasslands.

The Asian water buffalo *(Bubalus bubalis)* is quite different. Most are domesticated but the wild forms are creatures of swamps and flooded grasslands rather than forests, although they rest in forests during the day. Related forest oxen are the anoa *(Bubalus depressicornis)* of Sulawesi, the smallest of all cattle, and the only slightly bigger tamarau *(B. mindorensis)*, on Mindoro, in the Philippines. (Small size is typical of island species of other mammals as well as of birds.)

Other forest cattle in Asia include the gaur *(Bos gauri)*, the largest of all cattle and which is widely distributed from India to Malaysia, and the more restricted banteng *(B. javanicus)*, found mainly in Java.

Several species of deer live in the forests of Asia. One of the commonest is the sambar *(Cervus unicolor)*, which is similar to but larger than the red deer *(C. elaphus)*. Another is the barasingha or swamp deer *(C. duvauceli)*, which tends to prefer wetter forests. The chital or axis deer *(C. axis)*, and the several species of the small muntjac deer *(Muntiacus* spp.) are other widely dispersed deer of the tropical Asian forests.

2.2 Mammalian Carnivores

Those large tropical ungulates support a range of forest carnivores. The largest is the tiger *(Pathera tigris)*, which is confined to Asia although not to the tropics, and the most ubiquitous is the leopard *(P. pardus)*, whose range extends from Africa to Asia. Its niche in the South American forests is filled by the jaguar *(P. onca)*. A slightly smaller cat, the clouded leopard *(Neofelis nebulosa)*, inhabitats the forests of Southeast Asia. Some cats, such as the leopard, occur outside the forests and others, such as the puma *(Felis concolor)*, may occur in forests, although their more usual habitats are in open country.

Two species of bear are found in the tropical Asian forests. They do not prey on the large ungulates but feed mainly on plant food, invertebrates or small mammals and birds. One, the Malayan or sun bear *(Helarctos malayanus)* is the smallest of all bears. Despite its name it is not confined to Malaya, being widespread throughout Southeast Asia. The other is the sloth bear *(Melursus ursinus)*, which occurs in mountains as well as in the lowland forests of India and Sri Lanka.

Numerous small carnivores inhabit tropical forests. South American examples include several cats, such as the jaguarundi *(Felis yagouaroundi)*, margay *(F. weidi)* and tiger cat *(F. tigrina)* and several members of the racoon family (Procyonidae), such as the coatis (*Nasua* spp.) and the vegetarian kinkajou *(Potos flavus)*. The weasel family (Mustelidae) is represented by the metre-long tayra *(Eira barbara)*, which has been known to kill deer. Small predators of the African rain forests include some primitive primates, the potto *(Perodicticus potto)* and several species of bush-baby (*Galago* spp.), as well as genets (*Genetta* spp.), civets (*Viverra* spp.) and palm civet *(Nandinia binota)*. There is even a forest mangoose, the kusimanse *(Crossarchus ansorgei)*. Viverrids and mustelids are also well represented in Asian forests, with such species as the spotted linsang *(Prionodon pardicolor)*, the oriental civet or rasse *(Viverricula indica)* and the formidable biturong *(Arctictis biturong)* the largest of all viverrids and capable of killing small deer, although it is mainly a vegetarian.

The mammalian fauna of the tropical forests is certainly rich and varied but those species considered so far are not significant in world trade. The density of the large herbivores is too low for them to be cropped on a commercial scale for meat even if the inpenetrable vegetation had not made this impracticable. They are, of course, hunted or trapped by indigenous peoples, and in this sense are a valuable biological resource. It would be possible to trap the smaller carnivores for their skins, but this has not been done commercially, again probably because the dense vegetation would severely hamper a trapping programme. In any case, animals would probably not last long in the trap given the wide variety of scavenging animals. Further, the fur of animals killed in traps would soon deteriorate in the hot, humid forest.

2.3 Primates

Of the rain forest-dwelling mammals it is the primate that are of economic significance, for in addition to being locally exploited for food there is an international trade in several species. Few reliable data are available on subsistence hunting, but exploitation is known to be heavy and has led to some species becoming rare. No attempt is made at rational harvesting. Large numbers are taken at a time and hunting is indiscriminate, with young animals as well as pregnant and nursing mothers being taken. West Africa and South America are the main regions where primates are hunted for food. If fewer are killed in Asia it is probably because they are less numerous although religious taboos may play a part in Buddhist or Hindu regions. Monkeys killed by hunters are often sold in local markets but there is no international trade in this meat. In parts of west Africa, Chimpanzees *(Pan troglodytes)* and gorilla *(Gorilla gorilla)* are also eaten.

Of great economic value is the use of primates as experimental animals in medical research. Again, the precise numbers involved in the trade are unknown but exports of some species have been so high that fears have been expressed for their survival. Even the once common rhesus monkey *(Macaca mulatta)* of Asia has been so over exploited in recent years that number has declined drastically and the Indian government has forbidden further trade in the species. This is likely to increase pressure on alternative species, particularly in Africa, where the vervet monkey *(Cercopithecus aethiops)* has been particularly favoured for research. Another popular species, less able to withstand such attrition, is the chimpanzee because its close relationship with man makes it a suitable subject in medical research. Although the chimpanzee is now protected under the CITES regulations, an international treaty for the control of animal trade, specimens continue to find their way to medical laboratories.

The trade in primates is undesirable for a number of reasons, other than those concerned with conservation. In particular, laboratories should breed their own stock so as to produce standard animals. Few researchers would use wild-caught rats in order to assay a new drug and the heterogenous antecedants of laboratory primates must detract from the value of the experiments. But wild-caught primates continue to be used because breeding primates in captivity is usually difficult and expensive. There are also important ethical considerations.

A reprehensible business has also developed in the ornament use of primate products and in the use of live animals for tourist purposes. In Rwanda, in the 1970s, for example, a particularly hideous trade in the mummified heads and hands of gorillas developed to satisfy the ornamental tastes of Belgian expatriates. Although the numbers involved were few, the survival of the population, which numbered less than 300, was seriously threatened. Fortunately this trade had ceased.

An equally reprehensible trade in chimpanzees has recently developed in Spain where young animals are kept by beach photographers so that tourists may be photographed holding one (Templer, 1983). The trade is a lucrative one but, apart from other considerations, it represents a serious drain on the wild populations, since the young animal soon outgrows its attractiveness and has to be replaced. As with all captures of apes from the wild, several animals are killed or die for every one delivered to the overseas customer.

The demand for primates by zoos has resulted in a widespread trade. Zoos are now rather more circumspect than they were formerly in advertising for specimens, since the trade has undoubtedly helped to reduce some populations to endangered levels and zoos nowadays like to cultivate a conservation image. All four types of great ape were captured for the zoo trade, with chimpanzees being particularly popular. Few apes breed readily in captivity and only recently have the prospects of managing self-sustaining populations in captivity seemed realistic. Previously, replacements for zoo species that died were simply taken from the wild, usually by shooting the adults and catching the youngsters. The populations of chimpanzees, gibbons *(Hylobates* spp.) and siamangs *(Symphalangus* spp.) have usually been large enough to withstand such attrition, but those of the gorilla and orang-utan *(Pongo pygmaeus)* have not. The orang-utan is particularly specialised for life in the trees and cannot easily adapt to loss of forests. The latter has certainly been of greater consequence in the ape's decline but the added burden of supporting a zoo trade may have tipped the balance against its survival. The change in attitudes may have come in time, however, and there is hope for a recovery in numbers, particularly since an active rehabilitation programme now exists.

2.4 Rodents

Although they do not figure in international trade, rodents are an important wildlife resource for the inhabitants of forested regions. Where the original forests have been destroyed or broken up, the influx of grassland rodents, such as the cane rat or grasscutter *(Thryonomys),* which can attain a size of 8–9 kg, has greatly enhanced the productivity of "bush meat" (Asibey, 1974). The giant rat *(Cricetomys),* which reaches some 3 kg, is another widely-eaten rodent. Both are of a suitable size for consumption by a family at a single meal. Some West African peoples keep the animals in captivity, and laboratory studies have shown that both species are suitable as domestic animals (Ajayi & Tewe, 1980; Ajayi *et al.* 1978). Other rodents eaten in West Africa include all forms of squirrels, porcupines and various rats and mice. Elsewhere an equally diverse assortment of rodents is eaten, particularly among subsistence level societies (Eltringham, 1984).

2.5 Bats

Bats, particularly the large fruit bats, are consumed throughout the tropics. Much of the meat is preserved by smoking and sold. Fruit bats are not confined to forests and most are caught in inhabited regions, particularly where bananas and other fruit trees are grown.

2.6 Birds

The South American rain forests are particularly rich in brids, with several endemic groups that include humming birds, toucans and woodcreepers. In African and Asian forests hornbills take the place of toucans and sun birds take over the role of humming birds. The ancestor of the common chicken, the jungle fowl *(Gallus gallus)*, lives on the floor of Asian forests as does the peacock *(Pavo cristatus)*. Australian forests are the homes of the birds of paradise and their close relatives, the bower birds. There also are found the cassowaries *(Casuarius* spp.). These are ratites and unusually large for forest birds. Parrots are restricted mainly to the tropics, and most are forest dwellers. The group reaches its greatest taxonomic diversity in the Australasian region, but Central and South America has the greater number of species. Parrots are of little economic importance, except as cage birds and, to a certain extent, as pests of fruit and grain crops.

Birds are eaten everywhere in the tropics, and not only in forests. Most species are taken for the pot, including such unlikely examples as birds of prey and song birds. In certain places, such as Papua New Guinea, feathers are used for body ornaments but the effect on bird populations is negligible compared with the impact of the international trade in live birds and in their products, which involves mostly forest or woodland species.

The numbers of birds involved in the trade are unknown but estimates are in the millions (Bruggers 1982) with Senegal alone exporting an average of 1.4 million a year since 1955 at least. The importing countries are principally in western Europe where some 84% of the birds are sold. Since the demand is for pets the most colourful species are preferred and most of these are tropical. The mortality rate is high after capture, with perhaps more than twice as many birds dying as are sold. Certain species, particularly the seed-eating finches, do well in captivity but the insectivorous birds rarely survive under the unsuitable conditions in which they are kept by the catcher or dealer. Those from Africa are common. Some such as the queleas, weavers and sparrows are agricultural pests so there can be no objections from conservationists to their capture although there may be some on ethical grounds.

The case is different in India, the other tropical region from which large numbers of live birds are exported, for some of the species involved are rare and their survival is threatened by the trade. The hill mynah *(Gracula religiosa)*, for example, was once common but its popularity as a mimic has caused it to become rare. Not only are numbers reduced by direct capture, but their replacement is endangered owing to the common practice of felling trees to catch the nestlings. The consequent loss of nesting holes has severely limited the species' ability to breed successfully.

Parrots and macaws in particular have suffered from the live bird trade and in many cases the offtake is much greater than the capacity of the birds to replace themselves. The most popular species as a pet is the grey parrot *(Psittacus erithacus)* from the rain forests of central Africa. Although that species is probably numerous enough to withstand the exploitation, this is by no means the case with others. Some of the South

American macaws are particularly vulnerable, often owing to a limited range and low population size.

Feathers still feature in trade but they are no longer an important biological resource. When plumes were used to decorate women's hats, large numbers of birds were caught. A few, such as ostriches *(Struthio camelus)* and marabou storks *(Leptoptilus crumeniferus),* are not forest species, but many are, particularly the birds of paradise. Some certainly became rare because of the millinery trade, but there has been little demand for many years. Stuffed birds of paradise and other species, as well as their skins, continue to be traded, and as late as 1976 some 2,000 birds skins were being exported from Indonesia (Inskipp & Wells 1979).

2.7 Reptiles

Reptiles occur throughout the forests and many, including some poisonous snakes, are eaten. Lizards are more commonly taken because they are harmless and easy to catch. The varanid lizards are widely distributed throughout the tropics, except in the Americas, although they are never found far from water. They are perhaps the most favoured of all lizards because of their large size and succulent flesh. Whenever possible pythons, boas, anacondas and similar large constricting snakes are caught for food in all parts of the tropics. These snakes are frequently found near rivers inside the tropical forests. Crocodilians are also taken in forests, one of their many habitats. Some species are, dangerous and are not easy to kill, but they are nevertheless hunted for meat.

Although reptiles are an important source of local subsistence food supply, there is little international trade in reptile meat but there is a wide demand for their skins.

Crocodile leather is of high quality and is used for making expensive footwear and accessories. The uncontrolled shooting of crocodiles, alligators and caimans almost led to the extinction of some species. Although now adequately protected, the survival of many species is still not assured. Particularly endangered are the South American caimans. The Mississippi alligator *(Alligator mississipiensis)* almost became extinct but now appears to be safe from hunters at least, although it is still threatened by habitat destruction through drainage schemes.

Crocodile skins still appear in trade, especially in France, but as a biological resource the animals are no longer particularly important. Crocodiles are easily kept in captivity and "crocodile farms" are widespread in the tropics. These are mainly crocodile rearing stations since the reptiles do not normally breed in captivity. Eggs or hatchlings are taken from the wild and reared on the farms. This is good conservation practice since mortality among young crocodiles is high and as the farms are obliged to return some of the reared crocodiles to the wild, the annual increment of the wild population is greater than it would be were the nests left undisturbed. The existence of crocodile farms also ensures that the wild animals continue to constitute a biological resource.

Snakes and lizards are much more important than crocodiles in international trade.

They are exploited principally for their leather and, as with crocodiles, their history in trade is one of gross over-exploitation. The principal exporters are India, Bangladesh, Pakistan and Singapore and the principal importers are France, Italy, West Germany and Japan. Although now strictly controlled under the CITES regulations, large quantities of reptile skins continue to be traded. India banned the export of reptile skins in 1976 as a conservation measure. Trade within the country is permitted but large quantities continue to be exported, since at least some western imports originate from India (Inskipp 1981).

There is still no concerted attempt to manage reptiles rationally. Dealers and customers are concerned that the supply should not dry up but probably feel that there is little they can do as individuals. There is no central body to oversee the harvesting of reptiles and most are caught speculatively by local people hoping to sell them to dealers in towns. This is a pity since a profitable harvest could certainly be taken without endangering the species concerned, although admittedly it would be difficult to assess the sustainable yield since nothing is known of the sizes of the populations from which the harvests are taken. It is certain that some populations have been heavily overcropped, judging from the decline in numbers apparent from the extra effort required to catch further specimens. Other species, such as the monitors (*Varanus* spp.), seem better able to withstand the drain on their numbers and it is likely that worthwhile sustainable harvests could be taken because reptiles are prolific breeders and their low metabolic rates enable them to survive on limited food resources. The only control that can be applied is on the retail trade, but much of the effort, particularly of conservationists, is directed towards abolishing the trade rather than controlling it.

2.8 Amphibians

Amphibians are an important food resource in many parts of the tropics. But the species most closely adapted to forest life, the tree frogs, are not edible and are rarely a source of useful products. An unusual exception is the arrow poison derived from South American frogs (*Dendrobates* spp.). The giant toad *(Bufo marinus)* is also a source of such poisons. Many of the tree frogs secrete poisonous mucus as may be deduced from their bright warning coloration.

The tropics are an important source of frog meat for export to Europe, Asia and North America. The bull frog *(Rana adspersa)* and goliath frog *(Rana goliath)*, the largest of its kind, are eaten locally in West Africa in large numbers. Many amphibians used for food occur in forest streams and pools, although their distributions often extend into rivers and wet areas of the savannas.

2.9 Invertebrates

A great many invertebrates are taken for food by indigenous forest peoples, particularly by hunter-gatherers. Such societies are now becoming rare but even people fully integrated into modern ways still collect grasshoppers, flying termites and other edible insects when the opportunity allows. A feature of forest invertebrates is gigantism, so that finding even a single insect can be worthwhile. Most invertebrates collected for food in tropical forests are insects for shortage of calcium does not favour a diversity of snails, the other principal edible group. However, giant snails (*Achatina* spp.) in West Africa are important food items (Orraca-Tetteh, 1963).

2.10 The Future of Tropical Forests

The tropical rain forests are undoubtedly of importance to local societies, and not only because of the food resources provided by the animal inhabitants. Rain forest soils are poor and are soon eroded once the tree cover is removed. The destruction of forests for their timber and their replacement with agricultural systems has usually been disastrous in the humid tropics, yet it is continuing unabated at about 245,000 km^2/yr (Myers, 1979). Rain forests are the most varied and productive of all terrestrial ecosystems yet they are also among the most fragile. Many animal species have already became extinct and many more will follow unless the widespread clearance of the forests is halted. Unfortunately there is little evidence that this will happen.

3. SAVANNAS

3.1 African Savannas

1a. Mammals In the more arid regions of the tropics, loss of the forest leads not to disaster but to a wooded grassland known as derived savanna. The Guinea Savanna, a band of rangeland across West and Central Africa south of the Sahel, is probably such a derived savanna and is kept from reverting to forest by grass fires. It is the wettest of all African savannas, with an annual rainfall between 1000 and 1500 mm. The large mammal fauna of this region is varied, but total numbers are not large so that the area is not suitable for cropping schemes. Most of the species occur elsewhere in Africa but characteristic examples are the roan antelope *(Hippotragus equinus)*, bubal hartebeest *(Alcelaphus buselaphus)*, kob *(Kobus kob)*, waterbuck *(Kobus defassa)* and the giant or Lord Derby's eland *(Taurotragus derbianus)*, the largest of all antelopes. Its much smaller relative, the bushbuck, is found wherever the vegetation is thick enough. Elephants move into the Guinea Savanna in places but there are now no rhinoceros. There is the typical range of African carnivores of lions, leopards and hyaenas. The bird fauna is particularly characterised by members of the weaver family, but the region is also important as a dry season retreat for migrant birds from within Africa.

As the climate becomes drier northwards, the Guinea Savanna grades into the Sudan Savannas and finally into the Sahel Savanna. The fauna shows an increase in desert-adapted species, such as the addax *(Addax nasomaculatus)*, a close relative of the oryxs. There is a similar belt of savanna running across Africa south of the equator. This is the miombo woodland, which is dry and provides few animal resources other than honey. The mammalian fauna is rich but densities are low.

The East African savannas are by far the most important of any tropical rangeland in terms of large mammal biomass. Most of the famous African national parks occur in these savannas, which are, however, far from uniform in their vegetation. Those in the west, in Uganda, are often thickly bushed and support large numbers of elephants, hippos and buffaloes, but relatively few antelopes. The savannas of northern Tanzania and southern Kenya are open, sometimes treeless, and contain vast numbers of ungulates, particularly giraffe, wildebeest *(Connochaetes taurinus)*, topi *(Damaliscus korrigum)*, gazelles, buffaloes and zebra *(Equus burchelli)*. The wildebeest and buffalo have shown remarkable increases in numbers over the past twenty years or so (Sinclair, 1979). Wildebeest increased five-fold between 1961 and 1977 and buffalo by between two and three times their original number. These increases are believed to be due to a recovery from the effects of rinderpest, aided by improved rainfall, which increased primary production. The numbers of zebra, a species unaffected by rinderpest, did not rise during this period.

These huge concentrations of large mammals in the Serengeti, numbering some 1,300,000 in the case of wildebeest, has raised the idea that they could be cropped for meat and skins both for local food and exports. Such schemes have sometimes been promoted over-enthusiastically and without consideration of practical difficulties (Eltringham 1984).

One of the principal problems is to harvest the crop efficiently. It is no easy matter to slaughter wild animals by the thousand while ensuring that the meat obtained is produced under acceptable conditions of hygiene. The migratory habits of some populations is a particular difficulty. Thus a scheme to crop zebra in the Loliondo District, to the east of the Serengeti National Park in Tanzania, failed because the appearance of the migrating zebra in the cropping area was not easily predictable and their stay was too short. Processing of the carcass for meat presents many problems because of its perishability under humid tropical conditions. Unless the meat is to be disposed of cheaply to local people, the very stringent standards for the production of meat from domestic animals must be observed. Those require the use of refrigerated storage vans and mobile abattoirs with adequate supplies of chlorinated water. Owing to parasite infection, which occurs at a much higher level in wild than in domestic animals, many carcasses fail to pass government inspection. Since meat is not a particularily valuable product, financially successful cropping schemes depend on the sale of skins, which are cheaper to process and much less perishable than meat. In purely financial terms, an ungulate cropping scheme should be based on removal of the skins, the meat being left to rot.

The building of a permanent abattoir in the cropping area avoids many of the problems of processing carcasses but, as experience of elephant cropping in the Luangwa Valley, Zambia, has shown, the system is too inflexible. Stocks in the vicinity of the abattoir became exhausted and elephants had to be cropped at progressively further from the base. The resulting problems of increased haulage costs and risk of meat spoilage jeopardised the economic success of the enterprise.

Even if meat can be harvested and processed economically, there remains the problem of marketing. Large populations of wild mammals tend not to occur near centres of population, thus transport costs are excessive. Competition with the domestic meat industry is another formidable problem and several schemes have collapsed either because the meat trade has withdrawn co-operation or has blackmailed retail butchers into refusing to handle the game meat. Wildlife croppers also have to face the opposition of conservationists, who often object to wildlife being slaughtered for any reason. Although such opposition has no legal backing, through lobbying and public campaigns it can have a severely depressing effect on marketing.

The problems of wildlife cropping are logistic and political rather than biological and the vast herds of ungulates in the tropical savanna of East Africa have a great potential for meat and skin production. As such, they remain a wildlife resource of the highest importance. In many cases, they occur in regions where conventional meat production from domestic animals is impossible, either through lack of water or the presence of disease organisms. Much of the miombo woodland is of little or no use for agriculture or pastoralism owing to low or irregular rainfall and because of the tsetse fly, the vector of *trypanosomiasis,* a disease fatal to cattle.

An alternative to cropping is wildlife ranching or domestication. Experiments on the Galana ranch in Kenya (King *et al.* 1977) and elsewhere have shown that many wild ungulate species adapt readily to captivity and become, in effect, domestic animals. Unfortunately, many of the advantages that wild animals have over domestic stock through their better adaptations to the environment are lost under conditions of captivity and, in terms of animal production, the new domesticates are generally no better than the old.

The value of African savannas as a biological resource is not solely dependent on meat production, since they can also be exploited for tourism. The national parks, which were originally intended only for the conservation of threatened species and their habitats, have proved to be great tourist attractions and foriegn exchange earners. Kenya, in particular, has benefitted from wildlife-based tourism with income exceeding expenditure by a factor of three (Myers, 1975). Other countries, such as Zambia, have been less successful, for the outlay on the infra-structure of a tourist industry is heavy and the hoped-for influx of tourists has not occurred.

Sport-hunting is a tourist-based industry similar to game-viewing and it can be profitable, although the sums involved are not large relative to the G.N.P. in most countries. There has been little activity in this field in recent years since, at one time or another, all three East African countries have banned hunting as a conservation measure, Tanzania has reintroduced sport-hunting and Kenya seems set to do so shortly, since the hunting of birds is again permitted. Hunting showed a marked increase in Zambia between 1973 and 1977 but the absolute numbers of hunters remained too low for it to be a significant economic activity.

The fauna of the East African savannas comprises much more than just the large herbivores. There is a full range of predators, which undoubtedly contribute to the touristic potential and big game hunting, ranging from lions, leopards and cheetah *(Acinonyx jubatus)* to jackals, hunting dog *(Lycaon pictus)* and hyaenas, both spotted *(Crocuta crocuta)* and striped *(Hyaena hyaena)*. Although these predators are of significance in maintaining the health of the ungulates, in that they tend to weed out weak or unhealthy animals, they would be a problem in cropping schemes because of their involvement with the life histories of ungulate parasites, of which they are often the primary host.

Mention should also be made of the wealth of small ungulates on the tropical savannas and of the many small cats, civets, genets and mongooses. The weasel family is also well-represented, including the powerful ratel or honey badger *(Mellivora capensis)*, which has been known to drive a lion off its kill.

1b. Birds Bird life is extremely rich in the African grasslands and visible, particularly in the case of the ostrich *(Struthio camelus)*, the largest of all living birds. Other large birds include the secretary bird *(Sagittarius serpentarius)*, a terrestrial 'eagle', the kori bustard *(Ardeotis kori)*, the heaviest flying bird and the martial eagle (*Polemaetus bellicosus*), which, if not the biggest eagle in the world, is in the top two or three. Birds are taken for food and their eggs are popular items in the diet, particularly of people living on tropical islands where large colonies of sea birds nest (Feare, 1976).

1c. Reptiles and Amphibians Reptiles and amphibians are also common and diverse in savannas although amphibians are confined to rivers, swamps or other wetlands. Lizards and snakes are particularly abundant and where conditions are suitable, so, too, was the Nile crocodile, which has been relentlessly persecuted. The crocodile could be a valuable resource, for its skin makes a fine leather, but its numbers are now too depleted for it to be commercially exploited on a sustainable yield basis.

3.2 South American Savannas

There is little savanna in other parts of the tropics apart from rangeland that has replaced the indigenous forests. There are, however, savannas north and south of the equatorial rain forests of South America, which are probably natural, although it is possible that fire has played a part in their development. There is now little in the way of wildlife on these grasslands, for most regions have been developed for cattle, but there are still pampa deer *(Ozotoceras bezoarticus)* to represent the large mammals. There are no bovids in South America and the equivalents, the cameloids, tend to be mountain animals so that large mammals are not typical of the American tropical grasslands. Those that are present are, however, of great interest and include the great ant eater *(Myrmecophaga tridactyla)*, the nine-banded armadillo *(Dasypus novemcinctus)* and the long-legged maned wolf *(Chrysocyon brachyurus)*, which preys on rodents or other small mammals and does not hunt in packs like the true wolves. It is clear that these grasslands do not support wildlife of much economic importance.

3.3 Asian Savannas

There are stretches of savanna in tropical Asia although those in India are probably derived savanna and consequently did not originally contain any grassland species. They have, however, been colonised by antelopes from the surrounding temperate grasslands, including gazelles (*Gazella* spp.) in the west. Some originally forest animals, such as the nilgai antelope *(Boselaphus tragocamelus)* and the four-horned antelope *(Tetracerus quadricornis)*, have become adapted to life on grassland. There is also the blackbuck *(Antilope cervicapra)*, which appears to be a genuine savanna ungulate and not a forest relic. Its place of origin, remains, however, a mystery. There are natural semi-arid grasslands further east in Burma and the Indochinese Peninsula but the rare kouprey *(Bos sauveli)* is the only indigenous large ungulate. Although called the forest ox, it appears to be a true savanna animal living in the glades. As in the case of South America, the ungulates of the tropical Asian savannas are too sparse to be exploited commercially.

3.4 Australasian Savannas

The northern half of Australia lies within the tropics and although much of the central area is desert with little wildlife, the savannas in the north are well-populated with kangaroos and wallabies as well as with many introduced large mammals such as water buffalo, horses and cattle, which are now feral. Of the kangaroo family, there are the grey kangaroos and their relatives, the wallaroos (*Macropus* spp.) as well as the red kangaroo *(Megaleia rufa)*. Their smaller relatives include the agile wallaby *(Wallabia agilis)* and several species of rat kangaroos. The kangaroos are of economic importance both for their products of meat and skins, and as agricultural pests. The latter problem is probably greatly exaggerated but under certain conditions, as during drought, kangaroos may well compete with sheep for grazing. That they are a wildlife resource of some potential is obvious from the export of meat and hides, which according to official statistics amounted to nearly 1600 t of meat and over 1,500,000 skins in the 1981/82 season. Many more kangaroos were shot as pests or for consumption within the country, but not all came from tropical regions of Australia.

4. MOUNTAINS

There are numerous mountains and mountain ranges within the tropics and although their main biological potential is in timber products, some support wild animals of commercial value. One example is the vicuña *(Vicugna vicugna)* of the High Andes in Peru. The great value of this small relative of the camel lies in its wool, which is the finest mammalian fibre in the world. Vast herds existed in Inca times and they were regularly rounded up and sheared. This practice was presumably too tedious for the conquering Spaniards, who simply shot the animals in order to obtain the wool. Subsequent slaughter of the vicuña over the following centuries brought the species close to extinction and it was not until a conservation programme was established on the Pampa Galeras, the last stronghold, that numbers recovered to a relatively safe level. A scheme was introduced in 1972 to manage the vicuñas and exploit them for their meat and wool on a sustainable yield basis. The project was probably premature, since the vicuña is still fully protected under the CITES regulations and its products cannot be traded. There were other political and ecological controversies and the project has now been abandoned. Nevertheless, the vicuña represents a most valuable wildlife resource that might one day play an important role in the economy of the High Andes.

5. THE FUTURE

This brief survey of wildlife resources in the tropics has revealed a potential that has not yet been exploited satisfactorily. The nearest to success is probably the profitable operation of national parks in Kenya. The many failures in attempts to exploit wildlife commercially probably result, not from biological factors, but rather from unwieldy administration. Exploitation at the local level, as for example a licensing system to allow villagers to crop wildlife for themselves, is more likely to be successful. As in many other areas, the greatest threat to wildlife resources in the increasing human population. In the tropics, this has led especially to the destruction of forests either for timber or to make way for agricultural settlements. The latter have sometimes been successful but often not, particularly in the case of rain forests, where soils are unsuitable for agriculture and soon washed away. Whatever the outcome, it is always disastrous for much of the wildlife.

REFERENCES

1. Ajayi, S S and Tewe, O O, 1980. Food preference and carcas composition of the grasscutter *(Thryonomys swinderianus)* in captivity. *African Journal of Ecology*, 18, 133–140.
2. Ayaji, S S, Tewe, O O and Faturoti, E O, 1978. Behavioural changes in the African giant rat (*Cricetomys gambianus* Waterhouse) under domestication. *East African Wildlife Journal*, 16, 137–143.
3. Asibey, E O A, 1974. Wildlife as a source of protein in Africa south of the Sahara. *Biological Conservation*, 6, 32–39.
4. Bourlière, F (Ed), 1983. Tropical Savannas. 730 pp Amsterdam: Elsevier Publishing Company.
5. Bourlière, F and Hadley, M, 1970. The ecology of tropical savannas. *Annual Revue of Ecology and Systematics*, 1, 125–152.
6. Bruggers, R L, 1982. The exportation of cage birds from Senegal. TRAFFIC Bulletin, 4, 12–22.

7. Eltringham, S K, 1984. Wildlife Resources and Economic Development. 329 pp Chichester: John Wiley and Sons.
8. Feare, C J, 1976. The exploitation of sooty tern eggs in the Seychelles. *Biological Conservation*, 10, 169–181.
9. Grubb, P, 1972. Variation and incipient speciation in the African buffalo. *Zeitschrift für Saugetierkunde*, 37, 121–144.
10. Inskipp, T, 1981. Indian Trade in Reptile Skins. 13 pp Cambridge: Conservation Monitoring Centre.
11. Inskipp, T and Wells, S, 1979. International Trade in Wildlife. 104 pp London: Earthscan.
12. Jeffrey, S, 1877. How Liberia uses its wildlife. *Oryx*, 14, 168–173.
13. King, J M, Heath, B R and Hill, R E, 1977. Game domestication for animal production in Kenya: theory and practice. *The Journal of Agricultural Science*, 89, 445–457.
14. Myers, N, 1979. The Sinking Ark. 307 pp Oxford: Pergamon Press.
15. Myers, N, 1975. The tourist as an agent for development and wildlife conservation: the case of Kenya. *International Journal of Social Economics*, 2, 26–42.
16. Orraca-Tetteh, R, 1963. The giant African snail as a source of food. *Symposia of the Institute of Biology*, 11, 53–61.
17. Pratt, D J, Greenway, P J and Gwynne, M D, 1966. A classification of East African rangeland, with an appendix on terminology. *The Journal of Applied Ecology*, 3, 369–382.
18. Sinclair, A R E, 1979. The eruption of ruminants. 82–103 pp in: *Serengeti. Dynamics of an Ecosystem* ed. by A R E Sinclair and M Norton-Griffiths.
19. Templer, P, 1983. Survey of the use of chimpanzees as photographic models on Spanish beaches. *World Wildlife Fund Monthly Report*, February 1983.

Resource Management and Optimization
1990, Volume 7(1–4), pp. 115–140
Reprints available directly from the publisher.
Photocopying permitted by license only.
© 1990 Harwood Academic Publishers GmbH
Printed in the United States of America

TIDAL WETLAND RESOURCES IN THE TROPICS

PETER R. BURBRIDGE
House of Ross Comrie, Perthshire Scotland PH6 2JS

CONTENTS

This paper examines tidal wetlands and associated resource systems including mangrove, peat swamps, estuaries and lagoons. Bio-physical features of wetland ecosystems are discussed in relation to the influence of tides, functions provided by wetlands and other coastal ecosystems. Tidal wetland soils are also discussed in terms of their potential for agricultural use and management concerns which also influence their suitability for other resource uses.

Development pressures on wetlands are discussed in conjunction with principal uses of wetlands and management issues which have a bearing upon sustainable use of wetland resources. Emphasis is placed upon two wetland types—forest dominated swamps and mangrove and associated nipah. Uses discussed include: forestry, agriculture, irrigation, aquaculture and mining.

The work concludes with a discussion of the evaluation of wetland development alternatives and the role of economic analysis.

1. INTRODUCTION

Throughout the world the meeting place for land and sea—the costal zone—represents one of the most complex areas in which man can attempt to manage his environment. This is especially true of coastal areas in the tropics where the richness and diversity of resources available has stimulated complex and diverse patterns of social, cultural and economic development which are dependent on the continuing functions of fragile ecosystems.

There are, however, marked contrasts in the historical traditions of coastal resource use. For example, in Asia coastal areas have long been the focus of settlement, commerce and often highly intensive forms of resource exploitation. In contrast, areas of South America are only now reaching a level of coastal development comparable to long established regions in Asia. There is mounting development pressure on tropical coastal land and water systems including estuaries, lagoons and shallow seas. Their associated tidal wetlands formed by mangrove, freshwater swamp forests and grass dominated swamplands are also under pressure and they rank amongst the most ecologically critical and threatened areas of the globe (IUCN, 1980, Salm and Clark, 1984).

The management of these dynamic yet fragile systems is not solely a question of ecology. Tidal wetlands provide man with a wide array of goods and services ranging from commercial timber to the control of coastal erosion. Many of these goods and services cannot be technically or economically replicated by man.

The current state of our management expertise has advanced to the point where we can manipulate these ecosystems in order to increase the production of specific resources such as mangrove timber or convert them to produce rice using salt tolerant varieties. However, we have not reached a stage where we can place sufficient emphasis upon the management of tidal wetlands for sustained, multiple use.

In attempting to maximize the production of one or even a small part of the total resources offered by wetlands there are corresponding environmental and economic impacts upon other wetland dependent activities. For example, in exploiting the maximum sustainable yield of mangrove fuel wood it is highly possible that the production of *Chanos chanos* (milkfish) fry, which are dependent upon the mangrove, may be reduced. Are we achieving optimal levels of sustainable resource production through managing complex wetlands for only a small number of the resources they offer? The answer is elusive because most people do not stop to consider the full economic or socio-cultural costs associated with the maximization of but a part of the spectrum of potential resources wetlands offer. In reality we are at an early stage in the definition of the range of goods and services wetlands provide and in the establishment of truly sustainable forms of exploitation of those resources. However, there are strong indications that the potential that tidal wetlands offer for meeting the development needs and objectives of many coastal nations lies in the maintenance of the functional integrity of these natural systems, *not* in their conversion or management for single uses.

Major efforts to harness the more tangible assets of wetlands, such as the development of tidally irrigated rice cultivation in the peat swamp forests of Indonesia represent experiments whose long-term viability are by no means certain.

While the basic spirit of experimentation is to be applauded, great caution should be exercised in initiating the large-scale conversion of tidal wetlands *before* their functions and their ecologic, economic and social significance are clearly illustrated to decision makers.

If this is not done, there can be a loss of development options, natural hazards such as flooding and salt water intrusion can be increased and irreversible changes can be initiated which can have a "knock-on" effect on surrounding resource systems and development projects.

Here we will explore some of the major ecological and economic features of tidal wetlands and will discuss factors which condition the sustainable utilisation of wetlands resources.

2. TIDAL WETLANDS, ESTUARIES AND LAGOONS

2.1 Tidal Wetlands

The term "tidal wetland" is normally used to refer to land areas periodically inundated by salt or brackish water and includes grass dominated swamps, nipah, mangrove and vegetated tide flats. However, there are also extensive freshwater swamp forests in many areas of the tropics which are influenced by tidal movements.

The distribution and extent of tidal wetlands depends upon a variety of factors including topography, bathemetry of the coastal waters, coastwise transport of sediments, availability of sediment from upland areas, and the volume and rate of freshwater discharge by major rivers and streams. Tidal wetlands are normally found in association with coastal landforms characterized by shallow coastal waters, broad coastal plains often with rapidly accreting shorelines, embayments and major rivers and streams providing regular supplies of sediments and nutrients carried from upstream areas.

2.2 Estuaries and Lagoons

Where major rivers and streams are found in association with embayments with continuously free connections with the sea, the term "Estaury" is used (see Pritchard, 1967; Caspers, 1967; Clark, 1977). The term lagoon is used to denote embayments which are seasonally closed to the sea by sandbars or coral reefs or which do not have significant freshwater inflows to dilute the sea water (see Salm and Clark, 1984).

Where estuaries and lagoons are closely associated with extensive tidal wetlands, they provide habitats of great biological productivity which are generally recognised as playing a critical role in supporting major fisheries (see IUCN, 1980). Much of this productivity is attributed to mangrove (for example see Odum and Heald, 1972; MacNae, 1974; Hamilton and Snedaker, 1984).

3. BIO-PHYSICAL FEATURES OF TIDAL WETLANDS

3.1 The Influence of Tides

Tidal influence can vary considerably between areas and between seasons. The amplitude (range) of tides in the tropics, although generally smaller then in cooler climates to the North and South, can influence areas far inland from the coast.

Both the areal extent and the duration of tides has an influence on morphology of mineral soils affecting the chemistry of the deposited sediments and the flora and fauna colonizing these materials (Driessen and Ismangun, 1972). Distinct zones can often be differentiated in different wetlands based upon vegetation (for example see the study of New Guinea mangrove by Percival and Womersley, 1975) and can act as an indicator of tidal influence and related soil conditions, salinity, and other edaphic factors such as water quality and exposure to strong winds, waves or currents. The influence of tides on soil chemistry, drainage and other factors is sometimes used to help determine the suitability of wetland soils for agriculture (see for example the work of Kevie and Yenmanas, 1972; de Glopper and Poels, 1972 in Thailand).

Silvicultural management units have also been devised based upon the frequency and depth of inundation by tidal waters (for example see the tidal inundation units devised by Watson, 1928 in Malaysia and later modified by Versteegh, 1951; 1952 in Indonesia.

Tidal amplitude also has a bearing upon the viability of aquaculture in wetlands. While mangrove are not well suited to shrimp or fishpond development, areas with a tidal range of between 1 and 3 metres are considered more favourable in terms of water exchange in the ponds (Jamandre and Rabanal, 1975; Pedini 1973).

The transition between saline and freshwater conditions does not accurately reflect the extent of tidal influence. For example, van Wijk, (1951) recorded tidal influences extending more than 100 Km inland in swamplands in Southern Kalimantan, Indonesia. The rise and fall of tides is harnessed in Indonesia to help pump freshwater to form a source of irrigation water in extensive areas of former peat swamp forest which have been converted to agricultural use for rice production. Although the soils of the freshwater swamps are not normally saline, the influence of tidal forces requires extremely careful management of both irrigation and drainage if the agricultural viability of these soils is to be maintained and hazards such as salt intrusion are to be avoided. Due to the influence of tides on those swamplands, they are considered here as part of the broader definition of tidal wetlands.

3.2 The Role of Wetlands in the Productivity of Estuarine and Coastal Waters

The mangrove and other vegetation helps to convert inorganic compounds and sunlight into materials which, in the form of detritus, act as sources of food in the waters of estuaries and adjacent coastal waters. There is, however, some dispute over the contribution to estuarine and nearshore waters made by mangrove and other tidal vegetation, including sea-grass beds. Although it is generally recognised that the tidal wetlands in the lower reaches of major estuaries make a very significant contribution to biological productivity and to commercial fish stocks, attention must also be given to the contribution of other tidally influenced wetlands such as swampforests to the productivity of riverine waters feeding into estuaries and coastal waters. The vegetation in these more inland wetlands often decomposes within the wetland itself and, rather than contributing detritus to the rivers and estuaries, they supply dissolved basic nutrients (Clark, 1977).

3.3 Linkages Between Wetlands and Coastal Waters

Apart from the contribution to primary productivity, there are other factors which are important to consider in terms of the linkages between different forms of tidal wetland. Water is the major element linking wetland systems. Apart from the work done by water in transporting nutrients, sediments and other materials to downstream areas, the volume and timing of fresh water flows also have a major influence on tidal wetlands.

While annual rainfall and the vegetation cover within watersheds influences the total amount and rate of run-off, wetlands act as components of natural flood plains. In this capacity they absorb the energy of floodwater, they trap sediments and following the flood peak, they control the discharge of water into river systems. These wetlands also act as a storage mechanism for water and act as an important source of ground water seepage which helps to supply rivers and estuaries with fresh water during dry periods.

Changes in the natural vegetation and drainage characteristics of riverine and estuarine wetlands can therefore have a significant influence on the amount and rate of water discharge into the coastal environment. A major reduction in the vegetation cover of these wetlands can lead to increased flooding downstream. Conversely, increased drainage can reduce their ability to store water and can contribute to increased seasonability of water flows in adjacent rivers and streams. This, in turn, can lead to the increased penetration of saline waters inland during dry periods with a consequent shift in the numbers and mix of aquatic organisms.

In addition to changes in salinity in the water column and the possible salinization of soils adjoining tidally influenced streams and creeks in inland areas, changes in freshwater dilution of saline waters will have an impact on organisms which normally live in freshwater or marine waters but which migrate to brackish water environments for part of their life cycle. The giant freshwater prawn *Machobracium* is one example.

While normally inhabiting freshwater rivers, *Machobracium* migrate to brackish water areas and, it is believed, to tidal wetlands to breed. Major changes in the salinity, vegetation and drainage of tidal wetland areas could have a significant negative impact on breeding conditions for these commercially very valuable organisms.

Linkages between lower estuarine wetlands and more inland systems are also very important. The role of mangrove in the absorption of stormwave and wind forces helps to reduce the erosion of other tidal lands. There is therefore a strong case for the development of wetland management approaches which incorporate the linkages between different wetland types and their corresponding interdependence. Management of individual types, such as mangrove, in isolation from factors which are acting upon other wetlands can result in weak plans which often fail.

3.4 Wetland Functions

Table I outlines some of the natural wetland functions identified in the United States and Europe. Although these are not tropical countries, the wetland functions identified can also apply to the tropics. Unfortunately there is not a correspondingly rich literature pertaining to freshwater tropical wetlands. Far more attention has been given to mangrove. Table II presents information on the major functions and uses of estuaries, tidal swamp forests, mangrove and sea-grass beds.

A very significant point to consider is that the functions served by wetlands offer a wide array of goods and services which are provided at no cost to man. Table III, relating specifically to mangrove, identifies some of the more significant uses of the goods and services provided by a tidal wetland in its natural functioning state.

Many activities can utilize wetland resources without conflicting with other users. Some competing uses, however, result in irreversible change which alters the functioning of the natural system. Relatively simple systems of resource development such as filling of wetlands for urban development, or conversion to agricultural use, are poorly adapted to exploiting the wide variety of wetland goods and services normally available to a diffuse body of beneficiaries. We have yet to evolve multiple-use strategies for developing the full resource attributes of wetlands.

3.5 Tidal Wetland Soils

Two broad categories of wetland soils can be identified, namely organic (peat and peaty soils) and mineral soils. The alluvial soils commonly found in wetlands range from clays and silty clays with low organic content, to peaty mineral soils and deep organic peats. Peat soils have been largely ignored historically, as it was thought that the high temperatures of the tropics, and corresponding rapid decomposition of organic matter precluded the accumulation of tropical peat (Soepraptohardjo and Driessen, 1976). There are, however, large areas of lowland peat in the tropics, for example

TABLE I
Natural Wetland Functions. United States and Europe.

Function	Wetland Type	Reference
Hydrologic functions		
Conveyance of flood flows	Riverine	
Flood storage	Inland, Riverine	Niering, 1968; Dewey and Klopper, 1964; Mass Water Res. Comm., 1971; U.S.Army Corps or Engineers, 1975
Water supply	Inland	Motts and Haley, 1973
Barrier to waves and erosion	Coastal, Riverine, Inland	Teal and Teal, 1969
Ecologic		
Pollution control	Coastal, Inland	Grant and Patrick, 1970; Janota and Loucks, 1975
Production of oxygen	Coastal, Riverine, Inland	Grant and Patrick, 1970
Source of nutrients and habitat for fish and shellfish	Coastal	Teal and Teal, 1969; Darnell et al., 1976
Habitat for waterfowl and other wildlife	Coastal, Inland	Golet, 1972; Niering, 1968; Barske, 1966
Habitat for endangered species	Coastal, Inland	Bureu of Sport Fisheries and wildlife, 1966
Resource functions		
Food production	Coastal, Inland	Teal and Teal, 1969; Clark, 1974; Johnson, 1969;
Timber production	Inland	Wharton, 1970
Socio-economic		
Historic, archeological values	Coastal, Inland	Smardon, 1972
Education and research	Inland, Coastal	Niering, 1961
Open space and esthetic values	Coastal, Inland	Smardon, 1972; Gosselink et al., 1973; Larson, 1976

Source: Adapted from Kusler and Harwood, 1977.

Indonesia is thought to have 17 million hectares, 5 million of which are influenced by tides.

The mosaic of soil types in tidal wetlands is complex and definitions of mineral or organic soils differ from country to country. Driessen and Sudjadi (1984) present a very detailed analysis of the morphology, characteristics and problems of mineral and organic soils and we summarise some of the more pertinent management issues presented in their paper below:

A. Mineral River Basin soils: for rice cultivation on mineral soils the most pressing management problem is water control, including flooding hazards and low water availability during dry periods. Adequate water control is particularly important where shallow, base-poor fluvial sediments occur on top of pyritic (iron sulphide) marine

TABLE II
Tidal Wetlands and Related Ecosystems

Broad System Description	Functions and Uses	Detrimental Uses and Practices	Adverse Environmental Consequences
Estuaries (including associated mud flats and embayments)	nutrient influx to coastal waters fisheries production nursery & spawning areas for many coastal fish links to mangroves, sea-grasses, pelagic & demersal fisheries	Urban pollution (sewage, thermal) industrial pollution hydrologic modifications (upland irrigation and water withdrawal) conversion to tambak & dry land over exploitation of resources	reduced fishery production reduced habitat for adults and fry of fishery species infilling & sedimentation reduced estuarine habitat degradation of water quality
Tidal swamp forests	habitat for fish, wildlife & plants flood storage links to mangroves timber and fuel links to rice culture fisheries production	excessive logging conversion to tambak conversion to dry land reclaimation & irrigation destruction of mangrove buffer inappropriate channelization transmigration sites	reduced fishery yields reduced timber & fuel degradation of habitat loss of wildlife & plants reduced rice yields disruption of hydrologic regime, acid soils, loss of peat
Mangroves	sediment filter nutrient filter fishery resources (fin & shellfish) net transfer of production to coastal fisheries breeding & spawning grounds for many coastal species nursery ground for coastal & estuarine species links to seagrass, coral reefs shoreline protection buffer for tidal swamps timber fuel tanning & other chemicals	transmigration sites conversion to dry land excessive upland soil erosion overexploitation of wood overexploitation of fishery resources upland irrigation & water withdrawal oil pollution	degraded coastal water quality loss of most values, functions & uses loss & degradation of habitat due to sediment infilling reduced fishery production reduced fry production reduced nursery habitat secondary impacts to reefs, seagrasses, swamps
Seagrass beds	nutrient filter net transfer of production to coastal fisheries feeding habitat for green turtles, dugongs, nursery grounds for coastal fisheries links to mangroves, coral reefs fishery production, esp. finfish	coastal urban pollution—thermal & domestic sewage industrial pollution coral mining (excessive) excessive upland soil erosion overexploitation of fisheries inappropriate coastal development construction & dredging oil pollution	degradation of habitat loss of habitat due to infilling loss of habitat due to hydraulic changes displacement of seagrasses reduced fishery production loss of fry & breeding habitat

Adapted from Burbridge & Maragos, 1985

TABLE III
Uses of Mangrove Goods

Direct Products from Mangrove Forests	
Uses	**Products**
Fuel	Firewood for cooking, heating
	Firewood for smoking fish
	Firewood for smoking sheet rubber
	Firewood for burning bricks
	Charcoal
	Alcohol
Construction	Timber for scaffolds
	Timber for heavy construction (e.g., bridges)
	Railroad ties
	Mining pit props
	Deck pilings
	Beams and poles for buildings
	Flooring, panelling
	Boat building materials
	Fence posts
	Water pipes
	Chipboards
	Glues
Fishing	Poles for fish traps
	Fishing floats
	Fish poison
	Tannins for net preservation
	Fish attracting shelters
Agriculture	Fodder
	Green manure
Paper Production	Paper of various kinds
Foods, Drugs, and Beverages	Sugar
	Alcohol
	Cooking oil
	Vinegar
	Tea substitutes
	Fermented drinks
	Dessert topping
	Condiments from bark
	Sweetmeats from propagules
	Vegetables from propagules, fruits, or leaves
	Cigarette wrappers
	Medicines from bark, leaves, and fruits
Household Items	Furniture
	Glue
	Hairdressing oil

continued

TABLE III (continued)
Uses of Mangrove Goods

Direct Products from Mangrove Forests	
Uses	**Products**
	Tool handles
	Rice mortar
	Toys
	Matchsticks
	Incense
Textile and Leather Production	Synthetic fibers
	Dye for cloth
	Tannins for leather preservation
Other	Packing boxes

Indirect Products from Mangrove Forests	
Source	**Product**
Finfish (many species)	Food
	Fertilizer
Crustaceans (prawns, shrimp, crabs)	Food
Molluscs (oysters, mussels, cockles)	Food
Bees	Honey
	Wax
Birds	Food
	Feathers
	Recreation (watching, hunting)
Mammals	Food
	Fur
	Recreation (watching, hunting)
Reptiles	Skins
	Food
	Recreation
Other Fauna (e.g., amphibians, insects)	Food
	Recreation

materials. If water levels drop too low, the resulting aeration of the soil causes acids to form and creates associated problems of aluminum toxicity and low levels of phosphorous which inhibit plant growth (Pons, 1980). Soils of Marine plains: Marine plains (bare tidal flats, mangrove swamps and marshes inland) consist of silty or clayey material with low oxygen levels (reduced) and varying amounts of raw vegetal material. The tidal flats and mangrove are generally saturated and show little soil development. Where siltation has reached levels above the mean high water mark (upper elevation of the mangrove and adjacent marshland) soil development is more advanced.

B. Organic soils Most tropical peat soils are made up of ombrogenous raised bogs which are not influenced by tides. Topogenous peats (those influenced by tides and or ground water) are less problematic, they are however less extensive and the majority have already been colonized and are under agricultural production in countries such as Indonesia, Malaysia and Thailand.

Major problems associated with the raised bog or ombrogenous peats are related to their low mineral content and extremely high porosity. Drainage can lead to severe subsidence and the disruption of irrigation or drainage works. The shallower peats found at the edges of the raised bogs are slightly better in terms of their potential for agriculture. However, the peats over 1m in depth are marginal agricultural soils and their sustained use and profitability are highly uncertain. Driessen and Sudjadi (1984) suggest that their agricultural use is beyond the resources of the average settler.

3.6 Acid Sulphate Soils

An important phenomenon in coastal wetlands is the development of highly acid soil conditions following drainage and exposure of the soil. The soils that develop are collectively termed acid sulphate soils, as the active mechanism is the formation of sulphuric acid from the oxidation of pyrite. The sulphates produced may be partly neutralized by compounds such as calcium carbonate. The remaining acid breaks down clay minerals in the soil, often causing excessive amounts of aluminum to be liberated into solution which is toxic to plant roots and microorganisms.

Coastal wetland soils which are maintained in a water-logged condition will not have developed acid sulphate conditions, but have a high probability of doing so if drained and aerated. Such potential acid sulphate soils are often viewed as prime lands for large development schemes.

There are no universally suitable indicators of the potential for acid sulphate conditions to develop; indeed, such soils have been known to develop adjacent to very productive areas (Coulter, 1973). Thus the importance of recognising potential acid sulphate conditions in advance of major investments and population movement cannot be underestimated.

The major management tool of use in controlling acid formation is to restrict the drainage of potential acid sediments to a minimum. Other techniques include moderate application of lime accompanied by leaching through mounding the soil and deepening drainage ditches. This can yield good coconut groves but the process takes several years. This system is combined with rice cultivation by the Banjarese in Kalimantan, Indonesia (Collier, 1979).

3.7 The Need for Water Control

Water control is essential to the management of marine soils found in coastal lowlands for a number of reasons including subsidence of the soil following drainage. Inundation with saline or brackishwater helps to remove noxious materials from the root zone. Flooding is also a problem and can result from greater than normal rainfall or extremely high tides. The coincidence of both excessive rainfall and high tides can extend the area of flooding to very large areas of the coastal lowlands. This problem is compounded by poor upland management which increases erosion and increases surface water runoff during the wet season and reduces stream flows during the dry season.

3.8 Inadequate Information for Agricultural Development of Peat Soil

Part of the difficulty in managing peat soils in the tropics is the basic lack of information concerning the morphology, chemistry, physical properties, microbiology and other biophysical factors which condition the potential uses of these soils (Driessen and Rochimah, 1976; Tie and Kueh, 1979). In addition to the lack of basic data, the methods employed to analyse the peat soils, and the criteria used for classifying them, have been criticised. Several workers have argued that it is the physical properties of peat which pose the major problems in determining their suitability for reclamation (Driessen and Rochimah, 1976; Soepraptohardjo and Driessen, 1976, Pons and Driessen, 1975). Assessments of peat based primarily on chemical properties, and expressed on a weight per unit weight basis (concentration), are considered misleading due to the wide variation in the bulk density and wood volume of peats (Driessen, 1977). The significance of such physical characteristics, in conjunction with chemical characteristics, is better expressed as a weight of chemical constituents per unit volume (Van Wijk, 1951; Andriesse, 1964; Boelter and Blake, 1964; Driessen and Suhardjo, 1975; Driessen, 1977; Tie and Kueh, 1979).

Driessen (1977) is particularly forthright in his criticism of current approaches to assessment of peat soils:

> ". . The often heard but utterly wrong opinion that peats are quite uniform in their physical properties is just as difficult to eradicate as the equally false contention that these properties are 'favourable'. Our failure to appreciate the complexity of the matter and the deplorable habit of evaluating peat soils with mineral soils as a reference, account for numerous unjustified simplifications, and for a high incidence of 'unforeseen' events in past attempts to reclaim and use these tropical peats for agriculture" (Driessen, 1977, p12).

A basic issue which remains unresolved is whether agricultural development of much of the tidal swampland soils can be sustained. There is little quantitative information on the long term effects of practices such as fertilizer of lime application. There are also major unanswered questions concerning the effects of mechanical forest clearing and controlled burning of peat.

The complexity of the morphology of tidal swamps and their ecological linkages with other coastal ecosystems suggest that large scale and poorly managed agricultural development can have serious ecological effects. However, little money or scientific expertise is being invested in answering such questions. A basic question which must be considered is whether the reclamation of deep peats is justified in terms of the marginal gains to agriculture and the potential loss of estuarine and coastal fisheries.

4. DEVELOPMENT PRESSURES AND PRINCIPAL USES OF TIDAL WETLANDS

4.1 Wetlands Perceived as Underutilized Resources

Tidally influenced lands, principally mangrove and swamp forests, are often viewed as underutilized resources. The development of those lands, commonly termed "reclamation", is seen as a means of coping with a range of social and economic issues including the need to:

1. increase food production;
2. replace lost offshore fisheries;
3. resettle people displaced by development projects;
4. redistribute population;
5. stimulate regional development; and
6. improve the security of a nation's coastline.
   (Burbridge and Stanturf, 1984).

Tidal lands are generally not densely populated, and seemingly support low levels of resource production. These factors, together with the perceived potential for conversion of wetlands to more productive use ("reclamation") have led government officials and planners to formulate large-scale, often single purpose schemes for tidal wetland development.

These schemes involve major investments in infrastructure and reclamation works to transform tidal lands into agricultural fields or aquaculture ponds. Tidal wetlands, however, are a prime example of complex land and water systems whose resource attributes are neither fully understood from an ecological perspective nor valued comprehensively in economic terms.

Similarly, the social aspects of the use of these resource systems by indigenous cultures are poorly understood. Our lack of understanding of the ecologic, economic and social parameters of tidal wetlands development should cause us to proceed cautiously in making major changes in the nature, scale, or rate of exploitation of these resources. Major changes are, however, taking place in the tidal wetlands of Asia and there is mounting concern about the long-term impact of current policies and programmes which commit wetlands to uncontrolled development and irreversible change.

Several considerations, vitally important in properly evaluating the merits and disbenefits of altering these lands, are generally overlooked. These include the:

(i) ecological role of tidal wetlands in supporting diverse resource development activities and in maintaining the functional integrity of coastal ecosystems;
(ii) economic and social significance of the environmental and pecuniary goods and services provide at little or no cost to man by tidal lands in their natural state;
(iii) low chemical fertility and physical unsuitability of many of the mineral soils which limit their development potential;
(iv) high levels of risk and uncertainty associated with the agricultural development of organic soils found in tidal lands; and
(v) the question of whether such development can be sustained in the face of the high levels of capital and management skills required to ensure that these lands produce agricultural yields comparable to dryland areas, or yields from aquaculture that compensate for loss of natural fisheries resulting from the alteration of freshwater swamplands and mangrove.

4.2 Principal Uses and Management Issues

A. Tidal Swamp Forests and Grass Dominated Swamps

i) Forestry In Asia tidal swamp forests contain Dipterocarp species of commercial value. In Indonesia the annual production of such commercially valuable trees is estimated as 25 m3 per hectare per year for tidal forests in Sumatera (Anderson, 1981). This compares favourably with upland forest in Sumatera where 30 m3 per hectare per year can be harvested (Anderson, 1981). There are, however, considerable differences in production between the quality of tidal forests on deep, umbrogenous peat and those on the shallow peat/mineral swamp soil areas. Anderson, (1981) differentiates three classes of swamp forest according to the depth of the peat soil and their corresponding forestry potential. These are:

1. forest on shallow peat which generally has a low potential for economic forestry;
2. forest on peats of moderate depth located on the margins of the deeper umbrogenous peat bogs. This has a high potential for sustainable economic forest production including valuable species such as Ramin, Meranti, Shorea spp and Durio; and

3. the "Padang" forest located on deep peat areas. This forest type has a low potential for economic forestry.

It is significant that the better forest is produced on peat soils of moderate depth (1–2m). These same soils are often the sites for agricultural reclamation because they have greater agricultural potential than the more problematic deep peats.

While sustained yield forestry is feasible in tidal swamplands, the value of such resource production is often discounted in project appraisals favouring agricultural conversion of the forest. Burbridge et al (1981) evaluated the potential net benefit from sustained forest production versus agricultural conversion. Their study compared options for agricultural development of tidal swamp forest, *Imperata cylindrica* grasslands and mature upland forest. The value of the sustained production of forest products from both the tidal and upland forests were comparable and clearly indicated that the rehabilitation of former forest sites which were subsequently dominated by Imperata was a viable option which did not entail the net loss of forestry resource production.

ii) Agriculture The agricultural potential of the mineral and peat soils commonly found in grass or forest covered tidal swamps is generally low. Nevertheless, the huge expanses of tidal forests in areas such as Asia appear to offer a solution to problems such as landlessness, food shortages or the need to redistribute population.

Indonesia is a good example of a nation attempting to fulfill several national development objectives through a vigorous, large scale agricultural programme designed to convert tidal swamps for the production of rice and other food crops. Because a major objective of the programme is the redistribution of population from the crowded islands of Java, Madura and Bali to the less populous "Outer Islands", it is known as the "Transmigration" programme.

Detailed soil surveys are not complete for large areas of Indonesia, but it appears that areas of swamplands left for development are made up mainly of the less agriculturally suitable oligotrophic peats or the more shallow peats overlying potential acid sulphate clays (Soepraptohardjo and Driessen, 1976; Pons and Driessen, 1975; Driessen, 1977; Hanson and Koesoebiono, 1979). The low chemical fertility and potential acid sulphate condition of the peat soils which have not already been developed reduce their agricultural suitability and will require high levels of management to allow them to respond on a sustained basis to agricultural development. These issues will be taken up in greater details in subsequent sections.

Evidence is accumulating that transmigration in the deeper peats is leading to subsidence, erosion (Chambers, 1979) and declining rice yields (Collier et al., 1984). Collier et al., (1984) suggest that the poor performance for some of the official transmigration projects is due to poor site selection and inadequate land clearance. At best these "deep peats" have marginal agricultural potential and reclamation projects appear very expensive in relation to the benefits they are likely to produce. Driessen and Sudjadi (1984 p.147) in their review of tidal swamp soils state that "it is imperative that any research on specific aspects of tidal swampland utilization be preceded by proper land evaluation''. They also give priority to (a) hydrological research (b) the protection of farmland against floods, and (c) a tidal irrigation and drainage.

Experience in the Management of Tropical Wetland Peat Soils Traditional agricultural practices have evolved over centuries, and do not appear to create major problems in the richer but limited areas of eutrophic peat soils. Investigations of the response of the infertile oligotrophic peats to traditional cultivation practices in Indonesia have shown that even minor changes in the ecosystem can cause rapid deterioration of these deep peats (Driessen and Suhardjo, 1975; Driessen and Subago 1975; Driessen et al., 1975). Past attempts at reclaiming oligotrophic peats, and areas underlain by acid sulphate potential soil material, have usually failed. Pons and Driessen (1975) estimated that several hundred thousand hectares of waste land had been created through such failures in Indonesia.

As a result of government sponsored transmigration, spontaneous migration by the Banjarese and Buginese, and to a lesser extent the Javanese and Balinese, Indonesia has accumulated considerable experience in the development of tidal wetlands for agriculture. The reclamation of tidal swamplands has not always proven successful. Most coastal peats are infertile, even when acid sulphate potential of the subsoil is not of concern.

Soepraptohardjo and Driessen (1976) summarized the major problems encountered in reclaiming these lands. Although the physical properties of the oligotrophic coastal peats are generally considered favourable for plant growth, studies have shown that these soils are only marginal in a greater part of the coastal swamps, even after reclamation, due to the following conditions.

1. high subsidence after drainage and removal of the vegetation;
2. extremely rapid horizontal, or slow vertical, hydraulic conductivity, which is also locally variable;
3. high heat storage capacity and low thermal conductivity of many peats, resulting in variable but often lethal temperatures to seedlings at the soil surface;
4. locally slight decomposition of the organic matter, often due to a high percentage of woody material;
5. low bearing capacity of the soil and poor rooting conditions for top heavy crops;
6. rapid oxidation (decomposition) of organic matter after drainage;
7. irreversible shrinkage of the soil profile, causing adverse water retention and increased sensitivity to erosion.

iii) Irrigation Due to the high water table in tidally influenced freshwater wetlands, their exploitation as a source of irrigation water has been considered. In southern Thailand, attempts to pump water from such wetlands to irrigate adjacent agricultural schemes has led to the overdrainage of acid sulphate potential soils with the subsequent acidification of the ground water and consequent damage to crops from the acid irrigation waters.

B. Mangrove and associated Nipah Table III illustrates the wide use made of mangrove and mangrove related products. A significant proportion of the goods and services provided by mangrove are not directly related to forestry. Nevertheless, because mangrove are trees, mangrove are generally managed as a forest system primarily by foresters for the production of forest products. In many countries sustained yield forestry has been achieved for mangrove, however this does not necessarily mean that many mangrove goods and services and the economic and social benefits they provide are sustained. For example, the capture of shrimps in coastal waters which may be dependent upon mangrove for part of their life cycle can be reduced due to forestry operations. These fisheries' questions normally lie outside the interest of foresters, however, increasing attention is being given to the role of mangrove in fisheries and other activities. A case in point is, Indonesia which has instituted a "greenbelt" policy where mangrove considered of value in supporting fisheries are protected within 200–400 metres of the coastline and some 50 metres from rivers.

There are indications that pressures to increase the yield of mangrove wood products involving large scale clear felling are not sustainable (see Section II Hamilton and Snedaker, 1984; Gong and Ong, 1983; Burbridge and Koesoebiono, 1984). Apart from sustained yield forestry, mangrove are often allocated to other uses. Two major uses which entail major disturbance to the mangrove system are aquaculture and mining.

i) Aquaculture In many countries such as Ecuador, Indonesia and Thailand, mangrove have been reclaimed for fish or shrimp ponds. Although traditionally the use of mangrove for fish rearing has been practised for a very long time, few species have been involved and the extent of the practice is thought not to have caused any major impact on the coastal environment (Hamilton and Snedaker, 1984). In recent years this has changed with the rapid increase in the area of mangrove used for the culture of high-value crops such as penaeid shrimp. The export revenues which can be earned from the culture of shrimp give rise to few controls over the conversion of mangrove and, as in the case of Indonesia. Major loans from agencies such as the World and Asian Banks are available to expand shrimp and fish pond operations. Many of these loans do not entail adequate assessments of the impact of the loan supported projects or of the sustainability of the resulting development.

Mangrove environments are not necessarily ideal locations for aquaculture ponds. Site conditions which vary greatly between areas and major factors which influence the success of ponds are tidal range, soils, availability of freshwater, fresh and salt water quality and the supply of fry. Natural hazards such as flooding and storm surges represent major factors influencing the risk of operations in these highly dynamic coastal systems.

The Handbook for Mangrove Area Management (Hamilton and Snedaker, 1984 p75) states that ".. siting ponds in mangrove areas can be a costly mistake and should be resorted to only in the absence of other options." Such options include the siting of ponds above the intertidal area and improving the efficiency of existing ponds *before* sanctioning the use of further mangrove. However, these alternatives entail improved management, increased inputs of fertilizers and other products and investment in improved water exchange systems which are often expensive and beyond the financial and management resources of the operators. Location within the intertidal area provides tidal exchange of water and the opportunity to stock the ponds with wild fry, carried by the incoming tide. Therefore, most of these shrimp and fish ponds are located in mangrove.

The disadvantages of siting ponds in mangrove are: high land clearing and construction costs; difficulty of maintaining water quality and other factors within the pond's environment which are conducive to rapid growth and low mortality of the cultured species; and problems of drainage and harvesting (see Section II Mangrove Area Management Handbook, Hamilton and Snedaker, 1984).

There are a series of related issues which also make the siting of ponds in mangrove unfavourable, including the reduction of nutrient exchange, breeding grounds and nursery areas of importance to coastal fish stocks. A further consequence of disturbance to the mangrove system is the reduction of sources of wild fry. The fry of the Milkfish (*Chanos-chanos*) and *Penaeus* species of shrimp are thought to be dependent upon mangrove and tidal creeks during part of their life cycle. Clearance of mangrove to build ponds creates the risk of reducing the supply of the post larvae which are the prime source of seed stock for the ponds.

Pond operation in mangrove can be sustained due to limitations imposed by their location and the costs of overcoming these limitations are less favourable than sites for intensive cultivation practices. While the World market for high value cash crops remains bouyant, these more marginal ponds will continue to offer their operators a profit. However, with many tropical countries allocating extensive areas to shrimp pond operations, the question of supply and demand will have a bearing upon the long-term viability of ponds with low levels of productivity and high management costs. Little attention is currently being given to the prospect of over supply and the impact of falling prices on marginal ponds. There is a real danger that the areas of extensive yet low productivity ponds will become increasingly less profitable and many will be abandoned. The loss, meantime, to other fisheries resulting from the damage to the mangrove may be substantial in terms of employment, income and protein with which to feed growing populations.

Apart from questions of equity related to who benefits most from pond operations, the long-term role of mangrove and the fisheries and other resource based activities they support in fulfilling national development objectives need to be weighed before decisions to alter mangrove are sanctioned.

Ong (1982) in comparing the productivity of pond operations versus the natural fishing support functions of mangrove observes that "unless the artificial ponds can very significantly surpass the natural ecosystem, the establishment of aquaculture ponds may be a case of robbing Peter to pay Paul—with the possible added cost of having to compensate Peter later". Furthermore, many pond sites are abandoned which raises the question of the sustainability of many mangrove aquaculture operations.

Gong and Ong (1983) examined three uses of mangrove—traditional charcoal production using long term rotation, woodchip production and conversion to aquaculture in Malaysia. They concluded that only the traditional charcoal production was sustainable in the majority of the sites studied.

The weight of evidence suggests that the social and economic advantages of managing mangrove for multiple- function, multiple use purposes outweigh the maximization of any single product or their conversion to alternative use. Short term economic gains, however still appear to rule the day. Major mangrove areas are due to be converted to ponds in Indonesia and many other coastal nations.

An alternative to the conversion of wetlands to form pond sites is the use of estuarine and nearshore waters for mariculture. Two basic operations in waters enriched by mangrove, other wetlands and waters from upland areas are:

1. Bottom culture of cockles, oysters and mussels and other organisms such as seaweeds; and
2. Open water culture using floating cages or rafts and pens for fish molluscs and crabs.

Although shrimp are not easily cultivated using these methods, milkfish (*Chanos-chanos*) are reared successfully in embayments such as Lagunà de Bay in the Philippines.

ii) Mining The extraction of tin through dredging in mangrove is common in Thailand. An estimated 26,189 hectares of mangrove have been mined for tin (Kongsangchai, 1984). The value of the ore and the government royalties are thought to far exceed the cost of rehabilitating the mangrove following mining operations (National Research Council, 1977).

Apart from direct destruction of mangrove through dredging, the discharge of dredge spoil into coastal waters, or directly onto mangrove sites, can destroy young seedlings and the fauna of the mangrove and nearby waters. Sediment from the dredge spoil can also interfere with spawning and feeding of aquatic fauna (Aksornkoae et al, 1980).

iii) Rehabilitation Rehabilitation of mangrove disturbed by mining is meeting with some success in Thailand. Growth rates of mangrove species on such sites are low in comparison with other mangrove areas, and mortality is higher (Kongsangchai, 1984). The long term effects of mining on mangrove and related ecosystems are not yet evident, but concern over degradation and the difficulties of rehabilitation are causing Thai scientists to recommend that mining in mangrove areas be severely curtailed (Kongsangchai, 1984).

4.3 The Question of Sustainability

There are clear indications that the development of aquaculture and agricultural uses in many tidal wetlands is not sustainable. In the limited areas where such developments can be sustained, they remain subject to the natural and, where bad management prevails, man induced hazards or increased vulnerability to natural hazards.

Even where such developments can be sustained, there remains a basic question as to whether their reclamation produces sufficient benefits to compensaten for the loss of environmental and economic goods and services. This question can only be addressed by individual nations, there is no standard answer—yes or no.

Our present management expertise allows us to maintain the integrity of the ecological systems of different wetlands. However, we seldom choose to do this. Instead we continue to take a narrow view of wetland functions and set development targets which aim to maximize specific forms of resource production and largely ignore the value of the goods and services which may be lost.

This is partly due to a general lack of knowledge on the part of officials and politicians. It is also due to inadequate project evaluation where financial rather than more broadly based economic analyses are utilized.

Management frameworks are also at fault. Tidal wetlands are complex systems which have strong functional linkages to rivers, estuaries, lagoons, seagrass beds and other coastal ecosystems. Such linkages are not generally reflected in management arrangements. Estuaries, with their attendant wetlands are seldom managed as units. Instead we allocate the forests—whether mangrove or fresh water swamp forest—to a forestry department. The rivers may be under the jurisdiction of a multitude of bodies ranging from the Navy to the Department of Fisheries. Such agencies, unless they are brought together in a clearly defined management framework with clearly worked out multiple use objectives for the estuarine resources, are not capable of independently managing resources outside their specific concern.

Perhaps the most logical framework to adopt and to adapt to individual regions is the "Coastal Zone Management" approach pioneered in America but now widely used in tropical countries. Using this approach, a zone or area sufficiently large to reflect the unique features of tidal wetlands and other coastal systems can be defined. Relevant management units which reflect the functions of various ecosystems can then be drawn up as for example for coral reefs, estuaries, lagoons etc. Within these management units, special arrangements can be made to bring the various agencies involved in specific activities together to work out common management objectives, plans and procedures.

5. THE EVALUATION OF WETLAND DEVELOPMENT ALTERNATIVES

The common use of the term reclamation when applied to tidal wetlands implies that these lands are unproductive and will be upgraded to some more beneficial use. The trade-off between maintaining tidal lands in their natural functioning state, and their alteration to some alternative use needs careful examination *before* irreversible changes are made.

The evaluation of tidal lands is not easy due to the complexity of the ecological systems and the subtlety and complexity of many of the resource uses and the economic web of activities they stimulate. A fundamental problem in demonstrating the economic value of wetlands and their natural functions is that many of the goods and services they provide are never directly possessed, utilized, or traded by man, although they may be important to production of other goods and services which are easily exchanged. For example, wetlands can provide nursery and spawning areas that are important in the life cycle of wild fish species. These functions support commercial fish stocks which any fisherman can capture and sell for a profit. Without the wetland functions, commercial fish stocks may decline, as will the economic livelihood of the fisherman. Although goods and services such as wetland spawning areas are not directly utilized by the fisherman, they do contribute to the size and value of the fishery.

Other goods from the wetlands can of course be exchanged, such as mangrove timber. Economists have therefore made the distinction between environmental goods and services, which are not normally captured or exchanged in markets, and pecuniary goods and services which can be possessed, valued, and exchanged using market mechanisms. This distinction is helpful in giving economic recognition to environmental goods and services, however the task of defining their value in a comprehensive manner does not easily follow.*

There are two facets of wetland resources which are difficult to deal with in terms of conventional economic analysis. Firstly, many of the goods and services provided are not exchanged in the market place and therefore monetary values are difficult to establish. Secondly, the benefits of many of the goods and services are realised away from the wetland itself. These "off-site" economic benefits are external to the physical boundary of the wetland and have to be treated as economic externalities. An example would be reduction of damage from coastal storms provided by mangrove acting as a buffer to wind and waves.

A very logical way of categorizing wetland resources in terms of whether they are marketed or not and whether they occur within the wetland or at some off-site location appears in Section IV of the Handbook for Mangrove Management (Hamilton & Snedaker, 1984) and is reproduced below as Figure 1. This figure helps to identify explicitly all the potential economic resources of wetlands so that they can be entered into an economic evaluation of the benefits and costs associated with alternative management options.

The term *economic* evaluation is stressed because, unlike a purely *financial* evaluation, it permits a critical examination of the benefits and disbenefits to society which may result from an individual or sectoral proposal.

The following six-step analysis based upon the work of Hufschmidt et al., (1983); Hufschmidt and Hyman, (1982) and Dixon et al., (1983), provides a clear methodol-

---

*For a more full discussion of the economic nature of the wetlands see Burbridge, 1982.

		Location of Goods and Services	
		On - Site	Off - Site
Valuation of Goods and Services	Marketed	1 Usually included in an ecomomic analysis (poles, charcoal, woodchips, mangrove crabs)	2 May be included (fish or shellfish caught in adjacent waters)
	Nonmarketed	3 Seldom included (medicinal uses of mangrove, fuelwood, food in times of famine, nursery areas for juvenile fish, feeding areas for estuarine fish and shrimp, viewing areas for wildlife)	4 Usually ignored (nutrient flows to estuaries, buffers to storm damage)

FIGURE 1. Relation between location and type of mangrove goods and services and traditional economic analysis.
Source: Figure 42, Hamilton & Snedaker (1984)

ogy for valuing a natural system such as a wetland and the effects of alternative development projects/programs:

1) Identify project to be analysed as clearly as possible. — For discrete projects (e.g. a port or a fishpond) this is straightforward. For more extensive projects (e.g. wood chip harvesting in mangrove) this requires more complex analysis.

2) Define physical boundaries of analysis. — The boundaries should be broad enough to incorporate most of the major "externalities" generated by the project would have boundaries for analysis which include the expected benefits and costs which occur offsite (e.g. changes in coastal fish catches, increased storm damage inland, etc.).

3) Identify all the physical inputs and outputs of system. — These include all important inputs/output factors (many of which may be hard to measure). Social surveys are frequently used to obtain information on how natural resource systems are utilized. Net energy analysis may assist in analysing flows.

4) Quantify the physical — Changes in production and yields or other factors previously identified are now quantified to the extent possible (e.g. wood-chip production or charcoal yields from mangrove forests).

5) Place monetary values on the physical quantities measured or alternative costs or benefits. — This is frequently the hardest step. The goal is to explicitly include all the benefits and costs in the analysis. Where monetary issues cannot be determined, the data on physical flows can still be used in the decision-making process.

6) Perform economic evaluation of the proposed program or project (s). — The techniques for economic evaluation are quite straightforward and well-defined. They rely on measurements made earlier (Steps 3 and 5). Among the most commonly used approaches are cost-benefit analysis including net present value, internal rate of return and benefit-cost ratios. Other approaches, such as cost-effectiveness analysis are also used.

The six steps force the analyst to clearly state assumptions and to clearly identify factors that are included or excluded from the economic analysis. The resulting analysis is not an end in itself and provides no final answer, but it does aid the decision-maker by providing a more perceptive and realistic accounting of the economic and related social implications of projects.

By combining economic analysis with an environmental assessment of a wetland it is possible to present a very clear analysis of the ability of the resource system to

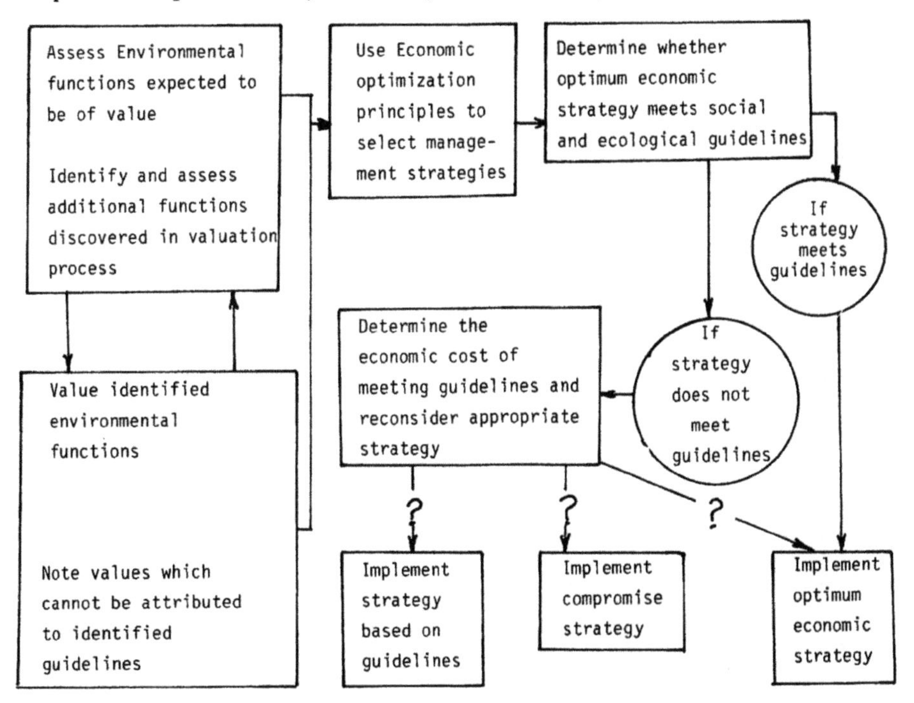

FIGURE 2. The Process of Assessment, Valuation and Selecting a Management Strategy

respond to alternative forms of development and the economic and social trade-offs involved. Figure 2 illustrates how economic and environmental analyses complement one another. This figure is taken from the Coastal Area Management and Planning Handbook produced by the participants in a Coastal Management training program in Thailand edited by Baker and Kaonian (1985). Such handbooks are helping to promote the improved utilization of coastal wetland resource systems and demonstrate the mounting attention given to what were formerly thought to be ''worthless swamps''.

REFERENCES

1. Aksornkoae, S., Kongsanchai, J. and S. Panithsuko, 1980. Mangrove forest in Thailand. 27th Conference of the Ecological Society of Japan, 18–20 July 1980, Hirosaki University.
2. Anderson, J.A.R. 1981. Forestry sector review in Riau Province, *in* Northern Sumatra Regional Planning Study. Jakarta, PADCO, Dept. Public Works and BAPPEDA, Riau.
3. Andriesse, J.P. 1964. The use of Sarawak peat for agriculture. Research Circ. No. 4, Dept. Agriculture, Sarawak.
4. Andriesse, J.P. 1974. Tropical lowland peats in South-East Asia. Comm. 63, Dept. Agric. Res., Royal Tropical Institute, Amsterdam.
5. Baker, I. and P, Kaeoniam (eds.), 1986. Manual of Coastal Development Planning and Management for Thailand. UNESCO and the Thailand Institute of Scientific and Technological Research, Bangkok.
6. Barske, P. 1961. Wildlife on the coastal marshes. Conn. Arboretum Bull. 12:13–15.
7. Boelter, D.H., and G.R. Blake, 1964. Importance of volumetric expression of water contents of organic soils. SSSA Proc. No. 28:176–178.
8. Burbridge, P.R., Dixon, J.A. and B. Soewardi, 1981. Land allocation for transmigration. Bull. Indonesian Economic Studies 17(1):108–113.
9. Burbridge, P.R., 1982. Valuation of tidal wetlands. *in* C. Soysa, L.S. Chia and W.L. Collier (eds.) Man Land and Sea, Agricultural Development Council, Bangkok, pp. 43–64.
10. Burbridge, P.R. and J.A. Stanturf, 1984. The development and management of tidal wetlands in Asia. Proc. MAB/COMAR Regional Seminar, Tokyo, pp. 157–160.
11. Burbridge, P.R. and Koesoebiono, 1984. Management of mangrove exploitation in Indonesia. *in* E. Soepadmo, A.N. Rao and D.J. Macintosh (eds.) Proc. Asian Symposium on Mangrove Environment: Research and Management. Kuala Lumpur 25–29 August 1980, Univ. Malaysia/UNESCO, pp. 740–760.
12. Burbridge, P.R. and J.E. Maragos, 1985. Coastal resources management and environmental assessment needs for aquatic resource development in Indonesia. International Institute for Environment and Development, Washington D.C..
13. Bureau of Sport Fisheries and Wildlife, 1966. Rare and endangered fish and wildlife in the United States. U.S. Dept. of the Interior, Resource Publ. No. 34, Washington D.C..
14. Caspers, H., 1967. Estuaries: Analysis of definitions and biological considerations. *in* G.H. Lauff (ed.), Estuaries (AAAS Publ. No. 83). Washington D.C.: American Association for the Advancement of Science. pp. 6–8.
15. Chambers, M.J., 1979. Rates of peat loss on the Upang transmigration project, South Sumatra. Proc. Third National Symp. Tidal Swampland Development, pp. 765–777.
16. Clark, J.R. 1974. Rookery Bay: Ecological constraints on coastal development. The Conservation Foundation, Washington D.C..
17. Clark, J.R., 1977. Coastal ecosystem management: A technical manual for the conservation of coastal zone resources. John Wiley & Sons, New York, 928 p.
18. Collier, W.L., 1979. Social and economic aspects of tidal swampland development in Indonesia. Occasional Paper No. 15, Development Studies Centre, Australian National University.
19. Collier, W.L., Rachman, B., Supardi, Ali, B., Rahmadi, and A.M. Jurindar, 1984. Cropping systems and marginal land development in the coastal wetlands of Indonesia. Workshop on Research Priorities in Tidal Swamp Rice, International Rice Research Institute, Los Banos, the Philippines.

20. Coulter, J.K. 1973. The management of acid sulphate and pseudo-acid sulphate soils for agriculture and other uses. ILRI Publ. No. 18:255–275, Wageningen, the Netherlands.
21. Darnell, R.M., Pequegnat, W.E., James, B.M., Benson, F.J., and R.A. Defenbaugh, 1976. Impacts of construction activities in wetlands of the United States. U.S. E.P.A., Corvallis, Oregon, Publ. No. EPA-600-3-76-045, 393 p.
22. DeGlopper, R.J. and R.L.H. Poels, 1972. A general study with tentative recommendations to the government on reclamation possibilities in the coastal area of the Central Plain of Thailand. FAO, Bangkok.
23. Dewey, D. and H. Klopper, 1964. Report of the effect of loss of valley storage due to encroachment-Connecticut River. Connecticut Water Resources Commission, Hartford, Connecticut.
24. Dixon, J.A., Hufschmidt, M.M. and S. Wattanavitukul (eds.) 1983. A case study workbook on economic valuation techniques for the environment. East-West Center, Honolulu, unpublished.
25. Driessen, P.M. and Ismangun, 1972. The influence of tidal fluctuations on soil formation and agriculture in the coastal region of Southern Kalimantan. Proc. Second ASEAN Soil Conf. Jakarta, Indonesia.
26. Driessen, P.M. and Subagjo, 1975. Growth and subsidence of tropical ombrogenous peats. A semi-quantitative approach. Third ASEAN Soil Conf., Kuala Lumpur.
27. Driessen, P.M., Soepraptohardjo, M. and L.J. Pons, 1975. Formation, properties, reclamation and agricultural potential of Indonesian ombrogenous lowland peats. Int. Peat Symp., Tel Aviv Israel.
28. Driessen, P.M. and H. Suhardjo, 1975. Reclamation and use of Indonesian lowland peats and their effects on soil conditions. Third ASEAN Soil Conf., Kuala Lumpur.
29. Driessen, P.M. and L. Rochimah, 1976. The physical properties of lowland peats from Kalimantan. *in* Peat and Podzolic Soils and their potential for agriculture in Indonesia, Bull no. 3 Soil Research Institute, Bogor, Indonesia, pp. 56–73.
30. Driessen, P.M. 1977. Peat Soils: Their formation, properties, reclamation and suitability for rice cultivation. Soil Research Institute, Bogor, Indonesia.
31. Driessen, P.M. and M. Sudjadi, 1984. Soils and specific soil problems of tidal swamps, *in* Workshop on Research Priorities in Tidal Swamp Rice. International Rice Research Institute, Los Banos, Philippines, pp. 143–160.
32. Golet, F.C. 1972. Classification and evaluation of freshwater wetlands as wildlife habitat in the glaciated Northeast. Ph.D Dissertation, University of Massachusetts, Amherst. 179 p.
33. Gong, W.K. and J.E. Ong, 1983. The environmental impact of the use of mangroves for aquaculture. University Sains Malaysia, Penang, Malaysia.
34. Gosselink, J.G., Odum, E.P. and R.M. Pope, 1973. The value of the tidal marsh. Urban and Regional Development Center, University of Florida, Gainesville.
35. Grant, R.R. and R. Patrick, 1970. Tinicum Marsh as a water purifier. *in* Two studies of Tinicum Marsh, Conservation Foundation, pp. 105–123.
36. Hamilton, L.S. and S.C. Snedaker, (eds.) 1984. Handbook for mangrove area management. East-West Center, Honolulu, Hawaii, 123 p.
37. Hanson, A.J. and Koesoebiono, 1979. Settling coastal swamplands in Sumatra: A case study for integrated resource management. *in* C. MacAndrews and L.S. Chia (eds.) Developing Economies and the Environment, McGraw-Hill, Singapore, pp. 121–175.
38. Hufschmidt, M.M. and E.L. Hyman, 1982. Economic Approaches to Natural Resource and Environmental Quality Analysis. Dublin, Tycooly International.
39. Hufschmidt, M.M., James, D.E., Meister, A.D., Bower, B.T. and J.A. Dixon, 1983. Environment, Natural Systems and Development: An Economic Assessment Guide. Baltimore, Johns Hopkins University Press.
40. IUCN, 1980. World conservation strategy: Living resource conservation for sustainable development. International Union for Conservation of Nature and Natural Resources, Switzerland.
41. Jamandre, T.J., Jr. and H.R. Rabanal, 1975. Engineering aspects of brackish water aquaculture in the South China Sea region. Manila, FAO/UNDP SCSP/75/WP/16: 96 17 annexes.
42. Janota, I. and O.L. Loucks. 1975. An analysis of the value of wetlands for holding inorganic phosphorous. University of Wisconsin, Center for Biotic Systems (mimeo)
43. Johnson, P.L. 1969. Wetlands preservation. Opens Space Institute, New York, New York. 45 p.
44. Kevie, W. van der and B. Yenmanas, 1972. Detailed reconnaissance soil survey of the Southern Central Plain area. S.S.R. 89, Soil Survey Division, Land Development Department, Bangkok.
45. Kongsangchai, J.H. 1984. Mining impacts upon mangrove forest in Thailand. *in* E. Soepadmo, A.N.

Rao and D.J. Macintosh (eds.) Proc. Asian Symposium on Mangrove Environment: Research and Management. Kuala Lumpur 25–29 August 1980, Univ. Malaysia/UNESCO, pp. 558–567.
46. Kusler, J.A. and C.C. Harwood, 1977. Wetland Protection: A guidebook for local governments. National Wetlands Inventory, U.S. Fish and Wildlife Service, Washington D.C.
47. Larson, 1976. Models for the assessment of freshwater wetlands. University of Massachusetts Water Resources Center, Amherst, 91 p.
48. MacNae, W. 1974. Mangrove forests and fisheries. Indian Ocean Fishery Commission, FAO. Rome, 35 p.
49. Massachusetts Water Resources Commission, 1971. Neponset River Basin Flood Plain and Wetland Encroachment Study. Boston, Massachusetts.
50. Motts, W.S. and R.W. Haley, 1973. Wetlands and groundwater. *in* J. Larson (ed.) A guide to important characteristics and values of freshwater wetlands in the Northeast. University of Massachusetts Water Resources Center, Publ. No. 31.
51. National Research Council of Thailand, 1977. Mangrove laws and regulations. Report Thailand National Task Force, Bangkok, 46 p.
52. Niering, W.A., 1961. Tidal Marshes: Their use in scientific research. Conn. Arboretum Bull. 12:3–7.
53. Niering, W.A., 1968. Wetlands and cities. Massachusetts Audubon Soc.
54. Odum, W.E. and E.J. Heald, 1972. Trophic analyses of an estuarine mangrove community. Bull. Mar. Sci. 22(3):671–738.
55. Ong, J.E. 1982. Mangroves and aquaculture in Malaysia. Ambio 11(5):252–257.
56. Ong, J.E. and W.K. Gong (eds.), 1984. Productivity of the mangrove ecosystem: management implications. Proc. UNDP/UNESCO Workshop, Penang, October 1983. 183 p.
57. Pedini, M. 1979. Penaeid shrimp culture in tropical developing countries. Committee for Inland Fisheries of Africa, Second Session of the Working Party on Aquaculture, Abidjan, 2–29 November 1979, 13 p.
58. Percival, M. and J.S. Womersley, 1975. Floristics and ecology of the mangrove vegetation of Papua New Guinea. Botany Bull. No. 8, Dept. of Forests, Div. of Botany, Lae, Papua New Guinea.
59. Pons, L.J. and P.M. Driessen, 1975. Waste land areas of ologotropic peat and acid sulphate soils in Indonesia, Symp. Wastelands, Jakarta.
60. Pons, L.J., 1988. Fine textured alluvial soils. Problem soils. ILRI Publ. No. 27:68–71, Wageningen, the Netherlands.
61. Pritchard, D.W., 1967. What is an estuary: physical viewpoint. *in* G.H. Lauff (ed.), Estuaries (AAAS Publ. No. 83). Washington D.C.: American Association for the Advancement of Science. pp. 3.5.
62. Salm, R.V. and J.R. Clark, 1984. Marine and Coastal Protected Areas. International Union for Conservation of Nature and Natural Resources, Switzerland, 302 p.
63. Smardon, R.C. 1972. Assessing visual-cultural values of inland wetlands in Massachusetts. MLA Thesis, University of Massachusetts. Amherst. 295 p.
64. Soepraptohardjo, M. and P.M. Driessen, 1976. The lowland peats of Indonesia, a challenge for the future. *in* Peat and Podzolic Soils and their potential for agriculture in Indonesia, Bull no. 3 Soil Research Institute, Bogor, Indonesia, pp. 11–29.
65. Teal, J.M. and M. Teal, 1969. Life and death of the salt marsh. Little. Brown and Co. Boston, 278 p.
66. Tie, Y.L. and H.S. Kueh, 1979. A review of the lowland organic soils of Sarawak. Technical Paper No. 4, Dept. Agriculture Sarawak.
67. U.S. Army Corps of Engineers, 1975. Natural valley Storages: A partnership with nature. Waltham, Massachusetts.
68. Versteegh, F., 1951. Proeve van een bedryfsregeling voor de vloedbossen van Bengkalis (Design for a working plan for the mangrove forests in Bengkalis, Sumatra). Tectona, 41:200–258.
69. Versteegh, F., 1952. Problems of silviculture and management of mangrove forests. Asia and Pacific Forestry Commission, Singapore.
70. Watson, J.G., 1928. Mangrove forests of the Malay Peninsula. Malayan Forest Rec. No. 6. Singapore: Fraser and Neave.
71. Wharton, C.H., 1970. The Southern River Swamp: A multiple use environment. Bureau of Business and Economics Research, School of Business Administration, Georgia State University, Atlanta, 48 p.
72. Wijk, C.L. van, 1951. Soil survey of the tidal swamps of South Borneo in connection with the agricultural possibilities. Contribution No. 123, General Agricultural Research Station, Bogor

Resource Management and Optimization
1990, Volume 7(1–4), pp. 141–169
Reprints available directly from the publisher.
Photocopying permitted by license only.
© 1990 Harwood Academic Publishers GmbH
Printed in the United States of America

MARINE RESOURCES IN THE TROPICS

S.Z. QASIM
Department of Ocean Development, 'Mahasagar Bhavan' Lodi Road, New Delhi—110003
and
M.V.M. WAFAR
National Institute of Oceanography, Dona Paula, Goa 403 004

CONTENTS

The biological productivity of five different tropical marine ecosystems, estuaries, coastal waters, ocean waters, coral reefs and mangroves (excluding seaweeds) determine their exploitable resources, mainly on the basis of primary productivity and trophic structure.

Based on their overall biological productivity, these ecosystems can be ranked, in terms of their exploitable resources: coastal upwelling areas, estuaries, coral reefs, shelf waters, mangroves and open ocean.

In the Mandovi-Zuari estuarine system, a prediction of fish yield, based on the production rates of phytoplankton, zooplankton and benthos compares well with its existing yield. Similar computations made for the coastal waters of India in 1976 indicated a scope for expansion of fishery along some coastal states and overfishing in one state. Fish yield for the latter years till 1981 did increase in all these states with the exception of one state for which overfishing was predicted. In these ecosystems, high primary production leading to a short food chain results in high fish yield. Conversely, in the open ocean, low primary production as a result of low nutrient availability and a long food chain ends up in low fish yield per unit area.

Coral reefs are the most productive of all tropical marine ecosystems where several producer-components contribute to primary productivity. Fish yield from this environment is generally higher than in many shelf waters with the exception of upwelling zones. Mangroves are also productive ecosystems, but these do not sustain important fisheries. Their influence, however, is manifested in the form of fisheries of adjacent waters and particularly in shrimp fisheries.

1. INTRODUCTION

Living resources of the sea are defined as the exploitable stocks which include herbivores and primary and secondary carnivores. These resources can be estimated in the form of "standing stocks", from which "potential yield" or "annual sustainable yield" can be determined. What could be regarded as the standing stock of a particular ecosystem and how best it can be measured quantitatively is a subject for discussion in this chapter. In recent years, a large number of measurements of primary production have indicated that the areas richest in marine resources are normally those which have a high rate of primary production, even if it is sustained for comparatively limited periods (Steele, 1965; Qasim, 1970). Leaving aside the higher latitudes, which have a high rate of daily production during a season of 4–6 months and are responsible for supporting large fisheries (Steele, 1965), the upwelling areas of the tropical and subtropical regions, which become exceptionally rich in nutrients for brief periods, in turn give rise to large fishery yield (Qasim, 1977). The commercially exploitable resources of the sea can be regarded as the final links in the food chain that begins with the sun's energy and its conversion by plant communities into organic matter and ending up with exploitable fish, crustaceans and molluscs.

During the last two decades considerable information has accumulated on the production mechanisms of tropical ecosystems. Hence it would be impossible to present a review of a large variety of different situations in this chapter. We have therefore chosen to confine ourselves to a consideration of a) estuaries, backwaters and lagoons; b) coastal waters; c) oceanic waters; d) coral reefs; e) mangroves; and f) seaweeds. Our main purpose is to evaluate the productive processes involved and to determine the overall picture of what and how much could possibly be exploited in the form of resources.

2. ESTUARIES, BACKWATERS AND LAGOONS

2.1 Ecology

An estuary is usually defined as a semi-enclosed coastal body of water which has a free connection with the open sea and within which the sea water is measurably diluted with freshwater derived from the rivers and land drainage (Pritchard 1967). Although this definition is generally adequate, a fuller understanding of an estuary's biological and chemical characteristics warrants a functional classification. The most commonly used criterion is based on salinity distribution. Accordingly, estuaries are classified as well-mixed, weakly or strongly stratified, fjords and arrested salt-wedge. Backwaters are, by definition, estuaries from which they commonly differ in that they consist of a system of inter-connected brackishwater lagoons and swamps, fed often by more than one river, and having more than one connection with the sea. A coastal lagoon can be defined as "a coastal zone depression below mean higher high water, having permanent or ephermeral communication with the sea, but protected from the sea by some type of barrier" (Lankford, 1976). Freshwater flow into coastal lagoons is often marginal, and evaporation generally exceeds dilution.

Tropical estuaries are generally coastal plain estuaries, with homogeneity in vertical salinity distribution during the dry season and weak to strong stratification in the wet season. In perennial river estuaries, which are filled with freshwater nearly up to the mouth during the period of heavy precipitation and run-off, a salt wedge formation occurs (Dyer and Ramamoorthy, 1969; Qasim and Sen Gupta, 1981).

Tides, freshwater flow and salinity are the most important environmental factors in a tropical estuary. Strength of tides and tidal currents in conjunction with the quantity of freshwater flow determines whether an estuary is well-mixed or stratified. In estuaries, typical tidal velocities range from 50 to 100 cm s^{-1} and the net river flow may range from 1 to 2 cm s^{-1} (Officer, 1983). When the river flow is very low, increased turbulent mixing induced by tidal currents renders a major section of an estuary vertically homogeneous. Conversely, during the period of high river run-off, stratification is produced. A typical example of this type of estuary is that of Cochin backwaters (Qasim *et al.*, 1969) which oscillates from the state of vertical homogeneity during the premonsoon season (February–May) to stratification in the monsoon months (June–September). Increased freshwater flow reduces the average salinity in the estuary significantly. In the tropical monsoonal estuaries, such as those along the west coast of India, the very high precipitation ($\approx$ 300 mm) during the monsoon months and the associated run-off have a profound effect on the salinity distribution. Salinity varies at the surface from near zero during the monsoon to about sea water in the premonsoonal months. In the deeper layers variations in salinity become less important.

2.2 Biological Productivity

Phytoplankton productivity of estuaries is controlled by the prevailing salinity regime, physical stability of the water column, turbidity and the consequent rapid attenuation of light, freshwater flow and nutrient addition, and cell losses during the estuarine flushing. Unlike temperate estuaries, water temperatures in tropical estuaries remain fairly high throughout the year and are not a major environmental factor. These physico-chemical features and phytoplankton productivity vary widely from among estuaries. However, even among those with widely varying physico-chemical attributes, there exist considerable similarities of primary production rates, which range from near zero to 2.5 gC m^{-2} d^{-1}, with an average of 0.58 gC m^{-2} d^{-1} (Boynton *et al.*, 1982). Under these conditions, a comparison of several estuaries will be of limited usefulness. Instead, we present here results on biological productivity studies carried out for several years in a tropical monsoonal estuary, the Mandovi-Zuari system. Similar extensive studies have been carried out on Cochin backwaters for nearly two decades and the results have been summarized occasionally (Qasim *et al.*, 1969; Qasim, 1970, 1973 and 1979).

The Mandovi-Zuari system is a tide-dominated coastal plain estuary which is homogeneous vertically but has lateral variations in salinity and also develops a salt wedge. The average tidal range in this estuary is about 2.0 m. The homogeneity of vertical salinity persists for about 8 months and moderate stratification sets in for about 3–4 months in the monsoon. Annual variations in the water temperature are small (5–6°C), whereas salinity variations are large (near zero–34%). Flushing-out (residence) time of water in the Mandovi is about 5–6 days during the monsoon and about 50 days during the dry season. In the Zuari, the residence time is somewhat longer. The addition of nutrients to the estuary peaks in the monsoon months. However, there is a further addition of nitrate to the estuaries in the post-monsoon months (October–January), as a direct result of application of nitrogenous fertilizers in the low-lying wetlands and there washing-out with river flow (Qasim and Sen Gupta, 1981). The depth of the euphotic zone varies seasonally from 0.75 to 6 m, being shallower during the monsoon and deeper at other times (Devassy, 1983).

2.2.1 Phytoplankton production: Annually, in the entire estuarine system, phytoplankton cell counts vary from 4.75×10^3 to 1370×10^3 cells l^{-1}. Phytoplankton abundance is generally high in the poast-monsoon period, moderately high in the pre-monsoon season and low in the monsoon months (Devassy, 1983). However, large phytoplankton blooms occur often in the low-salinity regions of the estuary in the monsoon (Bhattathiri *et al.*, 1976), presumably because of the increased addition of nutrients and the ability of several phytoplankton species in these waters to optimize photosynthesis at salinities ranging from 10 to 20% (Qasim *et al.*, 1972 *a*).

Maximum primary production occurs in the post-monsoon season when environmental conditions become favourable and vertical mixing releases nutrients to the entire water column. In the pre-monsoon season, although environmental conditions and available light are more favourable than in the post-monsoon period, primary production decreases, probably because of a nutrient limitation. In these months, regenerated nitrogen may become important (Verlencar, 1982). Low production in the monsoon months is associated with the low incident light and increased turbidity. By season primary production rates in the Mandovi estuary are: 570, 262 and 1077 mg C m^{-2} d^{-1} for pre-monsoon, monsoon and post-monsoon periods, respectively (Verlencar, 1982). Average primary production for the entire estuarine system is 510 mg C m^{-2} d^{-1} (Devassy, 1983).

2.2.2 Zooplankton production: The abundance and diversity of zooplankton in tropical estuaries is controlled primarily by the salinity regime and flushing time. In the Cochin backwaters, during the pre-monsoon season, the average high salinity and reduced flushing time leads to a sustained high zooplankton biomass. The lowest biomass values occur in the monsoon period, owing to rapid flushing and reduced salinity. In the post-monsoon season there is a partial recovery in abundance (Qasim *et al.*, 1969; Wellershaus, 1974).

Zooplankton production rates in the Mandovi-Zuari estuarine system were calculated from the biomass data and copepod regeneration time (Goswami, 1979; Bhattathiri *et al.*, 1976). Vast fluctuations in the production rates, from 1 to 83 mg C m^{-2} d^{-1} were observed. The average production rate was 22 mg C m^{-2} d^{-1} (Goswami, 1979). Similar estimations made by Bhattathiri *et al.* (1976) gave production rates ranging from 3 to 71 mg C m^{-2} d^{-1} during the monsoon period.

2.2.3 Benthic production: Parulekar *et al.* (1980) studied the macrobenthic biomass and production in the Mandovi-Zuari estuarine system. The annual mean biomass was 4.08 gC m^{-2} and a production estimate—based on Sanders' (1956) hypothesis that for macrobenthos dominated by polychaetes-bivalves it is about twice the standing stock—was 8.16 gC m^{-2} yr^{-1}. Low production occurs in the monsoon months, and high production rates were found either during the pre- or post-monsoon seasons, depending on the sampling station.

Characteristic seasonal cycles of the physico-chemical features and biological production rates in the Mandovi and Zuari estuaries are shown in Fig. 1.

2.3 Trophic Relationships and Fish Yield

The average primary production of Mandovi-Zuari estuarine system, in Goa, is 186 gC m^{-2} yr^{-1}; average secondary production (combining zooplankton and macrobenthos) is 16.19 gC m^{-2} yr^{-1}. The tertiary production calculated as 1% of primary production and 10% of secondary production, and then averaged, is equivalent to 1.74 gC m^{-2} yr^{-1} or 13 g live wt m^{-2} yr^{-1} (using a factor of 7.47). For the estuarine area of approximately 92 km^2, the estimated tertiary yield is 1200 ^{t} yr^{-1}. The present fish yield from the brackish and freshwater regions of Goa is about 1400 ^{t} yr^{-1}, of which more than 80% comes from the Mandovi-Zuari estuarine system. Thus the calculated tertiary yield predicts the actual fish yield with reasonable accuracy. Tertiary production is constituted by euryhaline resident fishes (pelagic) and shrimps, clams and to a smaller extent, crabs (demersal).

Thus, a prediction of the final biological yield as fish resources from an estuary is possible by an assessment of productivity rates at different trophic levels, as has been demonstrated above. However, for several other tropical estuaries such data are either not available or inadequate. Nor are there statistical data on fish catch, as fishing is mainly for sustenance and the yields in smaller estuaries are often less than a few hundred tonnes. Marten and Polovina (1982) summarize the fish yields from some tropical estuaries and coastal lagoons. Yields vary from 1 to $>$ 100 ^{t} km^{-2} yr^{-1}.

Although in many estuaries pelagic fishes form the principal component of the tertiary yield, demersal resources can be of considerable importance in others. For example, both the Hooghly–Maltah estuary on the east coast and the Cochin backwaters on the west coast of India, sustain nearly similar yields, at 14000–17000 ^{t} yr^{-1} (Jhingaran, 1982). In the former, however, the principal yield ($>$ 80%) is from pelagic

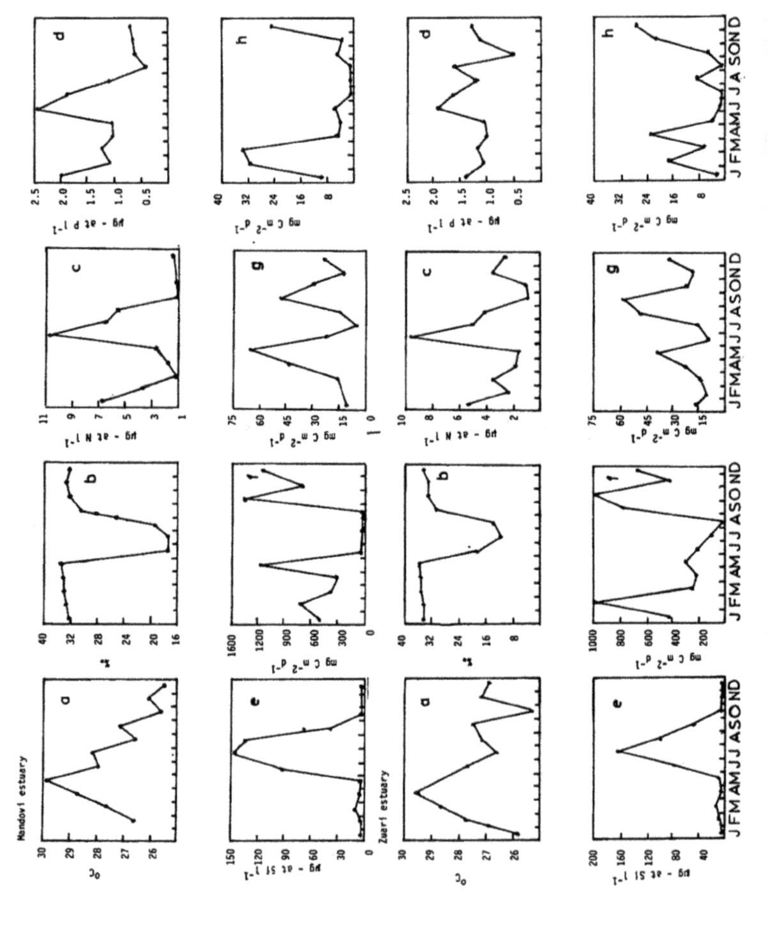

FIGURE 1. Characteristic seasonal cycles of some physicochemical parameters and biological productivity at different trophic levels in the Mandovi and Zuari estuaries. a) temperature; b) salinity; c) nitrate; d) phosphate; e) silicate; f) primary production; g) zooplankton production; h) benthos production.

fishes whereas in the latter, it is comprises up to 60–70% shrimps (juveniles of *Metapenaeus dobsoni, M. monoceros* and *Penaeus indicus*). The predominance of demersal fishery in the Cochin backwaters can be related directly to the quantitative importance of detritus transported to the bottom of the estuary, compared with the primary production. In the Cochin backwaters settled detritus production averages 14.96 gC m^{-2} d^{-1} (Qasim and Sankaranarayanan, 1972), whereas average primary production is only 0.77 gC m^{-2} d^{-1} (Qasim *et al.*, 1969). Thus the average food energy available for demersal forms is about 20 times greater than that for pelagic forms. The importance of detritus and detritus-based productivity is also reflected in other benthic resources of the Cochin backwaters; clams, next in importance to shrimps, are fished at a rate of about 88,000 ^{t} yr^{-1}.

3. COASTAL WATERS

3.1 Ecology

In terms of biological productivity coastal waters follow only estuaries in importance. The average primary production in coastal waters is about 164 g C m^{-2} yr^{-1} and the continental shelves of the world oceans, which account for only about 7.4% of the total ocean area (26.6×10^6 km^2), contribute about 18% (4.4×10^9 ^{t} C yr^{-1}) of the total marine primary production (Whittle, 1977). Continental shelf waters sustain primary production rates that on an average are three time higher than those of oceanic waters (Whittle, 1977). When the production rates are grouped near the coast, i.e. < 50 m depth, where the major fishing effort is concentrated, and off the coast (> 50 m depth), the average production at depths < 50 m is six times greater than in the offshore waters (Qasim, 1979). Such a high primary production is results primarily from two factors: 1) freshwater run-off into the coastal waters which adds a considerable amount of nutrients, and (2) wind-induced mixing in the relatively shallow coastal regions which brings the nutrients regenerated in the sediments into the entire water column.

Coastal upwelling zones in shelf areas are of particular interest from the perspective of biological productivity. These areas become exceptionally rich in nutrients for brief periods and give large fishery yields. The Peruvian upwelling system sustains primary production rates > 1000 gC m^{-2} yr^{-1}, and before it became overfished yielded about 20% of the world's annual fish catch (Walsh *et al.*, 1980). Ryther (1969) is of the opinion that the 0.1% of the oceans' areas where upwelling occurs support 50% of the world's fish catch. Though contested by Cushing (1971)—"production cycles in upwelling areas and temperate waters are essentially the same"—it nevertheless indicates the importance of upwelling zones in the biological productivity of coastal waters.

Along the west coast of India biological productivity of the shelf waters (< 50 m depth), at different major trophic levels, has been fairly well investigated and productivity rates have been correlated with the pelagic and demersal fisheries (Qasim *et al.*, 1978; Parulekar *et al.*, 1982). The results of these studies are presented below. We also discuss briefly the productivity of the Peruvian upwelling, the biologically richest coastal upwelling system in the world, and finally the fish yield of tropical coastal waters in general.

3.2 West Coast of India

3.2.1 Primary production Upwelling occurs along the west coast of India during the south-west monsoon and reaches its peak in July-August. During this process the nutrients from deeper waters are brought to the surface. In addition, in this period, large quantities of nutrients are added to the coastal waters through heavy precipitation and river run-off. Following this, the primary production during the late monsoon and post-monsoon periods attains its highest levels, which range from 0.48 to 2.45 gC m^{-2} d^{-1}, with an average of 1.19 gC m^{-2} d^{-1} (Nair *et al.*, 1973). However, during the pre-monsoon months the average primary production is about 3 times lower, at 0.33 gC m^{-2} d^{-1} (Qasim *et al.*, 1978). Combining these two sets of data, Qasim *et al.* (1978) calculated an average production rate of 0.76 g C m^{-2} d^{-1}. This value is about 10 times greater than the 0.070 g C m^{-2} d^{-1} reported for tropical oceanic waters (Koblentz-Mishke *et al.*, 1970). For most of the west coast of India and the southern part Tamil Nadu, on the east coast covering a total area of 42,525 km^2, primary production was equivalent to 11.80×10^6 t C yr^{-1}.

3.2.2 Zooplankton production Using zooplankton biomass data (dry weight and calorific values), species composition and copepod generation time, computations of secondary production were made. The average standing stock was 144 mg C m^{-2} and the average daily production was of the order of 125 mg C m^{-2}. For the estimation of total secondary production the production values in different ranges were first calculated separately and then summed. The production was found to be about 2.5×10^6 tonnes C yr^{-1} (Qasim *et al.*, 1978).

3.2.3 Prediction of tertiary and sustainable yields. The tertiary yield, calculated as 1% of primary production and 10% of secondary production, was averaged. The average value was 0.185×10^6 t C yr^{-1}. In terms of live weight, using a conversion factor of 10, the potential yield was calculated at about 2 million t yr^{-1}. From this, the sustainable (harvestable) yield was taken as 40%, since the fishery is dominated by pelagic species of short life-span, and thus a figure of 0.8 million t yr^{-1} was computed (Qasim *et al.*, 1978).

Those results are summarized in Table I by state for the four west coast and one east coast states studied. The ratio of harvestable to exploited yield was taken to indicate the state of fishery. Thus, when the ratio is one or close to one, fishing is at its optimum. Higher ratios indicate potential for further fishing whereas lower ratios reflect over-

TABLE I

Biological Productivity of the Coastal Waters of Some Indian States, Along with a Comparison of Calculated Harvestable Yields and the Exploited Yields

State	Coastal Area under Study (Km^2)	Primary Production (cal × 10^{14} yr^{-1})	Secondary Production (cal × 10^{14} yr^{-1})	Potential Fish Production (million t live wt yr^{-1})*	Harvestable Yield (million t live wt yr^{-1})	Exploited Yield (million t live wt yr^{-1}) Average of 1972–76	Ratio of Harvestable Yield to Exploited Yield	Exploited Yield (million t live wt yr^{-1}) Average of 1977–81
Southern Maharashtra	3600	104.2	23.68	0.181	0.072	0.033	2.2	0.037
Goa	3200	101.7	30.45	0.215	0.086	0.028	3.1	0.027
Karnataka	11725	364.5	71.31	0.571	0.228	0.089	2.6	0.129
Kerala	18000	621.2	54.20	0.616	0.246	0.383	0.6	0.321
Southern Tamil Nadu	6000	163.5	26.96	0.230	0.092	0.054	1.7	0.059

(after Qasim *et al.*, 1978)
*Calculated using '1 g dry wt of fish = 4718 cal', and 80% moisture.

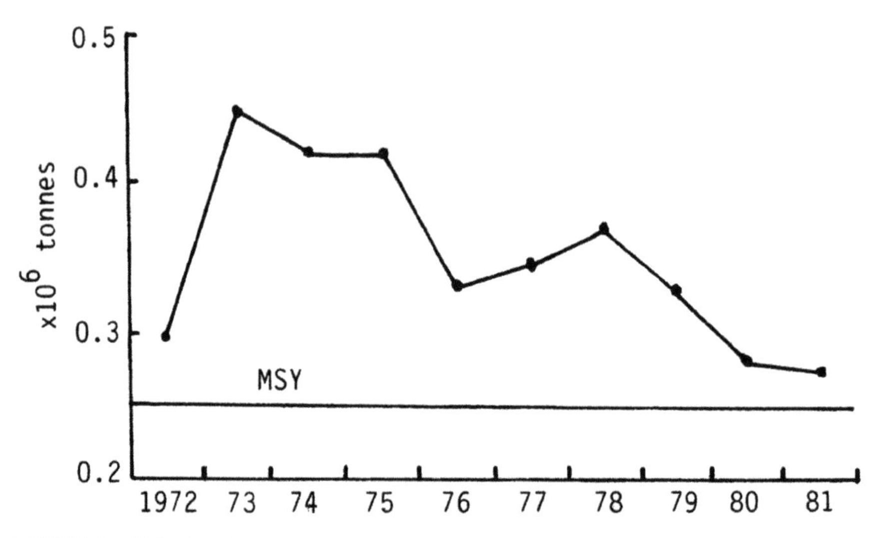

FIGURE 2. Fish yield of Kerala state, 1972–1981. The calculated maximum sustainable yield (MSY) is 0.25 million ᵗ.

fishing. From such an analysis it was concluded that there is scope for further expansion of fishery in all the states except Kerala, where the stocks seem to be overfished.

A comparison of the fish catch figures for the five years preceding the prediction and for the following five, should indicate the accuracy of the prediction. Although the average catch in Goa has remained almost stable, in Karnataka the yield has increased by about 45%, in southern Maharashtra by about 12% and in southern Tamil Nadu by nearly 10%. Of particular interest is the Kerala fishery for which overfishing was predicted. In this state, the average yield has declined by 16%. The yield for 1981 (0.345 million ᵗ) is about 40% lower than that of the peak year (1973) of 0.45 million tonnes. This is probably the results of over-exploitation during the preceding years. The decrease in the fish yield of Kerala became apparent as early as 1973–74 and is still continuing (Fig. 2).

3.2.4 Benthic production and demersal fisheries Parulekar *et al.* (1982) estimated the benthic production in the shelf waters of the west and east coasts of India. The average benthic production was 2.06 gC m^{-2} yr^{-1} for the Arabian Sea and 3.43 gC m^{-2} yr^{-1} for the Bay of Bengal. However, the production values in terms of wet weight in the two seas are nearly the same, being 2.68 g m^{-2} yr^{-1} in the former Sea and 2.76 g m^{-2} yr^{-1} in the latter. The potential demersal yield computed from these values was 0.75 million ᵗ for the Arabian Sea and 0.33 million ᵗ for the Bay of Bengal, totalling about 1.08 million ᵗ yr^{-1} for the Indian shelf waters. Of this, about 60% (0.65 million ᵗ) can be considered as exploitable yield (Parulekar, pers. comm.). On the other hand, the present yield is about 0.45 million ᵗ yr^{-1}, which suggests scope for increasing the demersal catch.

3.3 Peruvian Upwelling

Because of its high biological productivity, with a potential yield of 13 million t of fish (Gulland, 1971), the coastal upwelling along the Peruvian current is the most-studied such system in the world. Of this the anchoveta *Engraulis ringens* alone accounted for 12 million t in the peak fishery period of 1970 (Walsh, 1981). The upwelling zone extends along the Peru-Chile coast for about 2000 km (4–22°S), has an offshore width of about 50 km, and occupies an area of 1×10^5 km^2 (Walsh, 1981).

Estimates of primary production in this region vary widely, from 155 to about 1000 gC m^{-2} yr^{-1}. Such wide variations result from the varying assumptions made for the width and length of the upwelling area (Gulland, 1971) and also from the small-scale and intermittent localised variations in the upwelling process (Strickland *et al.*, 1969). However, production rates for 1966–1978 indicate that the mean production is about 4 g C m^{-2} d^{-1}, and at the rate of at least 3 g C m^{-2} d^{-1} the annual production can be well over 1000 g C m^{-2} (Walsh, 1981). Such a high primary production and the short food chain (phytoplankton → fish) results in fish yields of up to 100 t km^{-2} yr^{-1}.

The rise and collapse of the anchoveta fishery in the last three decades is a classic example of overfishing coupled with a significant decline in the survival of anchovy larvae during warm water intrusion (the *El Niño* phenomenon). The annual anchoveta harvest off the coast of Peru increased from 700,000 t (1958) to a maximum of about 12 million t in (1970). It then declined to about 2 million t (1973) and then to about 1 million t (1977–1979) (Walsh, 1981). This also illustrates how the energy flow in the ecosystem and the tertiary yields can readjust to new conditions.

The large range of the anchoveta yield before and after overfishing was accompanied by marked changes in trophodynamics (Walsh, 1981). A decline in the grazing pressure of anchoveta led to an increase in plankton biomass, a much greater detrital carbon production and the transfer of organic carbon to the sediments, leading to increased meio- and microbenthic production (Table II). Abundance of phyto- and zooplankton after the decline of the anchovy stocks resulted in an increase in sardine stocks, and the increased carbon flux to the benthos caused an increase in the demersal hake stock (Table II).

3.4 Other Tropical Coastal Waters

Marten and Polovina (1982) summarize the maximum sustainable yields (MSYs) of pelagic and multispecies demersal fisheries in tropical coastal waters (Table III). For demersal fisheries the MSYs were found to be correlated significantly with depth, higher MSYs appearing at depths <50 m. Conversely, the MSYs of pelagic fisheries were significantly correlated with primary productivity but not with depth. The authors explain the stronger correlation between pelagic fisheries and primary productivity as resulting from a direct connection between pelagic fish and the planktonic food chain. Demersal fish, on the other hand, were found to have a much less direct connection

TABLE II
Changes in Annual Carbon Fluxes of the Peru Upwelling Ecosystem Between 1966–69 and 1976–79

	1966–69 gC $m^{-2}yr^{-1}$	1976–79 gC $m^{-2}yr^{-1}$
Primary production	1570* (a,b)	1351 (b,c,d)
Copepod production	100^+	100^+
Euphausiid production	2.5^+	5^+
Anchovy yield	60^{*+}	6^{*+}
Sardine yield	0.4*	4*
Hake yield	0.05	0.7 (e)
Bacterioplankton production	215^+ (f)	22^+
Microbenthic production		9^{*+} (h)
Meiobenthic production	2^{*+} (g)	2.5^{*+} (h)
Macrobenthic production	0.5^{*+} (i)	0.1^+
Detrital carbon production	320^+ (a)	720^+
Sinking loss	105^+	698^+
Sediment storage/export	82^+	591^+

After Walsh (1981)
*Measured; $^+$calculated; (a) Walsh, 1975 *a*; (b) Barber and Smith, 1981; (c) Walsh *et al.*, 1980; (d) Doe, 1978; (e) Bruzhinin and Pshenichry (n.d.); (f) Walsh, 1975 *b*; (g) Pamatmat, 1971; (h) Rowe, (n.d.); (i) Rowe, 1971.

with the pelagic food chain. Noteworthy, therefore, is that even if the demersal fisheries have little direct relationship with the pelagic food chain, their MSYs still vary as a function of the primary productivity of the superadjacent water column. Marten and Polovina (1982) draw attention to the findings of Qasim (1979) that the average primary productivity in Indian coastal waters shallower than 50 m is about six times higher than in waters deeper than 50 m. The MSYs of multi-species demersal fisheries above and below 50 m, listed by them (Table III), vary in about the same proportion.

4. OCEANIC WATERS

Oceanic waters are the least productive of all marine ecosystems. Their average chl *a* concentration of 0.03 g m^{-2} is about 10 times lower than that in upwelling zones, and about 6 times lower than in shelf waters. Their average productivity rate of 60 gC m^{-2} yr^{-1} is about 3–5 times lower than in the shelf or upwelling areas (Whittle, 1977). This low production is, however, compensated to a certain extent by their vast area; about 92% of the total marine area. The total primary production for the entire oceanic realm is about 18.9×10^{9} ᵗ C yr^{-1}, which is requivalent to about 75% of the total marine productivity. Of this, oligotrophic tropical seas (excluding equatorial upwelling) occupying an area of 148×10^6 km^2, have a mean primary production rate of 70 mg C m^{-2} d^{-1} and a total production of 3.79×10^{9} ᵗ of C yr^{-1} (Koblentz-Mishke *et al.*, 1970).

TABLE III
Estimated Maximum Sustainable Yields of Tropical Pelagic and Demersal Marine Fisheries

Location	MSY ($mt/km^2/yr$)	Depth (m)	Primary Productivity ($gC/m^2/yr$)	Source
	Pelagic			
Java Sea (N. Coast of Java)	6.02	0–50	180	SCS (1979)
South Atlantic	4.43	0–550	135	Klima (1977)
Sumatra (West Coast	4.28	0–40	130	SCS (1979)
India (West Coast)	3.58	0–50	180	Anon. (1979)
Malaysia (West Coast)	3.38	0–100	130	Yesaki (unpub. data)
Gulf of Mexico (coast)	3.20	0–550	90	Klima (1977)
Atlantic (South America)	2.35	0–550	135	Klima (1977)
India (East Coast)	1.98	0–50	180	Anon. (1979)
India (West Coast)	1.03	0–200	135	Anon. (1979)
Thailand (West Coast)	1.02	0–100	55	SCS (1976)
South China Sea	0.81	0–500	45	SCS (1973)
India (East Coast)	0.70	0–200	90	Anon. (1979)
Philippines (offshore)	0.55	200 & more	110	Menasveta *et al.* (1973)
	Demersal			
North Coast Gulf of Mexico	6.7	0–110	135	Klima (1977)
U.S. Atlantic Coast (N. Carolina-Florida)	5.5	0–110	135	Klima (1977)
Gulf of Thailand	3.9	0–50	365	SCS (1978)
Philippines	2.8	0–200	135	Aoyama (1973)
North Coast of Java	2.6	0–50	180	SCS (1979)
Sunda Shelf—South	2.3	0–50	75	SCS (1978)
South China Sea	2.0	0–200	45	Aoyama (1973)
West Coast of Florida	1.8	0–110	135	Klima (1977)
Sunda Shelf—NW Borneo	1.7	0–50	75	SCS (1978)
South Coast of Kalimantan (Borneo)	1.5	0–50	135	SCS (1979)
Sunda Shelf—NW Borneo	1.1	50–200	75	SCS (1978)
Gulf of Thailand	1.1	50–200	365	SCS (1978)
Sunda Shelf	0.8	50–200	135	SCS (1978)
U.S. Atlantic Coast (N. Carolina-Florida)	0.5	110–548	45	Klima (1977)
North Gulf of Mexico	0.5	110–548	45	Klima (1977)
West Coast of Florida	0.4	110–548	45	Klima (1977)

After Marten & Polovina, 1982

Such low rates of primary production in oceanic waters are directly related to their macronutrients impoverishment, particularly of nitrogen. Wafar *et al.* (n.d.) studied the distribution of nitrogenous nutrients in relation to primary production in the oceanic waters of the Arabian Sea (8–10°N). Mean NO_3 concentrations in the euphotic zone were 0.1–0.5 μg-at l^{-1} and mean NH_4 concentrations were from traces to 0.35 μg-at

1^{-1}. Though the NO_3 concentrations were up to 20 μg-at 1^{-1} below the euphotic zone, transport into the euphotic zone was severely limited by the strong thermocline. The vertical eddy diffusion constant was 0.56 cm^2 sec^{-1} and the "new" nitrogen (NO_3) diffused into the euphotic zone accounted for only 37% of the primary production. "Regenerated" nitrogen (NH_4) supplied about 23% of the nitrogen requirements of phytoplankton. Thus the inorganic nitrogen was clearly a limiting factor for primary production. By contrast, the organic nitrogen was abundant (>8 μg-at N 1^{-1}) but its mode of utilization was not apparent. Urea could have been used but the quantity regenerated was not sufficient to supply the nitrogen required for the remaining quantum of photosynthesis.

Within the vast expanse oceanic waters with low productivity, however, equatorial upwelling zones are distinguished as regions of high productivity, as a result of the semi-permanent upwelling of intermediate waters of high nutrient content. These regions of equatorial upwelling, although relatively narrow in the western part of the oceans, widen to the east, where the phenomenon becomes well-marked (Wyrtki, 1966). In the Atlantic Ocean from 6–7°N to 7–8°S, upwelling is permanent east of 20–30°W; in the eastern part of the Pacific, equatorial upwelling is more constant and may be distinctly traced east of 160°–180°E, widening eastwards until it extends from 8–12°N to 6–8°S; in the Indian Ocean, strong upwelling at the Equator is sometimes observed in the eastern part of the ocean (Vinogradov, 1981).

The intensity of upwelling increases from west to east (Vinogradov, 1981). Correspondingly, phytoplankton biomass and productivity also increase from west to east. In the equatorial divergence of the eastern Pacific, from 155°W to 97°W, phytoplankton biomass increases from 3 to 49.5 g wet wt m^{-2}, cell counts from 26,000 to 1.742 $\times$ 10^6 cells ml^{-1} and productivity, from 0.48 to 3.15 gC M^{-2} (Sorokin *et al.*, 1975). According to Koblentz-Mishke *et al.* (1970), equatorial upwelling zones have an average primary production of about 200 mg C m^{-2} d^{-1}, which is 3 times higher than that for tropical oligotrophic waters. Equatorial divergence and the temperate oceanic waters have the same range of primary productivity, and together spread over 86.5 $\times$ 10^6 km^2 (Koblentz-Mishke *et al.*, 1970). If two thirds of this area is assumed to be a region of equatorial upwelling (Gulland, 1971), the productivity of this region will be about 4.2 $\times$ 10^9 ^{t} C yr^{-1}. The total productivity of tropical oceanic waters will then be equivalent to about 8 $\times$ 10^9 tonnes C yr^{-1}.

Though the oceanic waters account for 75% of the total marine primary productivity, fish yields from these waters, on a unit area basis, are several times lower than those of shelf waters. The tuna and billfish catches in the eastern tropical Pacific range from 0.002 to 0.04 ^{t} km^{-2} yr^{-1}, with an average of 0.024 ^{t} km^{-2} yr^{-1} (Calkins, 1975). In the tropical Atlantic off Africa they are as high as 0.05 ^{t} km^{-2} yr^{-1} (ICCAT, 1980; quoted in Marten and Polovina, 1982). The range of the existing yields in the tropical oceans is probably close to the MSYs (Marten and Polovina, 1982) and a comparison of these with the MSYs of the shelf waters (Table III) clearly shows a difference of an order of magnitude or more. Such a low yield in the open ocean is the result of a longer food chain, from phytoplankton to tertiary stage carnivores, encompassing five trophic levels, where much of the energy is lost at each level, as contrasted with a shorter food chain (phytoplankton → fish, or phytoplankton→zooplankton→fish) encountered in upwelling and several other coastal waters. Moreover, in oceanic environment, the food chain is also detritus-based, with detritus contributing more than 90% of the energy available at the primary levels and having a mean calorific value higher than in nearshore or estuarine environments (Qasim *et al.*, 1979). This might tend to lengthen the food chain even further.

5. CORAL REEFS

5.1 Ecology

In this chapter the term "coral" refers to coelentrates of the order Scleractinia (Class Anthozoa) that secrete massive calcareous skeletons and harbour endosymbiotic zooxanthellae. These are also known as hermatypic or reef-building corals. By contrast, ahermatypic corals are solitary and lack zooxanthellae.

Hermatypic corals are stenotypic and are limited to warm saline waters, essentially between the tropics of Cancer and Capricorn, where the water temperature does not fall lower than 18°C. Though stenotypic, requiring clear waters with uniform conditions of temperature and salinity, hermatypic corals have also been recorded from turbid waters with muddy bottoms (Goreau and Yonge, 1968), and from waters where temperatures fall to 10°C (Macintyre and Pilkey, 1969) and where salinity decreases to 20% for several months a year (Qasim and Wafar, 1979). The depth distribution of reef-building corals is restricted to the illuminated layers of the sea, a condition clearly associated with the endosymbiotic zooxanthellae, which require light for photosynthesis.

The concerted growth of hermatypic corals over several thousand years results in the formation of coral reefs. Basically, there are three major types of coral reefs: fringing or shore reefs which develop nearshore, barrier reefs separated from the shore by a lagoon, and atolls, annular reefs formed by the submergence of oceanic volcanic peaks. Minor types include table reefs, faros, micro-atolls, knolls and patch reefs.

The world's coral reefs cover an estimated 6×10^5 km^2, or 0.17% of the ocean area. More than half (54%) lie in the Asiatic Mediterranean and Indian Ocean. Of the remaining, Pacific reefs account for 25%, Atlantic for 6%, Caribbean for 9%, Red Sea for 4% and Persian Gulf reefs for 2% (Smith, 1978).

5.2 Primary Production

Coral reefs are among the most productive marine ecosystems, with daily gross production rates in the order of 2–12 gC m^{-2}. A summary of gross production and respiration rates (community metabolism) of several atolls and fringing reefs is provided by Qasim *et al.* (1972*b*), Sournia (1977) and Lewis (1981). Such high productivity of coral reefs is derived from several product-components: zooxanthellae, coral boring filamentous algae, attached benthic macroalgae, microalgae in sands, and phytoplankton.

5.2.1 Zooxanthellae The endosymbiont is the dinoflagellate *Gymnodinium microadriaticum,* which occurs in the coral polyps at densities of a million or more cells cm^{-2} of the coral surface. Chlorophyll *a* content of the zooxanthellae varies from 5 to 15 μg $million^{-1}$ cells (Titlyanov, 1981) and their productivity rate is about 0.9 $gC\ m^{-2}$ of reef day^{-1} (Scott and Jitts, 1977).

5.2.2 Boring filamentous algae A majority of these algae, living symbiotically with the corals, belong to the genus *Ostreobium.* Though their contribution to the total plant biomass in the corals is much higher than that of zooxanthellae (Odum and Odum, 1955), they contribute to only $< 10\%$ of the total productivity of the coral (Kanwisher and Wainwright, 1967), because of the low light energy reaching the level of these algae below the coral tissue and skeletal matter (Halldal, 1968).

5.2.3 Calcareous algae The calcareous algae (Rhodophyceae—*Lithothamnion, Porolithon* and other Corallinaeceae) are the most important secondary frame builders in a reef. Though their role in reef calcification is fairly well-known, studies on their organic production are very few. The available figures indicate that the net production rates of the coralline algae range between 0.66 $gC\ m^{-2}\ d^{-1}$ (Marsh, 1970) and 5.7 $gC\ m^{-2}\ d^{-1}$ (Littler, 1974).

5.2.4 Other benthic macroalgae The benthic macroalgae are another major group of primary producers in a reef environment. Qasim *et al.* (1972*b*) studied the abundance and productivity of benthic macroalgae on an atoll in the Lakshadweep Archipelago. In most of the algae, photosynthesis for 12 hours exceeded respiration for 24 hours, and the gross production of the macroalgae in the lagoon was 3.84 $gC\ m^{-2}\ d^{-1}$. By comparison, the reef had a gross production of 6.15 $gC\ m^{-2}\ d^{-1}$.

Seagrasses often form extensive beds in the lagoons of atolls, and contribute substantially to the overall productivity of coral reefs. Gross production of a seagrass bed on Kavaratti atoll, composed of *Thalassiosira hemprichii* and *Cymodocea isoetifoli,* was 11.97 $gC\ m^{-2}\ d^{-1}$, and respiration was 6.16 $gC\ m^{-2}\ d^{-1}$, with a P/R ratio of 1.94 (Qasim and Bhattathiri, 1971).

5.2.5 Benthic microalgae Benthic microalgae are increasingly known as important primary producers on a reef. In a Pacific atoll Sournia (1976) found that their biomass and productivity were 1000 and 50 times greater, respectively, than those of the phytoplankton per unit area. More recently, Sorokin (1981) has shown that the production of benthic microalgae (sand flora and periphyton) can be as high as 2–6 $gC\ m^{-2}\ d^{-1}$—a figure comparable to the gross production of many coral reefs.

5.2.6 Phytoplankton Primary production by phytoplankton and its contribution to reef productivity has been extensively studied. The consensus is that phytoplankton production in coral reefs is not particularly important and represents only the production rates of the surrounding waters, being often in the range of tens of mg C m^{-2} d^{-1}. On a unit area basis, phytoplankton production is only about 1% of total benthic production (Sournia, 1977).

5.3 Variety of Resources

5.3.1 Pelagic The pelagic resources of coral reefs are the endemic fish fauna. These can be further divided into food fishes and aquarium fishes.

5.3.1.1 Food fishes The high productivity of coral reefs indicates a high fisheries potential of reef fishes. Sevenson and Marshall (1974) summarize the studies on the abundance of fish aggregating on and around several coral reefs as well as their fishery. Their study shows that the fish biomass varies from 38 to 209 g m^{-2} and that the annual harvest ranges from 0.4 to 5 t km^{-2}. These rates compare favourably with the demersal fisheries of the North Sea and Georges Bank (< 5 t km^{-2} yr^{-1}). The yield of reef fish can vary far more widely, from <1 to 20 t km^{-2} yr^{-1}, with an average of 5.25 t km^{-2} yr^{-1} (Marten and Polovina, 1982). The annual fish yield of some Philippine coral islands is well above 10 t km^{-2} (Alcala and Luchavez, 1981). Smith (1978) while extrapolating the fish yield from Caribbean and North Atlantic reefs to the calculated world reef area, estimated a potential fisheries yield of 6×10^9 kg yr^{-1} (= 10 t km^{-2} yr^{-1}) for the world's reefs. This is roughly equivalent to 9% of the total oceanic fish landings and clearly shows that coral reefs, though located in nutrient-poor and plankton-impoverished oceanic waters, are as productive in terms of fishery yield as many shelf waters. The yields of some coral reef fisheries are shown in Table IV.

Recreational fishing is another important source of fish yield from reefs. However, it is well-developed only along certain areas, such as the Great Barrier Reef. Craik (1981) estimated that the recreational fishing yield (390,000 kg) in the Capricornia section of the Great Barrier Reef was three times higher than the commercial catch from the same area (130,000 kg).

5.3.1.2 Aquarium fishes Commercial exploitation of aquarium fishes from coral reefs has gained importance only recently, and accurate harvest and marketing data are lacking. The industry is important in the Philippines, where revenues from the export of aquarium fish increased from $120,000 (US) (1970) to $2.5 million (US) (1979) (Albaladejo and Corpuz, 1981). Hong Kong and Singapore are also important centres for the aquarium fish industry.

5.3.2. Demersal resources The demersal resources are principally molluscan forms exploited mostly for their shells and less for their meat. Corals themselves are exploited for the souvenir trade, and in some developing countries they are widely used as raw

TABLE IV
Yields from Coral Reef Fisheries

Location	Area (km²)	Catch (t/km²)	Source
Samoa	3	18[b]	Wass (1982)
Philippines	1	18[b]	Alcala (1981)
Samoa		8[b]	Hill (1978)
Ifaluk (Pacific)	6	5.1	Stevenson and Marshall (19
East Africa		5	Gulland (1979)
Mauritius	350	4.7[a]	Wheeler and Ommanney (1953)
Fiji		4.4[a]	Bayliss-Smith (pers. comm.)
Jamaica	2,860	4.1	Munro (1978)
Bahamas		2.4[a]	Gulland (1971)
Puerto Rico	2,300	0.8[b]	Juhl and Suarez-Caabro (197
Kapingamaringi (Pacific)	400	0.7[b]	Stevenson and Marshall (1974)
Cuba	55,000	0.5	Buesa Mas (1964)
Lamotrek (Pacific)	44	0.45[b]	Stevenson and Marshall (197
Bermuda	1,035	0.4	Bardach and Menzel (1957)
Raroia (Pacific)	400	0.09	Stevenson and Marshall (197

(After Marten and Polovina, 1982).
[a]MSY based on catch-effort relation over series of years.
[b]Probably near the MSY because of heavy fishing intensity.

materials for the cement and carbide industries with deplorable environmental consequences. On atolls coral boulders are commonly used for house construction.

The molluscan fishery of many reefs largely exports shell resources commercially. For example, the fishery for the coral reef snail *Trochus niloticus* in the Indo-Pacific has a dockside value of approximately $3 million (US) and an export value of roughly twice that amount (Heslinga, 1981). Similarly, the pearl oyster fisheries in the Gulf of Mannar yielded in 1961 a revenue of about $30,000 (US) (Mahadevan and Nayar, 1973). Fishing for pearl oysters is done at irregularly, often 10–20 years. However, the sacred chank (*Xancus pyrum*) fishery in the Gulf of Mannar is regular and the annual sale of shells earns about $20,000 (US). Further, several other molluscs are exploited commercially and traded for their ornamental value. Thus the annual trade in sea-shells runs to several million $ (US) (Wells, 1981).

Corals are also used as building-blocks, for road construction and in the lime and cement industries. Smith (1978) has estimated that the world reefs precipitate about 6 $\times 10^{12}$ moles of $CaCO_3$ yr^{-1}. This is equivalent to a $CaCO_3$ production of about 6 $\times$ $10^{8\ t}$ yr^{-1}.

The increasing exploitation of corals, particularly the such branching forms as *Acropora* spp. and *Pocillopora* spp., as decorative pieces, is a serious threat to the very existence of many reefs. Over the past 15 years the commercial trade in corals in the Kavaratti atoll of the Lakshadweep Archipelago has now reached such proportions that during a cruise in May 1984 the entire lagoon floor and the inner reef platform was found to have been completely denuded of all branching forms of corals. On this same atoll extensive removal of massive corals for construction purposes has led to severe erosion of the shoreline.

6. MANGROVES

6.1 Ecology

Mangrove is a formation of trees and shrubs along some coasts tropical and subtropical coasts. Their distribution is largely limited to the tidal zone, and usually to a narrow strip of shallow coastal waters, deltas, estuaries and lagoons. There is a total of 55 mangrove species worldwide (Chapman, 1970) and mangrove forests cover an estimated area of 0.3×10^6 km^2 (Ajtay *et al.*, 1979). Of that total, 44 occur in the Indian Ocean—western Pacific; 7 species are known from Pacific America, 9 from Atlantic America and 7 from the west Africa (Chapman, 1970). Mangrove vegetation may be divided geographically into two groups: that of the Indo-Pacific region and that of the western Africa and Americas (Walsh, 1974). By habitat, mangroves are classified into coastal, river bank, and creek-bordering mangroves.

Temperature is the main ecological factor that determines the development and survival of mangroves. Well-developed mangroves are found only along coasts where the average lowest temperature remains above 20°C and where the seasonal temperature range is less than 5°C (West, 1956). Salinity and tidal range influence the zonation of mangroves. Watson (1928) related mangrove species to tidal inundation conditions and described five classes: 1) inundated by all high tides; 2) inundated by medium high tides; 3) inundated by normal high tides; 4) inundated by spring tides, and 5) inundated occasionally by exceptional or equinox tides. Macnae (1968) related the mangrove vegetation of Southeast Asia to salinity. Zonation ranges from *Avicennia intermedia,* which prefers the coastal seawater, to *Sonnerati caseolaris,* which prefers low salinity upstream areas. Areas of intermediate salinity are occupied by a range of other mangroves. Substratum also plays a vital role in mangrove development. Extensive mangroves are found in the regions with a soft substratum composed of fine silt and clay rich in organic matter (Schuster, 1952).

6.2 Productivity

6.2.1 Productivity of mangrove vegetation Mangroves are a specialized tropical marine ecosystem that have productivity rates comparable to several other such ecosystems, like the coral reefs or the seagrass beds, which are also specialized. However, a major portion of this production is derived from the terresterial component, i.e. the mangrove vegetation, rather than from the aquatic component. Golley *et al.* (1962) made some of the first measurements on mangrove productivity in Puerto Rico. Production was estimated to be 8 gC m^{-2} d^{-1}, community respiration was of the same order but there was, nevertheless, an export of particulate matter of 1.1 gC m^{-2} d^{-1}. A reassessment of the data of Golley *et al.* (1962) gave a net production of 3.4 g organic matter m^{-2} d^{-1} (Miller, 1972). In the red mangroves of Florida, net production rates were 1 gC m^{-2} d^{-1} (Heald, 1971) and 2.8 g organic matter m^{-2} d^{-1} (Miller, 1972). Net production in, summarized for several mangroves varies from nil to 7.5 gC m^{-2} d^{-1} (Lugo and Snedekar, 1974). However, a net production of 1–2 gC m^{-2} d^{-1} appears to be a reasonable value for many mangroves.

6.2.2 Aquatic productivity As with several specialized tropical marine environments, phytoplankton production in mangroves represents only that of the surrounding water (i.e. the estuarine or the nearshore waters flowing in and out of the mangroves). Often this production is inadequate for the energy requirements of the whole community. Untawale *et al.* (1977) studied the community metabolism in a mangrove swamp and found that the gross production was 2.24 gC m^{-2} d^{-1} and respiration 2.90 gC m^{-2} d^{-1}, with a P/R ratio of 0.77. Pant *et al.* (1980) achieved similar results, with gross production varying from 0.8 to 2.06 gC m^{-2} d^{-1} but with P/R ratios of only 0.4 to 0.65.

The net phytoplankton production in mangroves is generally from 10 to 100 mgC m^{-3} hr^{-1}. Some observed values are: 14–86 mgC m^{-3} hr^{-1} in Goanese mangroves (Pant *et al.*, 1980); 11–91 mgC m^{-3} hr^{-1} (Teixeira *et al.*, 1969) and 11–79 mgC m^{-3} hr^{-1} (Tundusi, 1969) in Brazilian mangroves, and 2–8 mgC m^{-3} hr^{-1} in Coondapoor mangroves (India) (Untawale *et al.*, 1977). However, Krishnamurthy and Sundararaj (1973) reported an average net production of about 525 mgC m^{-3} hr^{-1} from the Pichavaram mangroves of India. This rate is unusually high, being several times greater than the averages from elsewhere.

In mangroves the lack of a substantial production of phytoplankton is offset by detritus production. Much of this detritus originates from the mangrove vegetation, since its production ultimately enters the aquatic food chain as detritus. The importance of detritus in mangroves can be illustrated as follows. In a mangrove at Goa leaf-fall adds about 40 g of material, which produces 25 gC m^{-2} d^{-1}, whereas phytoplankton production adds only 1–2 gC m^{-2} d^{-1} (Pant *et al.*, 1980). Thus the energy available as detritus is at least 10 times greater than the phytoplankton production. This is also reflected by the composition of secondary producers. For instance, although basically carnivorous, in mangroves shrimps subsist largely on detritus.

6.3 Resources

6.3.1 Mangroves Wood is the major mangrove resource. mangrove forests. Though the growth rates of mangrove trees remain for detailed study, some data on the wood yield from some managed mangrove forests are available. In Matang (Malaysia), the average yield over many years of the forest reserve is 8–10 m^3ha^{-1} yr^{-1}; in Thailand, plantations of *Rhizophora apiculata* have mean annual increments of 16 m^3 ha^{-1} yr^{-1}; in the Sunderbans of Bangladesh, however, the average yield during the last 20 years was only 1.9 m^3; ha^{-1} yr^{-1} (FAO, 1982).

Mangrove wood is used for several purposes. Its major use is as timber, which is put to a variety of uses. Mangrove wood is also used in charcoal production and as fuelwood. Tannin from the bark is used in leather-tanning, and the ink and plastic industries, as well as for boiler water, oil-well drilling and formaldehyde glues used in the manufacture of plywood and plastic board. The fruits and seedlings of some mangroves are edible, and the foliage of some are used as cattle fodder.

6.3.2 Mangrove-related fisheries Mangrove fisheries include those for the shrimps, of which the most important are *Acetes* spp. and *Macrobrachium* spp., crabs (*Scylla serrata*), molluscs (oysters, clams and snails) and fish. Fishing in mangroves is generally for subsistence only. However, the mangroves exert considerably indirect influence on other commercial fisheries, particularly for shrimps, since the post-larvae and juveniles of most shrimps depend on mangroves for shelter and food. A correlation between shrimp landings and mangrove areas has been established in Indonesia (Martosubroto and Naamin, 1977). Again, the well-known estuarine fisheries, such as those of the Sunderbans region of India and of the Gulf of Mexico, are fundamentally mangrove-based fisheries.

Mangrove areas are extensively used for brackish-water aquaculture. In 1973 about 0.4 million ha of mangrove forests in Indo-Pacific were under brackish-water pond culture (FAO, 1982) and by 1977, this had risen to 1.2 million ha (Saenger *et al.*, 1983). An estimated 2.5 million ha of mangrove forests and tidal flats might still be available for aquaculture (FAO, 1982).

7. SEAWEEDS

7.1 Ecology

Seaweeds are marine macrophytes that require light, water and mineral ions to produce organic matter. They colonize suitable areas along the coastal margins from near high water mark to a maximum depth of about 200 m (Lüning, 1981). As with marine phytoplankton, the growth and productivity of seaweeds are controlled by light, nutrients and a wide variety of other environmental factors, such as the type substratum or the level of dessication in the intertidal region (Lüing [1981], De Boer [1981] and Norton *et al.* [1981]).

7.2 Resources

Seaweeds form one of the major living resources of the sea. The potential yield of brown algae in the world oceans has been estimated at 14.6 million ᵗ (Table V) and that of red algae at 2.61 million ᵗ (Michanek, 1975). As shown in Table V, regions of maximum

TABLE V
Potential Harvest of Seaweeds in the World Oceans (in Thousands of Tons Wet Weight)

Area	Red Algae*	Brown Algae*
Arctic Sea	—	—
NW Atlantic	100	500
NE Atlantic	150	2000
WC Atlantic	(10)	1000
EC Atlantic	50	150
Mediterranean/ Black Sea	1000	50
SW Atlantic	100	2000
SE Atlantic	100	100
W Indian Ocean	120	150 (1000, Kerguelen)
E Indian Ocean	100	500
NW Pacific	650	1500
NE Pacific	10	1500
WC Pacific	50	50
EC Pacific	50	3500
SW Pacific	20	100
SE Pacific	100	1500

(After Michanek, 1975.)
*Broad indications of possible annual output
NW = North West; NE = North East; WC = Western Central;
EC = Eastern Central; SW = South West; SE = South East;
W = West; E = East.

potential seaweed resources are: the North East, West Central and South West Atlantic, and the North West, East Central and South East Pacific. Large potential resources in the Indian Ocean are relatively fewer. The average annual harvest of the world oceans for 1974–77 is about 1.36 million ^{t}, of which nearly 1 million ^{t} are brown algae (Fig. 3). The maximum yield is from Japan (0.6 million ^{t}), followed by South Korea (0.26 million ^{t}).

Seaweeds have a variety of uses; as food, fodder, fertilizer and in the pharmaceutical and chemicals industries. In Japan, South Korea, Taiwan and other countries, seaweeds—particularly *Porphyra*—form a staple component of human diet.

8. CONCLUSION

From the preceding discussion, it is evident that the fishery resources of any ecosystem correlate closely with the primary productivity and the type and length of the food chain in that ecosystem (see also Fig. 4). Given an understanding of the rates of production at different trophic levels and the nature of the food chain, it is possible to

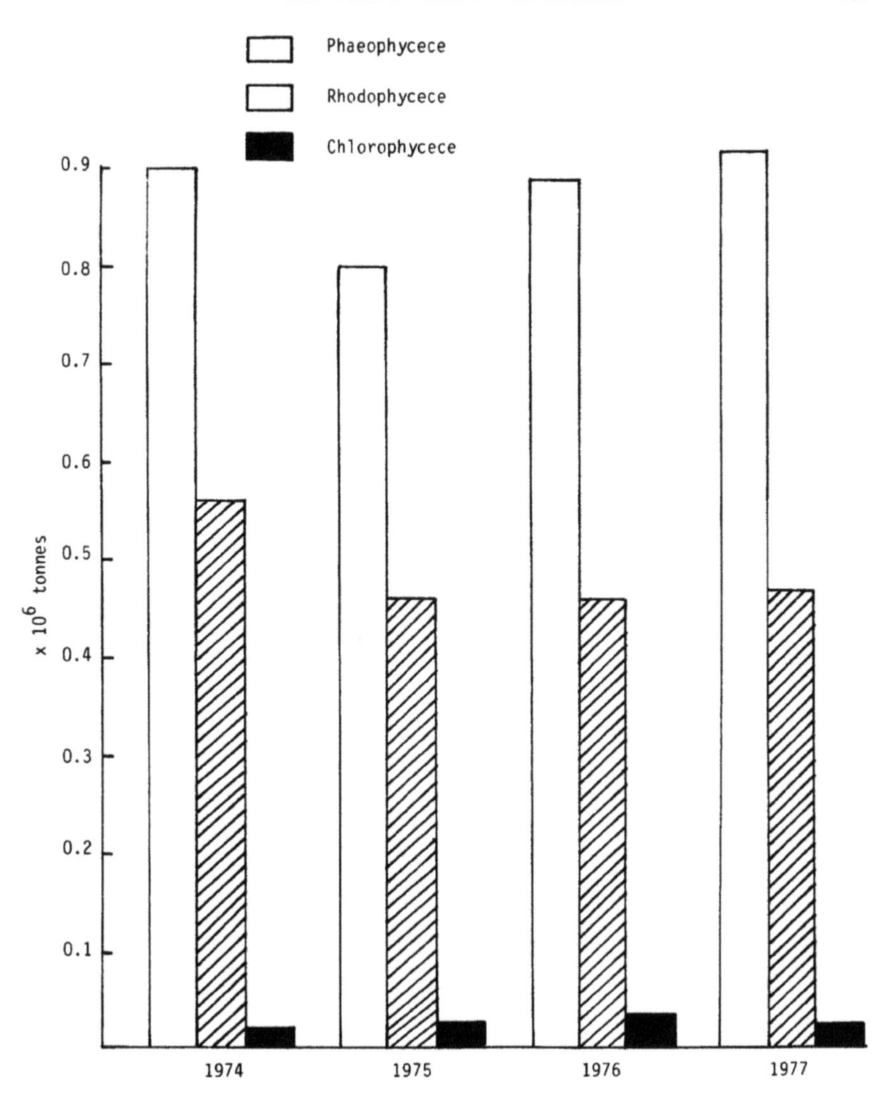

FIGURE 3. Harvested yield of seaweeds from the world oceans for the years 1974–1977 (Source: FAO Yearbook of Fishery Statistics, 1977).

predict reasonably the yield of harvestable fish resources, as we have demonstrated with reference to the Mandovi-Zuari estuarine system and the coastal waters of India. Given a knowledge of the complexity of the food web and trophodynamics of an ecosystem, it is also possible to envisage the type of fish yield. Examples are the shift

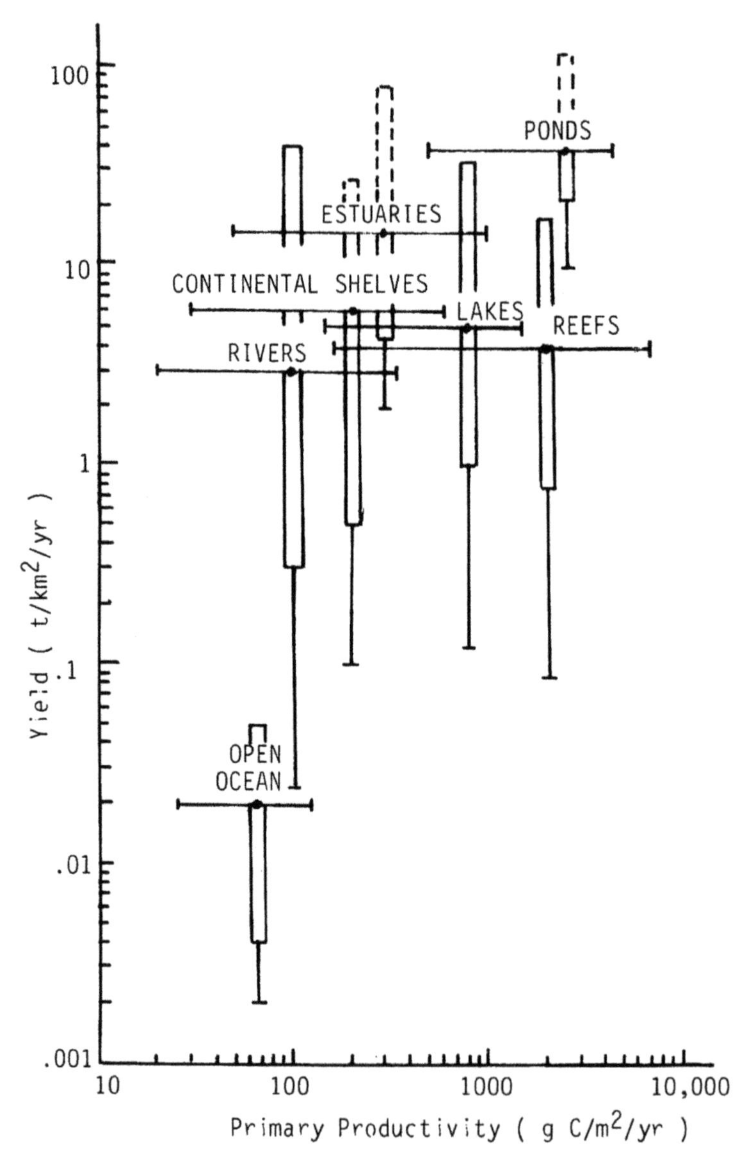

FIGURE 4. Range of fish yields and primary productivities in various tropical ecosystems. Dots at the intersection of ranges represent modal values. Thickened portions of the bars represent the range of maximum sustainable yields. Dashed projections at the top of the ranges for estuaries and ponds represent elevated yields from aquaculture with fertilization (but not supplemental feeding). The dashed projection for continental shelves represents higher yields which occur in areas of upwelling (from Marten and Polovina, 1982).

from ancoveta to sardine and hake in the Peruvian upwelling following the overfishing of anchoveta, or the relationship between mangroves and shrimp fisheries in adjacent waters.

Thus it is clear that biological oceanography provides useful concepts and ideas about living resources that can be used by fishery scientists to improve technologies for fish detection and exploitation. Until now our ability to predict stock sizes and set limits on exploitation have not been particularly fruitful, largely because fishery science has operated separately from related marine sciences. But during the last decade theoretical understanding of marine ecosystems has improved considerably and the influence of the major environmental factors on living resources is now beginning to be understood. With a greater interdisciplinary evaluation of resources it should be possible to predict with greater confidence stock sizes, as well as to recommend measures of exploitation and develop management models that will permit fisheries to be managed to provide maximum sustainable yields.

REFERENCES

1. Ajtay, G.L., Ketner, P. and Duvigneaud, P. (1979). 'Terresterial primary production and phytomass', in *The global carbon cycle*, SCOPE Report 13 (Eds. B. Bolin, E.T. Degens, S. Kempe and R. Ketner), pp. 129–181, John Wiley & Sons, Chichester.
2. Albaladejo, V.D. and Corpuz, V.T. (1981). 'A market study of the aquarium fish industry of the Philippines: an assessment of the growth and the mechanics of the trade', *Proc. Fourth Int. Coral reef Symp.*, Vol. 1, 75–81.
3. Alcala, A.C. (1981). 'Fish yield of coral reefs of Sumilon Island Central Philippines: implication for coral reef resources management in the Philippines', *Nat. Res. Counc. Philipp. Bull.*, 36, 1–7.
4. Alcala, A.C. and Luchavez, T.F. (1981). 'Fish yield of the coral reef surrounding Apo Island, Negros Oriental, Central Philippines', *Proc. Fourth Int. Coral reef symp.*, vol. 1, 69–73.
5. Anon. (1979). 'Trends in total marine fish production in India—1976', *Mar. Fish. Inf. Serv. T and E Ser.*, 9, 7–22.
6. Aoyama, T. (1973). *'The demersal fish stocks and fisheries of the South China Sea'*, SCS/DEV/73/3, pp. 46, FAO, Rom
7. Barber, R.T. and Smith, R.L. (1981). 'Coastal upwelling ecosystems in *Analysis of marine ecosystems* (Ed. A.R. Longhurst) pp. 31–68, Academic Press, London.
8. Bardach, J.E. and Menzel, D.W. (1957). 'Field and laboratory observations on the growth of certain Bermuda reef fisheries', *Proc. Gulf Caribb. Fish. Inst.*, 9, 106–11.
9. Bhattathiri, P.M.A., Devassy, V.P. and Bhargava, R.M.S. (1976). 'Production at different trophic levels in the estuarine system of Goa', *Indian J. Mar. Sci.*, 5, 83–86.
10. Boynton, W.R., Kemp, W.M. and Keefe, C.W. (1982). 'A comparative analysis of nutrients and other factors influencing estuarine phytoplankton production', in *Estuarine comparisons* (Ed. V.S. Kennedy), pp. 69–90, Academic Press, London.
11. Bruzhinin and Pshenichry (n.d.)
12. Buesa Mas, R.J. (1964). 'Las pesquerias cubanas', *Cent. Invest. Pesq. Contrib.*, 21, pp. 93.
13. Calkins, T.P. (1975). 'Geographical distribution of yellowfin and skipjack tuna catches in the eastern Pacific Ocean and fleet and total catch statistics, 1971–1974', *Inter-Amer. Trop. Tuna Comm. Bull.*, 17, 1–163.
14. Chapman, V.J. (1970). 'Mangrove phytosociology', *Trop. Ecol.*, 11, 1–19.
15. Craik, G.J.S. (1981). 'Recreational fishing on the Great Barrier Reef', *Proc. Fourth Int. Coral reef Symp.*, Vol. 1, 47–52.
16. Cushing, D.H. (1971). 'A comparison of production in temperate seas and the upwelling areas', *Trans. Roy. Soc. South Africa*, 40, 17–33.

17. DeBoer, J.A. (1981). 'Nutrients', in *The Biology of Seaweeds* (Eds. C.S. Lobban and M.J. Wynne), pp. 356–392, Blackwell, Oxford.
18. Devassy, V.P. (1983). '*Plankton ecology of some estuarine and marine regions of the west coast of India*', Ph.D. Thesis, University of Kerala, India, pp. 276.
19. Doe, L.A.E. (1978). 'Project ICANE. A progress and data report on a Canada-Peru study of the Peruvian anchovy and its ecosystem', *Bed. Inst. Oceanogr. Rep. Ser.*, BI-R-78- pp 211.
20. Dyer, K.R. and Ramamoorthy, K. (1969). 'Salinity and water circulation in the Vellar estuary', *Limnol. Oceanogr.*, 14, 4–15.
21. FAO. (1982). 'Management and utilization of mangroves in Asia and the Pacific', *FAO Environment Paper,* 3, pp. 160, FAO, Rome.
22. Golley, F., Odum, H.T. and Wilson, R.F. (1962). 'The structure and metabolism of a Puerto Rican red mangrove forest in May', *Ecology,* 43, 9–19.
23. Goreau, T.F. and Yonge, C.M. (1968). 'Coral community on muddy sand', *Nature,* 217, 421–423.
24. Goswami, S.C. (1979). '*Secondary production in the estuarine, inshore and adjacent waters of Goa*', Ph.D. Thesis, Panjab University, Chandigarh, India, pp. 248.
25. Gulland, J.A. (1971). '*The fish resources of the oceans*', Fishing News (Books) Ltd., Surrey, England.
26. Gulland, J.A. (1979). '*Report of the FAO/IOD workshop on the fishery resources of the western Indian Ocean south of the equator, Mahé, Seychelles, 23 October–4 November, 1978*', IOFC/DEV/79/45, FAO, Rome.
27. Halldal, P. (1968). 'Photosynthetic capacities and photosynthetic action spectra of endozoic algae of the massive coral *Favia*', *Biol. Bull.*, 134, 411–424.
28. Heald, E. (1971). 'The production of organic detritus in a south Florida estuary', *Sea Grant Tech. Bull.*, 6, pp 110, University of Miami.
29. Heslinga, G.A. (1981). 'Growth and maturity of *Trochus niloticus* in the laboratory', *Proc. fourth Int. Coral reef Symp,* Vol. 1, 39–45.
30. Hill, R.B. (1978). '*The use of nearshore marine life as a food resource by American Samoans*', Pacific Island Studies Program, pp 170, University of Hawaii, (Mimeo).
31. Jhingaran, V.G. (1982). '*Fish and fisheries of India*', Hindustan Publishing Corporation, Delhi.
32. Juhl, R. and Suárez-Caabro, J. (1972). 'La pesca en Puerto Rico', *Agro. Pesq.*, 4, 1–52.
33. Kanwisher, J.W. and Wainwright, S.A. (1967). 'Oxygen balance in some reef corals', *Biol. Bull.*, 133, 378–390.
34. Klima, E.F. (1977). 'An overview of the fishery resources of the West Central Atlantic region', *FAO Fish Rep.*, 200, 231–252.
35. Koblentz-Mishke, O.I., Volkovinsky, V.V. and Kabanova, J.G. (1970). 'Plankton primary production of the world ocean', in *Scientific exploration of the South Pacific* (Ed. W.S. Wooster), pp. 183–193, National Academy of Sciences, Washington.
36. Krishnamurthy, K. and Sundararaj, V. (1973). 'A survey of environmental features in a section of the Vellar-Coleroon estuarine system, South India', *Mar. Biol.*, 23, 229–37.
37. Lankford, R.R. (1976). 'Coastal lagoons of Mexico. Their origin and classification', in *Estuarine Processes* Vol. 2 (Ed. M. Wiley), pp. 182–215, Academic Press, London.
38. Lewis, J.B. (1981). 'Coral reef ecosystems', in *Analysis of marine ecosystems* (Ed. A.R. Longhurst), pp. 127–158, Academic Press, London.
39. Littler, M.M. (1974). 'The productivity of Hawaiian fringing reef crustose Corallinaeceae and an experimental evaluation of production methodology', *Limnol. Oceanogr.*, 18, 946–952.
40. Lugo, A.E. and Snedaker, S.C. (1974). 'The ecology of mangroves', *Ann. Rev. Ecol. and Syst.*, 5, 39–64.
41. Lüning, K. (1981). 'Light', in *The Biology of Seaweeds* (Eds. C.S. Lobban and M.J. Wynne), pp. 326–355, Blackwell, Oxford.
42. Macintyre, I.G. and Pilkey, O.H. (1969). 'Tropical reef corals: tolerance of low temperatures on the North Carolina continental shelf', *Science,* 166, 374–375.
43. Macnae, W. (1968). 'A general account of the fauna and flora of mangrove swamps and forests in the Indo-West-Pacific region', *Adv. Mar. Biol.*, 6, 73–270.
44. Mahadevan, S. and Nayar, K.N. (1973). 'Pearl oyster resources of India', *Proc. Symp. Living Resources of the Seas around India,* pp. 659–671, CMFRI, Cochin.
45. Marsh, J.A. Jr. (1970). 'Primary productivity of reef-building calcareous red algae', *Ecology,* 51, 255–263.

46. Marten, G.G. and Polovina, J.J. (1982). 'A comparative study of fish yields from various tropical ecosystems', in *Theory and management of tropical fisheries* (Eds. D. Pauly and G.I. Murphy), pp. 255–289, International Centre for Living Aquatic Resources Management, Manila, Philippines and Division of Fisheries Research, Commonwealth Scientific and Industrial Research Organization, Cronulla, Australia.
47. Martosubroto, P. and Naamin, N. (1977). 'Relationship between tidal forests (mangroves) and commercial shrimp production in Indonesia', *Mar. Res. Indonesia,* 18, 81–88.
48. Menasveta, D., Shinde, S. and Chullasam, S. (1973). 'Pelagic *fishery resources of the South China Sea and prospect for their development*', SCS/DEV/73/6, FAO, Rome.
49. Michanek, G. (1975). 'Seaweed resources of the ocean', *FAO Fish Tech. Paper,* 138, pp. 127, FAO, Rome.
50. Miller, P.C. (1972). 'Bioclimate, leaf temperature, and primary production in red mangrove canopies in south Florida' *Ecology,* 53, 22–45.
51. Munro, J.L. (1978). 'Actual and potential fish production from the coralline shelves of the Caribbean Sea', *FAO Fish Rep.,* 200, 301–321.
52. Nair, P.V.R., Samuel, S., Joseph, K. and Balachandran, V.K. (1973). 'Primary production and potential fishery resources in the seas around India', *Proc. Symp. Living Resources of the seas around India,* pp. 184–198, CMFRI, Cochin.
53. Norton, T.A., Mathieson, A.C. and Neushul, M. (1981). 'Morphology and environment', in *The Biology of Seaweeds* (Eds. C.S. Lobban and M.J. Wynne), pp. 421–451, Blackwell, Oxford.
54. Odum, H.T. and Odum, E.P. (1955). 'Trophic structure and productivity of a windward coral reef community on Eniwetok atoll', *Ecol. Monogr.,* 25, 291–320.
55. Officer, C.B. (1983). 'Physics of estuarine circulation', in *Estuaries and enclosed seas (Ecosystems of the world 26)* (Ed. B.H. Ketchum), pp. 15–41, Elsevier, Amsterdam.
56. Pamatmat, M.M. (1971). 'Oxygen consumption by the sea bed IV. Shipboard and laboratory experiments', *Limnol. Oceanogr.,* 16, 536–550.
57. Pant, A., Dhargalkar, V.K., Bhosale, N.B. and Untawale, A.G. (1980) 'Contribution of phytoplankton photosynthesis to a mangrove ecosystem', *Mahasagar-Bull. Natn. Inst. Oceanogr.,* 13, 225–234.
58. Parulekar, A.H., Dhargalkar, V.K. and Singbal, S.Y.S. (1980). 'Benthic studies in Goa estuaries: Part III—Annual cycle of macrofaunal distribution, production & trophic relations', *Indian J. mar. Sci.,* 9, 189–200.
59. Parulekar, A.H., Harkantra, S.N. and Ansari, Z.A. (1982). 'Benthic production & assessment of demersal fishery resources of the Indian seas', *Indian J. mar. Sci.,* 11, 107–114.
60. Pritchard, D.W. (1967). 'Observations of circulation in coastal plain estuaries', in *Estuaries* (Ed. G.H. Lauff), pp. 37–44, American Association for the Advancement of Science, Washington.
61. Qasim, S.Z. (1970). 'Some problems related to the food chain in a tropical estuary', in *Marine food chains* (Ed. J.H. Steele), pp. 45–51, Oliver & Boyd, Edinburgh.
62. Qasim, S.Z. (1973). 'Productivity of backwaters and estuaries', in *The Biology of the Indian Ocean* (Ed. B. Zeitzschel), pp. 143–154, Springer-Verlag, Berlin.
63. Qasim, S.Z. (1977). 'Biological productivity of the Indian Ocean', *Indian J. mar. Sci.,* 6, 122–137.
64. Qasim, S.Z. (1979). 'Primary production in some tropical environments', in *Marine production mechanisms* (Ed. M.J. Dunbar), pp. 31–69, Cambridge University Press, Cambridge.
65. Qasim, S.Z. and Bhattathiri, P.M.A. (1971). 'Primary production of a seagrass bed on Kavaratti atoll (Laccadives)', *Hydrobiologia,* 38, 29–38.
66. Qasim, S.Z. and Sankaranarayanan, V.N. (1972). 'Organic detritus of a tropical estuary', *Mar. Biol.,* 15, 193–199.
67. Qasim, S.Z. and Wafar, M.V.M. (1979). 'Occurrence of living corals at several places along the west coast of India', *Mahasagar-Bull. Natn. Inst. Oceanogr.,* 12, 53–58.
68. Qasim, S.Z. and Sen Gupta, R. (1981). 'Environmental characteristics of the Mandovi-Zuari estuarine system in Goa', *Estuarine, Coastal and Shelf Sci.,* 13, 557–578.
69. Qasim, S.Z., Wellershaus, S., Bhattathiri, P.M.A. and Abidi, S.A.H. (1969). 'Organic production in a tropical estuary', *Proc. Indian Acad. Sci.,* 69, 51–94.
70. Qasim, S.Z., Bhattathiri, P.M.A. and Devassy, V.P. (1972 *a*). 'The influence of salinity on the rate of photosynthesis and abundance of some tropical phytoplankton', *Mar. Biol.,* 12, 200–206.
71. Qasim, S.Z., Bhattathiri, P.M.A. and Reddy, C.V.G. (1972 *b*). 'Primary production of an atoll in the Laccadives', *Int. Revue ges. Hydrobiol.,* 57, 207–226.
72. Qasim, S.Z., Wafar, M.V.M., Sumitra-Vijayaraghavan, Royan, J.P. and Krishna Kumari, L. (1978).

'Biological productivity of coastal waters of India-from Dabhol to Tuticorin', *Indian J. mar. Sci.*, 7, 84–93.

73. Qasim, S.Z., Wafar, M.V.M., Sumitra-Vijayaraghavan, Royan, J.P., and Krishna Kumari, L. (1979). 'Energy pathways in the Laccadive Sea (Lakshadweep)', *Indian J. mar. Sci.*, 8, 242–246.
74. Rowe, G.T. (1971). 'Benthic biomass in the Pisco, Peru upwelling', *Inv. Pesq.*, 35, 127–135.
75. Rowe, (n.d.)
76. Ryther, J.H. (1969). 'Photosynthesis and fish production in the sea', *Science,* 166, 72–76.
77. Saenger, P., Hegerl, E.J. and Davie, J.D.S. (1983). 'Global status of mangrove ecosystems', *Commission on Ecology Papers No. 3,* International Union for conservation of Nature and Natural Resources.
78. Sanders, H. (1956). 'Oceanography of Long Island Sound, 1952–54. X. The biology of marine bottom communities', *Bull. Bingham oceanogr. Coll.*, 15, 345–414.
79. Schuster, W.H. (1952). 'Fish culture in brackish water ponds of Java', *Indo-Pacific Fish. Counc. Spec. Publ.*, 1, 1–143.
80. Scott, B.D. and Jitts, H.R. (1977). 'Photosynthesis of phytoplankton and zooxanthellae on a coral reef', *Mar. Biol.*, 41, 307–315.
81. SCS. (1973). *'Pelagic fishery resources of the South China Sea and prospects for their development'*, SCS/DEV/73/6, pp 68, FAO, UNDP, Rome.
82. SCS. (1976). *'Report of the BFAR/SCSP workshop on the fishery resources of the Visayan and Sibuyan Sea areas, 18–22 October 1976. Iloilo, Philippines'*, SCS/GEN/76/7, pp 26, South China Sea Fisheries Development and Coordinating Programme, Manila.
83. SCS. (1978). *'Report of the workshop on the demersal resources of the Sunda shelf, Part II. November 7–11, 1977. Penang, Malaysia'*, SCS/GEN/77/13, pp. 120, South China Sea Fisheries Development and Coordinating Programme, Manila.
84. SCS. (1979). *'Report of the workshop on demersal and pelagic fish resources of the Java Sea, 5–9 December 1978, Semerang, Indonesia'*, SCS/GEN/79/20, pp. 60, South China Sea Fisheries Development and Coordinating Programme, Manila.
85. Smith, S.V. (1978). 'Coral reef area and the contributions of reefs to processes and resources of the world's oceans', *Nature,* 273, 225–226.
86. Sorokin, Yu. I. (1981). 'Periphytonic and benthic microflora on the reef: biomass and metabolic rates', *Proc. Fourth Int. Coral reef Symp.*, Vol. 2, 443–447.
87. Sorokin, Yu. I., Sukhanova, I.N., Konovalova, G.V. and Pavelyeva, E.B. (1975). 'Primary production and phytoplankton in the area of equatorial divergence in the Equatorial Pacific', *Trans. Inst. Oceanol.*, 102, 108–122.
88. Sournia, A. (1976). 'Primary production of sands in the lagoon of an atoll and the role of foraminiferan symbionts', *Mar. Biol.*, 37, 29–32.
89. Sournia, A. (1977). 'Analyse et bilan de la production primaire dans les récifs coralliens', *Ann. Inst. Océanogr. Paris,* 53, 47–73.
90. Steele, J.H. (1965). 'Some problems in the study of marine resources', *Spec. Publs. int. Commn. N.W. Atlantic Fish.*, 6, 463–476.
91. Stevenson, D.K. and Marshall, N. (1974). 'Generalisations on the fisheries potential of coral reefs and adjacent shallow water environments', *Proc. 2nd Int. Coral reef Symp.*, Vol 1, 147–156.
92. Strickland, J.D.H., Eppley, R.W. and Rojas De Mendiola, B. (1969). 'Phytoplankton populations, nutrients and photosynthesis in Peruvian coastal waters', *Boln. Inst. mar. Perú,* 2, 37–45.
93. Teixeira, C., Tundusi, J. and Santoro Ycaza, J. (1969). 'Plankton studies in a mangrove environment VI. Primary production, zooplankton standing stock and some environmental factors', *Int. Revue ges. Hydrobiol.*, 54, 289–301.
94. Titlyanov, E.A. (1981). 'Adaptation of reef-building corals to low light intensity', *Proc. Fourth Int. Coral reef Symp.*, Vol. 2, 39–43.
95. Tundusi, J. (1969). 'Plankton studies in a mangrove environment—its biology and primary production', in UNAM-UNESCO, *Lagunas Costeras, Un Simposio,* pp. 485–494.
96. Untawale, A.G., Balasubramanian, T. and Wafar, M.V.M. (1977). 'Structure and production in a detritus-rich estuarine mangrove sqamp', *Mahasagar-Bull. Natn. Inst. Oceanogr.*, 10, 173–177.
97. Verlencar, X.N. (1982). *'Nutrients and organic production in the tropical coastal environment'*, Ph.D. Thesis, University of Bombay, 205 pp.
98. Vinogradov, M.E. (1981). 'Ecosystems of equatorial upwelling', in *Analysis of marine ecosystems* (Ed. A.R. Longhurst), pp. 69–93, Academic Press, London.

99. Wafar, M., Wafar, S. and Devassy, V.P. (n.d.) 'Nitrogenous nutrients and primary production in a tropical oceanic environment'.
100. Walsh, G.E. (1974). 'Mangroves: A review' in *Ecology of Halophytes* (Eds. R.J. Reimold and W.H. Queen), pp. 51–174, Academic Press, New York.
101. Walsh, J.J. (1975 *a*). 'A spatial simulation model of the Peru upwelling ecosystem', *Deep-Sea Res.*, 22, 201–236.
102. Walsh, J.J. (1975 *b*). 'Utility of systems models: a consideration of some possible feedback loops of the Peruvian upwelling ecosystem', in *Estuarine Research* (Ed. L.E. Cronin), Vol. 1, pp 617–633, Academic Press, New York.
103. Walsh, J.J. (1981). 'A carbon budget for overfishing off Peru', *Nature,* 290, 300–304.
104. Walsh, J.J., Whitledge, T.E., Esaias, W.E., Smith, R.L., Huntsman, S.A., Santander, H. and De Mendiola, B.R. (1980). 'The spawning habitat of the Peruvian anchovy, *Engraulis ringens*', *Deep-Sea Res.*, 27, 1–27.
105. Wass, R.C. (1982). 'The shoreline fishery of American Samoa—past and present', in *Ecological aspects of coastal zone management. Proc. Seminar on Marine and Coastal Processes in the Pacific. Motupore Is. Res. Centre* (Ed. J.L. Munro), pp. 51–83, UNESCO, Jakarta.
106. Watson, J.G. (1928). 'Mangrove forests of the Malay Peninsula', *Malay. Forest Rec.*, 6, pp 275.
107. Wellershaus, S. (1974). 'Seasonal changes in the zooplankton population in the Cochin backwater (a South Indian estuar) *Hydrobiol. Bull.*, 8, 213–223.
108. Wells, S.M. (1981). 'International trade in ornamental corals and shells', *Proc. Fourth Int. Coral reef Symp.*, Vol. 1, 323–330.
109. West, R.C. (1956). 'Mangrove swamps of the Pacific coast of Colombia', *Ann. Ass. Amer. Geogr.*, 46, 98–121.
110. Wheeler, J.F.G. and Ommanney, F.D. (1953). 'Report on the Mauritiu Seychelles fisheries survey, 1948–49', *Colonial Off. Fish. Publ.*, H.M.S.O., London.
111. Whittle, K.J. (1977). 'Marine organisms and their contribution to organic matter in the ocean', *Mar. Chem.*, 5, 381–4.
112. Wyrtki, K. (1966). 'Oceanography of the eastern equatorial Pacific Ocean', *Oceanogr. Mar. Biol. Ann. Rev.*, 4, 33–68.

Resource Management and Optimization
1990, Volume 7(1–4), pp. 171–190
Reprints available directly from the publisher.
Photocopying permitted by license only.
© 1990 Harwood Academic Publishers GmbH
Printed in the United States of America

IMPACTS OF ECONOMIC DEVELOPMENT AND POPULATION CHANGE ON THAILAND'S FORESTS*

PETER KUNSTADTER
Institute for Health Policy Studies, University of California, San Francisco, Sanga Sabhasri, Ministry of Science, Technology and Energy, Bangkok, Sanit Aksornkoae, Kasem Chunkao, and Sathit Wacharakitti, Faculty of Forestry, Kasetsart University, Bangkhen, Bangkok

CONTENTS

The fate of a nation's forest is closely related to the country's socioeconomic and demographic environment. In 1963 over half of Thailand was forest-covered but by 1982 less than a third of the Kingdom was forested. In some regions over half the remaining forest was cleared between 1970 and 1982. This paper examines the socioeconomic and demographic conditions in Thailand associated with rapid loss of forest, and describes in detail the interaction between people, development and forests in the northern highlands.

---

*Manuscript prepared for publication 6 February 1985.

1. ECONOMIC DEVELOPMENT AND POPULATION CHANGE IN THAILAND

Economic development involves progressive changes of several kinds, including increase in infrastructural services and facilities (health, education, communication and transportation), increase in the application of science and technology to a wide range of problems, and increase in energy consumption, which usually implies a switch from an emphasis on biomass fuels to fossil fuels and hydropower. In a traditionally agricultural country such as Thailand, development includes an increase in the proportion of the national economy derived from commercial (versus subsistence) agriculture, wage work, commerce and industry, and, usually, a decrease in the proportion derived from extractive industries such as mining and forestry. In Thailand the proportion of G.D.P. derived from agriculture and forestry declined from 38.9% in 1960 to 25.9% in 1979, while the proportion derived from manufacturing doubled from 10.4% to 20.4% (Thailand (a) 1966, 1980). In general, development has been accompanied by increase in production, and in Thailand per capita G.N.P. increased from about $100 in 1960 to $176 in 1969, doubled to $346 in 1975 and doubled again to $792 in 1982, as the Kingdom's economy has been transformed.

One major corollary of development has been the alteration of demographic rates. Generally, death rates have fallen first, with consequent population growth, until fertility is brought under control. Thailand's population grew slowly (at about 0.7% annually), from 5–6 million in 1850 to 8.3 million in 1911, and then increased at an accelerating rate (averaging about 2.0% annually) to 17.3 million in 1947. Death rates dropped rapidly after the Second World War, and the surplus of births over deaths doubled the population between the late 1950s and 1984 (average annual growth rate about 2.7%), to a total of over 50 million (Thomlinson 1971; Thailand (a) various years). The national rate of population growth is now slowing (to under 2.0%) in association with widespread and effective use of family planning, but remains high among minority groups in the northern highlands, where much of the attention on problems of deforestation has been focused (Kunstadter 1984a).

Population growth led to increased demands for farm land, which were met largely by expansion into lands previously considered marginal. Increased demands for off-farm employment have been associated with migration to urban centres and increased seasonal agricultural migration. Growing demands for food, fuel and building materials have been met by increasing the land under cultivation, increased cutting of wood for local consumption, and the substitution of fossil fuels and non-traditional materials for forest products, as these became scarce and expensive.

Economic growth in the post-war period has proceeded much more rapidly in Thailand than did population growth. The economy of Thailand, which was traditionally based on irrigated rice produced on family-owned farms for subsistence and secondarily for sale, is rapidly being transformed to emphasise a variety of upland cash crops and wage labour. Growth in income has led to demands for raw materials and manufactured goods much greater than required to keep pace with population growth.

Although Thailand has produced and exported increasing amounts of food and of many raw materials, the import of fuel and manufactured goods has accelerated, with a consequent negative effect on balance of payments. This has been associated with increasing emphasis on production of export crops to earn foreign exchange. Thailand's balance of trade, which was based largely on rice exports, was positive until the early 1950s, but has been negative since then. Balance of trade reached a nadir of about (U.S.) $3.2 billion in 1981, despite rapid diversification and expansion of exports to include a wide variety of agricultural products, especially upland crops. While the total value of rice exports increased 13-fold in that period, rice dropped from over half of all agricultural exports in the early 1950s to about 20% in 1980. Major increases in share of export values were seen, especially in the upland crops, including fibre crops, maize, cassava and tobacco, animal products, as well as in fishery products. During the same period forestry products declined as a proportion of exports, reflecting both increased domestic demand and declining supply (Thailand b, various years).

Post-war timber production reached a peak of 3.34 million cubic meters in 1977, and then declined to 2.54 million cubic meters in 1980; teak production peaked at 326,671 cubic meters in 1969, and declined to 97,323 cubic meters in 1980 (Thailand b, various years). There is considerable annual variation in timber production, and, because of illegal cutting, timber production figures are notoriously incomplete. Nonetheless, they suggest a decline in production as well as in timber quality, in terms of species composition. The fully mature trees of the higher quality species are almost gone, and size limits are being relaxed. Lower quality species (e.g., pine) are being substituted for teak and other high value species, and lumber is often marketed before it is fully dry.**

Farm production has increased in recent years primarily as a consequence of the expansion of land under cultivation, rather than by intensification of agriculture and increase in productivity. Production per unit area has remained essentially constant, despite substantial increases in the amounts of agricultural inputs in the form of irrigation systems, agricultural chemicals and agricultural machinery. Rice production, for example, averaged 1800 kg/ha in 1907–1916, and 1781 kg/ha in 1980–1983; cassava production averaged 17,088 kg/ha in 1960/61 and 14,388 in 1982/83. Use of agricultural fertilizer increased from 273,686 tons in 1969 to 506,428 tons in 1975 and 780,000 tons in 1982; use of pesticides trebeled from 9,000 tons in 1975 to 26,000 tons in 1982; and use of farm machinery was also increasing rapidly during this period (2 wheel tractors up 260% from 90,0001 to 323,846 between 1975 and 1982; small 4 wheel tractors from 14,575 to 45,688, up 213%; large tractors up 364% from 13,338 to 61,840; water pumps from 251,288 to 780,610 up 211%; and rice threshing machines up 661% from 3955 to 30,091) (Thailand b, various years). Because most of

**Note added in proof. All legal teak logging was halted throughout the Kingdom in 1988, following disastrous floods in southern Thailand attributed to forest destruction associated with conversion of forest to rubber plantations. Consequently the government negotiated large increases in teak imports from Burma and Laos.

these inputs are imported, they contribute to Thailand's balance of payments problems at the same time as they contribute to their solution by increasing productivity.

Average farm size has remained about the same (4.36 ha. in 1975; 4.28 ha. in 1981) but the number of farms increased about 1.6% annually, and the amount of land being farmed increased at about 1.3% annually during this period. Because some of the land cleared for farming has been abandoned after a short time this has been associated with rapid destruction of forest. During the same period the balance has been shifting from irrigated rice to upland crops. The area devoted to rice increased 71% from 5.6 million ha. in 1960/61 to 9.6 million ha. in 1982/83, while the area in upland crops (maize, cassava, sugarcane, mungbean, sorghum, kenaf and tobacco) increased 475% from 0.8 million ha. (14% of the area in rice) to 4.6 million ha. (48% of the area in rice) (Thailand (b), various years).

United Nations figures show that per capita energy consumption tripled from 20 kg coal equivalents in 1950 to 60 kg in 1960, tripled again to 183 kg in 1970, and then doubled to 364 kg in 1982. Production of energy in Thailand increased exponentially from 45,000 m coal equivalents in 1960 to 2,761,000 in 1983. At the same time, consumption grew from 1,619,000 to 17,671,000. The difference between production and consumption was made up primarily by imports of liquid petroleum products (U.N. 1984: Table I), and this has been a major factor in Thailand's balance of payments problems. In the near future the consumption of gas will probably increase as a result of local production in the Gulf of Thailand. Use of wood-based fuel by households (fuelwood plus charcoal) appears to be declining, from 1.27 m^3 per capita in 1970 to 0.95 in 1980 (Aksornkoae 1984; de Backer and Openshaw 1972; Mungkorndin 1983; Sirivahanakul and Tadyu 1980; Urapeepatanapong and Mungkorndin 1983). Most of the wood-based fuel is used in rural areas. Apparently household fuels are being supplemented by electricity and gas, especially in urban areas, whereas most industrial and commercial energy is derived primarily from fossil fuels and secondarily from hydroelectric power. Although these figures probably underestimate the total amount of wood used for fuel (wood burned in swiddening is not included in the total used by households; wood burned for tobacco curing is not included in the total used in the industrial sector) they suggest that despite the increase in population and in energy consumption, cutting of the forest for household fuel is not as important a direct cause of deforestation as is clearing for agricultural use.

2. EFFECTS OF DEVELOPMENT AND POPULATION CHANGE ON FORESTS

Comparisons of aerial photographic and LANDSAT imagery suggest a decline in forested area in Thailand from 53.3% in 1963 to 30.5% in 1982 (a 43% decline in the remaining forest) (Table I). These are conservative figures. More rapid rates of deforestation are indicated if allowance is made for degradation in the quality of the forest. Rates would appear to be even more rapid in the East and South were it not for

TABLE I
Total Area and Forested Area in Thailand, 1963, 1973 and 1982*

	Total Area	Forested Area					
		1963		1973		1982	
Region	km^2	km^2	%	km^2	%	km^2	%
North	169,645	116,275	68.56	113,595	66.96	87,756	51.73
East	36,503	21,163	57.98	15,036	41.19	8,000	21.92
Northeast	168,854	70,906	41.99	50,671	30.01	25,886	15.33
Central	67,399	35,661	52.91	23,970	35.56	18,516	27.47
South	70,715	29,626	41.89	18,435	26.07	16,442	23.25
Total	513,115	273,629	53.33	221,707	43.21	156,600	30.52

*Source: Klankamsorn and Charuppat 1983; 1963 data from aerial photographs; 1973 data from LANDSAT-2; 1982 data from LANDSAT 3.

extensive replacement of the natural forest by rubber plantations. The most rapid change has occurred since 1970, during which about 50% of the remaining forest was cleared in the Northeastern and Eastern regions of the Kingdom (Klankamsorn and Charuppat 1983; Wacharakitti et al. 1979; Wacharakitti 1984). The loss of forest is also reflected in declining targets for forest land area in the national economic development plans from 50% of the nation's land area in the First Plan (1961–66) to 37% in the Fourth Plan (1977–81), which also called for extensive reforestation to meet this goal.

To set the scene for a discussion of recent changes, three stages in the process of development in Thailand will be described, along with their socioeconomic and environmental consequences: the traditional period, prior to about 1860; early development, between 1860 and 1960, and recent development in the past 25 years.

Prior to 1860, when the population was about 5–6 million, most people in Thailand depended on small-scale subsistence rice farming on rain fed or naturally flooded fields in the lowlands. Transportation in the interior was slow and expensive, and markets for cash crops were poorly developed.

In the Northern and Western highlands (which were only loosely integrated into the Kingdom at the time) sparse populations of Karen, and Lua' and other Mon-Khmer-speaking minorities, living in permanently settled villages, cleared coherent tracts of secondary forest for subsistence upland rice cultivation under a regular system of field rotation. The upland forest in the middle elevations where they lived, from about 600 to perhaps 1300–1500 m, was largely old secondary forest. Vast tracts of forests remained at higher elevations and in the poorly drained Central Plains, which were essentially uncleared and unsettled except along the rivers.

As the country began to open itself to the Western world, in the mid-19th century, major engineering works were undertaken, often with the aid of immigrant Chinese labourers, to drain and irrigate the Central Plains, which previously had been thinly settled. This became a vast cultivated area served by a network of canals for both irrigation and transportation, which produced a large surplus of rice, much of it for export. Because large, sparsely settled areas were available, control of the scarce labour supply was more important than control of land. Rights to the productivity of the populations living in large tracts were given to nobility and other important people in the *sakdina* or tax farming system, and in corvée labour obligations, which survived in one form or another until the 1930s. A railway system was built, starting about the turn of the century, which had reached Chiang Mai in the North, Ubon and Udorn in the Northeast, and the Malayan border in the South by 1941. The rail network had an immediate effect on the transport of rice, especially for export, and on the import of foreign goods, and was also used for the transportation of lumber (Ingram 1955).

At that time there was little concern with protecting the lowland forests; adequate supplies of forest resources were easily available in the hills, which were all considered to be part of the Royal Forest. Most of the lowland forest in the Central Plains was cleared during this period. Large concessions were granted, primarily to foreigners, for harvesting teak and other economically valuable species in the lower elevations of Northern Thailand. Little effort was devoted to replanting, even though problems of depletion of the more easily accessible teak forests were noted as early as 1895 (Smyth 1898 vol. I:104–107). Early attempts by the Royal Forest Department to require concession holders to reforest and to exclude farmers from forest land apparently could not be enforced, and were quickly abandoned in the 1890s (Smyth 1898 vol. II: 280–281).

Small-scale subsistence rotational swiddening continued in the highlands, where population grew slowly to fill in the unoccupied portions of the middle elevations. But for the most part the middle elevations were still covered with relatively mature secondary forest. Hmong (Meo), Yao, Lahu (Musser), Lisu (Lisaw), and Akha (Ikaw) began immigrating from Laos and Burma into the higher, unoccupied parts of the northern Thai highlands, clearing and burning the forest in small patches for a mixture of rice, maize and opium (for home consumption and as a cash crop), abandoning their fields, and moving to new locations when soil was exhausted or *Imperata* grass became too thick.

Following the Second World War, and especially since the 1960s, large- and medium-scale irrigation projects were built in the Central Plains and on major rivers in the North to improve water control in the relatively easily irrigated lowlands. This allowed year-round cultivation. Although these developments increased productivity per unit area per year in the areas they served, they do not seem to have absorbed much of the population increase which was taking place in the post-war period. As land values increased in the irrigation project areas, and with the development of a market for upland crops, there was rapid clearing of the forests on the margins of the Plains, in the East and in the Northeast (which prior to this time had been primarily a subsistence farming area). These uplands, which had previously been considered to be of low value for agriculture, both because of difficulties of irrigation and of cultivation with animal drawn ploughs, were rapidly put under cash crops with the use of motorized tractors.

Thailand was traditionally an agricultural country with the vast majority of the population working on family-owned farms. The traditional response of both lowland and highland people to population growth was migration to as yet uncleared forest areas. Expansion of the area under cultivation apparently kept pace with population growth until after the Second World War. As the rate of population growth increased, large numbers of rural people began to move temporarily or permanently to cities, and in recent years some migrants from rural areas have gone to the Middle East. Thailand's cities are having increasing problems of coping with rural migrants, and the Middle East market for unskilled labour is declining, as the price of oil falls, and as basic infrastructural projects are completed. The problem of rural unemployment appears to be growing, as indicated by the increase in number and proportion of landless farm workers. Thus demand for new agricultural land has remained high.

Highland populations benefited from the control of smallpox and malaria after the war. Population growth increased in the highlands, owing both to decline in mortality coupled with continued high fertility, and to continued migration from Laos and Burma. At the middle altitudes swidden rotation cycle times were reduced and the forest no longer approached maturity when cut. At the same time there has been rapid expansion of small-scale irrigation systems, mostly constructed by the farmers themselves. At the higher elevations much of the mature forest has been cleared and in many areas it has been replaced by grasses covering abandoned swiddens. In some areas of extreme land shortage at the higher elevations farmers have begun the more labour-intensive task of cultivating the grass covered fields.

Virtually no reforestation was carried out prior to the first half of the twentieth century (a total of only 8754 ha from 1899 to 1960). Although the rate and amount of forest being converted to agriculture were higher in the Northeast and East, attention to problems of deforestation focused first on the highlands of the North. This was perhaps because of the association of some swiddening with the cultivation of opium (which became illegal in 1957), and because the northern highlands form the watershed for the most important rivers of the nation. By the mid-1960s the threat of loss of forest resources was perceived primarily in relation to swidden farming of highland minority peoples. Large-scale reforestation projects were begun in the highlands, often in direct conflict with agricultural use of the land. The amount of reforestation increased to about 2000 ha/yr in the early 1960s, about 6500 ha/yr in 1968–1974, and about 39,500 ha/yr in 1975–1982 (Thaiutsa 1983). But even this expanded rate of reforestation has failed to keep pace with loss of forest.

3. IMPACT OF ECONOMIC DEVELOPMENT AND POPULATION CHANGE ON THAILAND'S FORESTS

As already noted, economic development and population growth have proceeded rapidly in Thailand during the past century, and especially in the past 25 years. Much of this growth has been at the expense of the forest. The major effect of economic development and population change on Thailand's forests has been the rapid conversion of forest to agricultural use, in two major stages. Most of the lowland forest in the Central Plains was cleared for irrigated agriculture in the last half ot he 19th century and the early 20th century. More rapid and extensive clearing of forests has taken place especially in the East and Northeast in the 1970s and 1980s, as upland areas were cleared for such cash crops as cassava and sugar cane, whereas clearing took place more slowly in the Northern highlands in association with population growth and the cultivation of opium as a cash crop. The remainder of this chapter examines in more detail the consequences of development and population growth in traditional swiddening areas.

4. EFFECTS OF GOVERNMENT DEVELOPMENT PROGRAMS IN THE HIGHLANDS

In addition to infrastructural developments in the highlands, which have introduced primary schools and health stations in highland villages, many different governmental programmes have attempted to reduce forest destruction by reducing amount of swiddening by highlanders. Most government projects seem to envision a relatively low level of cash crop farming as the future for the highland population. Until recently little attention had been given to reducing the rate of natural population increase through family planning, no attempts have been made to reducing migration from the lowlands into the highlands, and little attempt has been made to secure title to the lands traditionally cultivated by highlanders. Scant attention has been paid to facilitating migration of highlanders into other ecological zones with more intensive economic systems, or to introducing permanent non-agricultural employment opportunities for highlanders in the highlands (such opportunities as there are have gone largely to lowlanders).

Agricultural development projects have been directed at the introduction of new crops including coffee, fruit trees, sericulture (including planting of mulberry trees), beans and potatoes on hillsides, and dry season cash crops (e.g., soybeans) on the fallow irrigated fields. At the higher elevations government projects have emphasized the introduction of cash crops to substitute for opium. Highlanders have been employed temporarily in development projects which seem to have more effect in providing temporary employment opportunities than in increasing productivity. Agricultural development has often been poorly coordinated with efforts at reforestation. Overall the projects have generally been unsuccessful in increasing productivity to keep pace with population growth and with the increase in standard of living which has occurred in the lowlands.

Given the rate of growth of highland populations, especially those at the higher elevations, and the removal of land from swiddening both by reforestation and loss of forest cover or soil exhaustion, it seems unlikely that these projects will succeed in reducing the pressure on the land. The conflict between agricultural use of the highland areas and reforestation seems likely to increase. For the foreseeable future, a series of unplanned, uncontrolled, and sometimes unwanted consequences will probably continue to occur in the highlands as a reflection of Thailand's socioeconomic development.

5. EFFECTS OF TRADITIONAL LAND USE SYSTEMS ON THE FOREST

The traditional pattern of forest use was either for swiddening or for the taking of small amounts of lumber (either "high grading" for commercial use, or cutting for local use as fuel or building materials). Lumbering of this type apparently has little ecological effect, if the volume of wood removed is low; heavier selective cutting can reduce the ability of the selected species to reproduce itself and may result in a change in the species composition of the forest (Bethel and Turnbull 1974:II-8 ff.).

Under subsistence swiddening by populations of low density, the climax forest may be replaced by secondary forest, often including many fire-resistant species. The gathering of wood for fuel, and even cutting of lumber for construction by subsistence farmers, has minimal ecological effects on the secondary forest; much of the fuel which is used is unburned debris from swiddening. Soil erosion, death of seedlings and other ecological damage, including modification of the microenvironment to favour transmission of malaria, often result from commercial logging, especially if large openings are made in the forest and the removal of logs is associated with the building of roads or trails. Similarly, lumbering roads may offer access to farmers who otherwise would not invade the forest (Chunkao 1975; Corvanich and Boonkird 1975; Kartawinata 1975). Commercial use of fuelwood (e.g., for curing tobacco) may result in forest destruction apart from large-scale lumbering or clearing forests for farms.

Foresters usually consider swiddening to be a method of forest destruction. It may also be viewed as a system of forest clear cutting and spontaneous reforestation. The primary products are energy (the slash is burned in order to clear the field for cultivation) and fertilizer (fertility elements bound up in the above ground biomass are deposited on the soil surface in the course of the burn), and food and other useful products which are grown on the swidden field, rather than forest products. As such, it has been an important source of sustenance to large numbers of people in the highlands and in the lowlands of Northern Thailand.

6. TYPES OF SWIDDEN SYSTEMS IN NORTHERN THAILAND

Several traditional swidden systems with different ecological consequences can be distinguished in Northern Thailand (Kunstadter and Chapman 1978:6–13):

1. Short cultivation—short fallow, used by Northern Thais at the lower elevations for supplementing production on lowland irrigated fields. Cultivation is usually for one year, followed by 2 or more years of fallow, during which time the

vegetation recovers to low-growing bush. Upland farmers in the Northeast use a short cultivation—long fallow system because secondary regrowth to bush or forest is very slow;

2. Short cultivation—long fallow, used by Karen and Lua' in the middle to high elevations for primary production of subsistence rice. Cultivation is for one year followed by 7 or more years of fallow, during which the secondary vegetation, composed largely of trees, restores itself; and
3. Long cultivation—very long fallow, used by Hmong and other high elevation groups for production of rice, maize and opium. Cultivation is continued until fertility declines or grassy secondary vegetation makes further cropping impractical in terms of labour requirements, then the field is abandoned.

Differences and gradations between these systems and their association with various ethnic groups can be quantified in terms of the proportion of land abandoned, i.e., removed from a regular rotation system. The proportion of abandoned land ranges from 0.0% (Lua' and Tai Yai or Shan), 0.1% (Karen), 0.9% (Khamu'), 7.4% (Htin), and 16.7% (Northern Thai) among groups practicing short cultivation systems, to 36.6% (Lisu), 71.9% (Yao), 97.2% (Akha) and 97.9% (Hmong) among groups practicing long cultivation systems (Task Force Thailand 1980: Table II).

7. ECOLOGICAL CONSEQUENCES OF SWIDDENING

Ecological consequences of swiddening vary depending on the pattern of land tenure, method of cultivation, length of the cycle of cultivation and fallow, and the environment in which the swiddening is carried out. Permanent land tenure, or rotational swiddening, is associated with restoration of vegetation and soil fertility to approximately their levels prior to cultivation during a decade or so of fallow. When land is treated as a free good and abandoned following cultivation to the point where soil is exhausted or the area is overcome by weeds, the forest is replaced by grassland which may return to forest only slowly if undisturbed. Weeding techniques (which allow successional species to survive on the plot), size of plot (and distance from seed sources), control of fire during the regrowth period (and fire tolerance of the successional species) appear to be particularly important practices of settled rotational swiddeners in determining the extent and speed with which the vegetation, biomass and fertility will be restored to their preclearing condition. Soil disturbance during cultivation is probably the most important variable in the cultivation system with regard to control of erosion.

In many places these traditional systems are now under great pressure as a result of population growth and various forms of economic development. Thus the traditional systems have become modified or combined with more permanent forms of land use (Kunstadter 1979b).

8. LOWLAND SWIDDENING IN NORTHERN THAILAND (SHORT CULTIVATION—SHORT FALLOW)

Swiddening has traditionally been practiced by ethnic Thais in the North and the Northeast, in the uplands above irrigated valley floors, and on the gentle slopes of the foothills. Sometimes this is a prelude to more permanent use of land, but more often it is a supplement to wet rice cultivation. In some parts of the North this has traditionally been a bush fallow system in which farmers return to their land after a relatively short period of fallow. This has been a relatively stable but marginally productive system (Chapman 1978; Charley and McGarity 1978; Judd 1964). Forest (as contrasted with bush) regrowth at these lower elevations may require relatively long fallow periods, as shown in the Northeast, for example, where the initial succession is dominated by *Saccharum spontaneum*, *Imperata cylindrica*, and/or *Eupatorium odoratum* for 9 or more years, with woody species becoming dominant in biomass only after 20 years or so, unless the clearings are small and seed sources are readily available for tree species (Drew, Aksornkoae and Kaitpraneet 1978; Sabhasri et al. 1974). Restoration of fertility under this system may also be delayed, as compared with the Lua' and Karen system, until regrowth of the forest (Drew, Aksornkoae and Kaitpraneet 1979).

9. MIDDLE ELEVATION SWIDDENING (SHORT CULTIVATION—LONG FALLOW)

The traditional Karen and Lua' system of swiddening followed a regular round of cultivation for a single year with 9 or more years of fallow. Swidden activities, including the location of the site, timing of activities and control of burning, were tightly organized by village leaders, whose orders were backed by religious sanctions, and ultimately by the local princes to whom the villagers paid tribute in exchange for being allowed to govern themselves. Swiddens were cleared in relatively large coherent blocks surrounded by fairly well-advanced secondary forest. Fire was carefully controlled to keep it within the boundaries of the area to be cultivated. Trees were felled at about 75 cm. above the ground, or were severely pruned to avoid shading the growing crops. Cutting weeds at the surface of the soil rather than digging them out insured that the structure of the soil was not disturbed and that regrowth could occur promptly from the remaining stumps. There was little erosion, and by the end of the fallow period soil fertility was restored to approximately the same level as at the start of the cycle. Secondary forest also regenerated to about the same species composition and amount of biomass as at the start of the cycle. Erosion was minimized by lining contours of hillsides with poles to reduce soil slippage, by building pole frameworks and planting viney crops along temporary water courses, by refraining from cutting and burning on very steep slopes, on the crests of hills and along stream banks, by shallow cultivation techniques, and by the resultant rapid regrowth of vegetation following harvest. The villagers used a wide range of species in the various stages of regrowth which were maintained as a result of the operation of this system (Kunstadter 1978, 1979a; Kunstadter, Sabhasri and Smitinand 1978; Sabhasri 1978; Zinke, Sabhasri and Kunstadter 1978).

10. HIGH ELEVATION SWIDDENING (LONG CULTIVATION—VERY LONG FALLOW OR ABANDONMENT)

Hmong and other groups living at higher elevations have traditionally followed a less conservative system of swiddening. Swidden cutting is controlled by individual households rather than by village leaders. Relatively small clearings are made at first, which are later expanded or abandoned depending on soil fertility and the difficulty in clearing weeds at the start of a new growing season. Hmong farmers, or their hired workers, remove all the stumps from the fields, which they hoe to a depth of about 20 cm. They do not line the hillside contours with logs, nor do they control burning of areas not to be cultivated. The result is a fair amount of erosion while the fields are under cultivation, and regeneration of fire-tolerant grasses rather than secondary forests. In general, when moving into a new area, their clearings expand in ''soap-bubble'' fashion, as new fields are cleared and burned. They make no systematic effort to limit the burning to the area which will be cultivated. The result is large grass-covered areas at the higher elevations (Keen 1978). Pressure on land is increasing both because of the rapid increase in population of these groups and because they cultivate both subsistence crops and opium as a cash crop, for which there has been, until recently, an insatiable demand.

In addition to replacement of the diverse mature montane forest by a few grassy species and the loss of biomass, this form of cultivation increases runoff speed and is associated with considerable erosion. Secondary succession to grass, however, greatly reduces runoff and virtually halts erosion, even far below the level under undisturbed forest. Return of species diversity and quantity of biomass in the fallow period appears to be much slower than in the Lua' and Karen system (Chunkao et al 1983) probably as a result of an absence of seed sources and repeated burning.

11. POPULATION CHANGE IN NORTHERN THAILAND

Population growth, owing to both natural increase and migration from other parts of Thailand, has resulted in increased pressure on swidden land at all elevations in Northern Thailand. Death rates declined rapidly in the lowlands of Northern Thailand following the second world war, as a result of the control of malaria, smallpox and several other diseases. Health conditions in the highlands lag behind those of the lowlands, but death rates have fallen there as well. Birth rates in the Northern lowlands have fallen rapidly since the mid-1960s so that the annual rate of natural increase is now well below 2%. Much of the population increase in the rural lowland areas of Chiang Rai, Mae Hongson, Nan and other Northern provinces has been the result of internal migration. Birth rates and rates of natural increase remain high among the highland minority groups (3.0–3.5% among Lua and Karen in the middle elevations, and even higher among such higher elevations groups as Hmong) (Kunstadter 1984a; 1984b). Population in the highlands is also growing as a result of international migration, and movement of lowlanders into the highlands.

12. EFFECTS OF DEVELOPMENT IN LOWLAND SWIDDEN AREAS

Efforts at development in areas swiddened by lowland Northern Thai farmers have been directed at increasing the production per unit area by allowing annual cultivation of fields which were cultivated on a three or more year swidden-fallow cycle. This has involved allocation to the farmers of permanent title to the swidden land, clearing of stumps, and developing crop, fertlizer and tillage combinations which will sustain productivity (Chapman 1978; Charley and McGarity 1978). This pattern of development has apparently been successful in parts of Nan Province as a result of careful adaptation of the socioeconomic and ecological features of the development projects to local conditions.

13. EFFECTS OF DEVELOPMENT IN MIDDLE ELEVATION SWIDDEN AREAS

The Lua' and Karen system of forest fallow cultivation was a relatively stable method of subsistence farming in the middle elevations for hundreds of years when population density was relatively low, but in recent years it has been overwhelmed by a number of changes. In the nineteenth century use of the highlands was governed by the Northern Thai princes, and many highland villages received grants, inscribed on metal plates, giving them control over their lands and the rights to self-government, and exemption from corvée labour requirements and from military service in return for nominal payments of tribute. When the North was incorporated into the Kingdom, in the early 1900s, the highlands became, nominally, part of the Royal Forest, and swiddening became illegal. A head tax, payable in cash, was imposed, which pushed some of the highland men into the wage labour market since cash cropping was impractical given the difficulties of transport to market (Kunstadter 1978).

As already noted, population growth among middle elevation groups has been rapid, and little or no land remains unclaimed or unused at the middle elevations. In some places large amounts of land has been removed from the swidden cycle by reforestation, further reducing the amount of land available per capita for swiddening. This has resulted in reduction of the swidden cycle time below that necessary for adequate regeneration of the forest. This, in turn, seems to have been associated with a decline in swidden productivity, which may further increase the tendency to reduce the length of the fallow period (e.g., Hoare 1984).

14. RESPONSES TO POPULATION GROWTH

For many years one response by householders to overpopulation or other causes of economic hardship among middle elevation groups has been temporary or permanent migration. At least since the turn of the century and the imposition of a head tax payable in cash, Karen and Lua' men have taken temporary jobs in the lumber industry or mines or on lowland farms to supplement their earnings from subsistence farming. Karen households also hive off to establish daughter villages in nearby unoccupied land or to reduce travel time to fields. Lua' villages have rarely split, but Lua' villagers have been moving to the lowlands since early in the century, where they have gradually transformed themselves into irrigated rice cultivators. Until recently there was virtually no migration from lowlands to highlands. This was one mechanism which maintained the stability of the middle elevation upland villages and reduced the danger of overloading the swidden system.

Permanent migration from highlands to lowlands is continuing, but now involves both very poor people who have little access to land, and relatively wealthy highlanders who sell out in their home villages and buy irrigated land in the lowlands. Members of other ethnic groups, in addition to Lua', are now joining the migration to the lowlands, but migration is not enough to stem the population increase in the highlands. Thus pressure on land resources in the highlands is increasing.

Another response to population increase by many of the villagers living at middle elevations has been to expand areas under irrigation, by terracing in the small valleys. This removes land from the swidden cycle, and thus reduces the land under secondary forest. In some places these small-scale irrigation projects seem to have reached the limits of water which can be supplied by the characteristically relatively small watersheds.

15. CONSEQUENCES OF THE LEGAL VACUUM SURROUNDING SWIDDENING

Before the Bangkok government took control from the Northern Thai princes, the princes at least indirectly controlled land use of highland villages, and periodically lectured village leaders on ways to conserve land resources under swiddening. Under the Royal Forestry Department all swiddening is illegal, and there is no legal system for the settlement of swidden land disputes between villages. It is now impossible for village leaders to maintain forested boundaries between highland villages. Thus a fallow field previously used by one village may be cultivated before its time by members of another village. Similarly, the ability of animistic village religious leaders to control swidden burning (and thus to conserve the regrowing forest in fallow fields) has been diminished both by the religious conversion of some of the villagers, and by the addition of governmental officials outside of the local animistic religious system (Kunstadter 1980, 1982).

16. EFFECTS OF ROADS

A rapidly increasing network of roads is now spreading throughout the highlands, allowing motorized transport to areas which previously could be reached only after days or weeks of walking. Roads are essential for development, but they have also had detrimental effects on the forest. Roads have opened the highlands to the activities of government officials or concessionaires seeking areas in which they can meet their increasingly ambitious targets for reforestation. Fallow swiddens are the easiest place in which to carry out these projects, despite their having been cultivated for many generations and in disregard of the natural reforestation process. Reforestation of such areas removes the fields from further agricultural use by the villagers, decreases the amount of land available for farming, and further accelerates the shortening of swidden cycles without inputs of fertilizer, soil quality and swidden productivity decline.

Roads into the highlands have also opened the area for exploitation by outsiders who have no permanent interest in maintaining the productivity of the highland forests. Entrepreneurs seeking forest products (mostly illegally) have rapidly removed some species. With the growth of the lowland population and increase in the price of fossil fuels, lowlanders now truck fuelwood from the highlands, leading to a depletion of the standing crop of trees. Shortage of grazing areas in the lowlands and easier access into the highlands has also led to a large increase in the number of cattle grazed in the highlands, with a consequent reduction in grasses in the secondary vegetation and probably an increase in erosion from the traffic of cattle on the hill slopes. The roads, built by bulldozer and inadequately sealed and drained, are themselves associated with extensive erosion.

Roads have also made possible the transportation of bulky cash crops to market. Increasing amounts of land at the middle elevations are now being cleared, mostly by lowland entrepreneurs, for more or less permanent cultivation of such crops as cabbage and tomatoes. Again, fallow swiddens are the preferred location for such fields.

17. EFFECTS OF DEVELOPMENT IN THE HIGHER ELEVATIONS

The traditional response to local overpopulation (as indicated by soil exhaustion) among groups using the long cultivation—very long fallow swidden system was to seek previously uncut or unclaimed forest for cultivation. Thus villages often split up or moved to new locations, in general staying in the higher elevations as they moved in a generally southward direction. The limits of expansion are being reached, both because the southern limits of the mountains are being reached, and because of the increasing effectiveness of governmental controls on cutting in previously uncut forest. At the higher elevations some villagers have continued to migrate to areas of uncleared forest, while others have begun to cultivate in grasslands which were previously abandoned because of the amount of labour required. A few Hmong are now beginning to settle in the lowlands, either as farmers, or as merchants.***

Government projects aimed at replacing opium cultivation with other cash crops have generally been unsuccessful in reaching an equivalent level of income (Lee 1981). Opium is still being grown, at least as a supplement to the other cash crops, and government programmes are shifting more towards active suppression of opium cultivation. In some places the opium growers apparently respond by growing opium on smaller more isolated plots, and may use fertilizers pesticides, sprinklers and other modern technology to increase yields.

18. EFFECTS OF REFORESTATION

Very extensive reforestation has taken place in recent years, especially in the Northern highlands. The effect on the local economy at the middle elevations appears to have been negative in several ways: reforested areas are removed from land available for swidden rotation, wages paid for unskilled forestry work may be an important supplement but are not an adequate substitute for subsistence farming, and the jobs are often taken by temporary migrants from the lowlands. Thus, although reforestation may be important for the preservation of national resources, alone it does not decrease pressure on forest resources by addressing the causes of swiddening. It is not surprising, therefore, that even very extensive reforestation has failed to reverse the trend of forest destruction in swidden areas. The effects of reforestation on the national economy are also unclear, as most of the trees planted since the early 1960s at the middle and high elevations have been pine, for which no major market has been developed. Pine is unsuitable for fuel, no local paper industry has developed using pine chips, and pine logs are not favoured for construction.

The ecological effects of reforestation have not been widely studied. Species diversity of the natural or secondary forest is reduced by this form of monocropping, but reforestation has been successful in restoring forest cover and reducing erosion, as compared with active swiddening at the higher elevations (Chunkao et al. 1984).

***Note added in proof. Beginning in the late 1960s and continuing through the 1980s the government resettled large numbers of Hmong to middle and low elevations where some Hmong farmers now make their fields. Other resettled farmers return seasonally to higher elevations make fields. Opium production has declined substantially since the government began to destroy poppy fields to enforce laws against opium production in the mid-1980s. Simultaneously the government intensified controls on clearing forests for swiddens and the forested area cleared annually appears to have decreased.

CONCLUSIONS

Deforestation and loss of forest quality is proceeding very rapidly in Thailand in association with economic development and population growth. It is apparent that much of the recent growth of the Kingdom's economy has been made possible by the existence of unfarmed forested land into which the growing agricultural population could move, and on which upland crops could be grown for export. Traditional subsistence farming systems, practiced by relatively small populations, which were in a more or less self-sustaining balance with natural reforestation, have been overtaken by expanding commercial farming systems and rapidly growing populations. Reforestation, the traditional cure for deforestation, addresses the symptom, but not most of the underlying causes of non-sustainable exploitation of forest resources: population growth which outpaces the creation of non-farm jobs, low agricultural efficiency and productivity, and commercialization of agriculture (all of which lead to increased demand for farm land), plus increased demand for forest products, coupled with inadequate control of land use.

It is likely that deforestation will continue to outpace reforestation in the Northern Thai highlands as long as development efforts are aimed at an agricultural future for the highlanders, and as long as population growth remains unchecked. Introduction of cash crops and commercialization of subsistence farming seems likely to increase the demand for farm land, and this can be met only at the expense of the forest. The population of the Northern highlands is a relatively small proportion of the total population of Thailand. Further, while the economics of the that area play a relatively small part in the national economy, the watershed is of great national importance. This suggests that it would at least theoretically be possible to restore and conserve some of the forests by offering alternatives to subsistence plus low level cash crop farming. Small-scale industries, using local resources, might offer employment and economic stability to the highlanders while reducing their need to cut the forest in order to make a living.

The prospects for the forests in other parts of Thailand are even less bright under present circumstances. Demand for upland crops is increasing, and exports of upland crops play a major role in Thailand's economy. Despite massive infrastructural investments in irrigation systems and transport networks, and despite rapidly increasing expenditures for agricultural chemicals and farm machinery, there has been little or no overall increase in agricultural productivity in Thailand. Nor has much attention been paid to creation of non-farm employment opportunities for the rural population. This seems to be a realistic objective, especially since Thailand produces surpluses of agricultural products.

It is often argued that agricultural productivity in Thailand has been discouraged by the rice premium system which maintains a domestic rice price lower than on world markets. Figures given in the Agricultural Statistics of Thailand suggest that the situation is more complicated than this. Rice productivity on the best land (which is irrigated year round) is twice as high *per crop* as the national average: 3259 kg/ha in

1980–1983, versus 1781 kg/ha This suggests that agricultural inputs, in the form of irrigation systems, are successful in raising productivity locally, but not nationally, as a result of expansion into marginal farm lands. Upland crops are not covered by the rice premium system, but with the exception of tobacco, their productivity has failed to rise. Tobacco productivity tripled from 344 kg/ha in 1950–51 to 1063 kg/ha in 1980–1983. The increase was associated with an integrated system of agricultural extension (including provision of seed, fertilizer and pesticides and technical advice on how to use them), production and quality controls, and marketing (including guaranteed price to the farmer). No similar organization has been present with respect to other crops, nor has any analogous system been attempted in government development projects. This suggests that the organization of agriculture as well as its technology must change in order to increase productivity.

If present trends continue, the end of the forests in many regions of Thailand is in sight, perhaps by the year 2000. The removal of most of the natural forests from the Central, Northeastern and Eastern regions will have major socioeconomic implications in addition ecological and esthetic consequences. The growth of production in Thai agriculture (without some major change in agricultural methods and organization) will stop, and Thailand will be unable to absorb its increasing agricultural population by expanding the area under cultivation. The ranks of the unemployed will increase, and the balance of payments problem will become worse. The rapid loss of forests is a signal that economic development efforts must be redirected at increasing agricultural productivity, efficiency and international competitiveness, and at providing alternative economic futures for a rapidly increasing portion of the rural population.

REFERENCES

1. Aksornkoae, Sanit. 1984. Multiple use of mangrove ecosystems in Southeast Asia (Thailand, Malaysia and Indonesia). Bangkok. Kasetsart University, Faculty of Forestry.
2. Bethel, James S. and Kenneth J. Turnbull, eds. 1974. The history of human use of the forests of Indo-China. *In* Studies of the Inland Forest of South Vietnam and the Effects of Herbicides Upon those Forests, James S. Bethel and Kenneth J. Turnbull, eds., Part II, pp II-1–II-13. The Effects of Herbicides in South Vietnam, Part B, Working Papers. Washington, D.C., National Academy of Sciences.
3. Chapman, E. C. 1978. Shifting cultivation and economic development in the lowlands of Northern Thailand. Ch. 12 *in* Farmers in the Forest, P. Kunstadter, E. C. Chapman and S. Sabhasri, eds., pp. 222–235. Honolulu. The University Press of Hawaii.
4. Charley, J. L. and J. W. McGarity. 1978. Soil fertility problems in development of annual cropping on swiddened lowland terrain in Northern Thailand. Ch. 13 *in* Farmers in the Forest, P. Kunstadter, E. C. Chapman and S. Sabhasri, eds., pp. 236–254. Honolulu. The University Press of Hawaii.
5. Chunkao, Kasem. 1975. The effects of logging on soil erosion. Proceedings of the Symposium on the Long-Term Effects of Logging in Southeast Asia, R. S. Suparto et al., eds., pp. 65–72. Bogor, Indonesia. BIOTROP, SEAMEO Regional Center for Tropical Biology.
6. Chunkao, Kasem et al. 1983. Final Report: Research on Hydrological Evaluation of Land Use Factors Related to Water Yields in the Highlands as a Basis for Selecting Subsistitute Crops for Opium Poppy, July 1980–June 1983. Bangkok. Highland Agriculture Project, Kasetsart University.
7. Chunkao, Kasem, Nipon Tangtham, Samakkee Boonyawat, and Wicha Niyom. 1984. Watershed Management Research on Mountainous Land: 15-Year Tentative Report 1966–1980. Bangkok. Department of Conservation, Faculty of Forestry, Kasetsart University.

8. Corvanich, Amnuay and Sa-ard Boonkird. 1975. The effects of logging on the environment in Thailand. Proceedings of the Symposium on the Long-Term Effects of Logging in Southeast Asia, R. S. Suparto et al., eds., pp. 35–41. Bogor, Indonesia. BIOTROP, SEAMEO Regional Center for Tropical Biology.
9. de Becker, M., and K. Openshaw. 1972. Timber trends in Thailand: detailed description of surveys and results. Project Working Document FA:DP/THA/69/017. Rome. Food and Agriculture Organization.
10. Drew, William B., Sanit Aksornkoae, and Wasan Kaitpraneet. 1978. An assessment of productivity in successional stages from abandoned swidden (rai) to dry evergreen forest in Northeastern Thailand. Bangkok. Faculty of Forestry, Kasetsart University. Forest Research Bulletin No. 56.
11. Drew, William B., Sanit Aksornkoae, and Wasan Kaitpraneet. 1979. The inventory of nutrients in vegetation during secondary succession from swidden to forest in Thailand. Bangkok. Faculty of Forestry, Kasetsart University. Forest Research Bulletin No. 61.
12. Hoare, P. W. C. 1984. The declining productivity of traditional highland farming systems in northern Thailand. Thai Journal of Agricultural Science 17:189–219.
13. Ingram, James C. 1955. Economic Change in Thailand since 1850. Stanford, California. Stanford University Press.
14. Judd, Laurence C. 1964. Dry Rice Agriculture in Northern Thailand. Ithaca, New York. Cornell University, Southeast Asia Program, Data Paper no. 52.
15. Kartawinata, Kuswata. 1975. Biological changes after logging in lowland Dipterocarp forest. Proceedings of the Symposium on the Long-Term Effects of Logging in Southeast Asia, R. S. Suparto et al., eds., pp. 27–34. Bogor, Indonesia. BIOTROP, SEAMEO Regional Center for Tropical Biology.
16. Keen, F. G. B. 1978. Ecological relationships in a Hmong (Meo) economy. Chapter 11 *in* Farmers in the Forest, P. Kunstadter, E. C. Chapman and S. Sabhasri, eds., pp. 210–221. Honolulu. The University Press of Hawaii.
17. Klankamsorn, Boonchana and Thongchai Charuppat. 1983. The forest situation in Thailand. Chapter 1 *in* Papers Presented at National Forestry Conference, 1983. Bangkhen, Bangkok. Royal Forestry Department.
18. Kunstadter, Peter. 1978. Subsistence agricultural economies of Lua' and Karen hill farmers, Mae Sariang District, northwestern Thailand. Chapter 6 *in* Farmers in the Forest, P. Kunstadter, E. C. Chapman and S. Sabhasri, eds., pp. 74–133. Honolulu. The University Press of Hawaii.
19. Kunstadter, Peter. 1979a. Ecological modification and adaptation: an ethnobotanical view of Lua' swiddeners in northwestern Thailand. *In* The Nature and Status of Ethnobotany, R. I. Ford, ed. Ann Arbor. Anthropological Papers of the Museum of Anthropology, University of Michigan.
20. Kunstadter, Peter. 1979b. Implications of socioeconomic, demographic and cultural changes for regional development in northern Thailand. *In* Agroforestry Systems and Highland Lowland Interactions, S. Sabhasri, P. Voraurai, and J. D. Ives, eds. Tokyo. United Nations University, Natural Resources.
21. Kunstadter, Peter. 1980. The impact of economic development on Southeast Asian tropical forests. Proceedings of the Vth International Symposium of Tropical Ecology, ed. J. I. Furtado, Part I, pp. 65–72. Kuala Lumpur. The International Society of Tropical Ecology.
22. Kunstadter, Peter. 1982. Consequences of development in a traditional agroecosystem. Paper presented at Population Institute—Environment and Policy Institute Summer Seminar on Population. Honolulu. East-West Population Institute.
23. Kunstadter, Peter. 1984a. Highland population in northern Thailand. Ch. 1 *in* Highlanders of Thailand, J. McKinnon and W. Bhruksasri, eds., pp. 1–45. Kuala Lumpur. Oxford University Press.
24. Kunstadter, Peter. 1984b. Demographic differentials in a rapidly changing mixed ethnic population in northwestern Thailand. Tokyo. Nibon University Population Research Institute Research Paper Series No. 19.
25. Kunstadter, Peter and E. C. Chapman. 1978. Problems of shifting cultivation and economic development in Northern Thailand. Ch. 1 *in* Farmers in the Forest, P. Kunstadter, E. C. Chapman and S. Sabhasri, eds., pp. 3–23. Honolulu. The University Press of Hawaii.
26. Kunstadter, Peter, Sanga Sabhasri and Tem Smitinand. 1978. Flora of a forest fallow farming environment in northwestern Thailand. Journal of the National Research Council 10:1–46.
27. Lee, Gary Y. 1981. The Effect of Development Measures on the Socio-Economy of the White Hmong. Sydney, N.S.W. Department of Anthropology, Faculty of Arts, University of Sydney. Doctoral dissertation.
28. Mungkorndin, S. 1983. Rural and urban demand, supply, distribution and use of fuelwood and charcoal for domestic purposes in Thailand. Paper presented at ILO/DANIDA Regional Seminar on Fuelwood and Charcoal Preparation, 21–25 March. Lampang, Thailand.

29. Richter, H. V. and C. T. Edwards. 1973. Recent economic development in Thailand. Chapter 2 *in* Studies of Contemporary Thailand, Robert Ho and E. C. Chapman, eds. Canberra. Department of Human Geography, Research School of Pacific Studies, Australian National University, Publication BG/8 (1973).
30. Sabhasri, Sanga. 1978. Effects of forest fallow cultivation on forest production and soil. Chapter 7 *in* Farmers in the Forest, P. Kunstadter, E. C. Chapman and S. Sabhasri, eds., pp. 160–184. Honolulu. The University Press of Hawaii.
31. Sabhasri, Sanga et al. 1974. A study of succession from shifting cultivation to forest in a Dry Evergreen forest in Central Thailand. *In* Studies of the Inland Forest of South Vietnam and the Effects of Herbicides Upon those Forests, James S. Bethel and Kenneth J. Turnbull, eds., Appendix 1, pp II-14–II-34. The Effects of Herbicides in South Vietnam, Part B, Working Papers. Washington, D.C., National Academy of Sciences.
32. Sirivadhanakul, T. and S. Tadyu. 1980. Situation on fuelwood and charcoal in rural Thailand. Bangkok. National Energy Administration.
33. Smyth, H. Warrington. 1898. Five Years in Siam from 1891 to 1896. London. John Murray, Albemarle Street. 2 vols.
34. Task Force Thailand. 1980. Population in Forest Communities Practicing Shifting Cultivation (Thailand). Final Report. [Bangkhen, Bangkok. Faculty of Forestry, Kasetsart University.]
35. Thailand (b). Ministry of Agriculture and Cooperatives. Various Agricultural Statistics of Thailand 1967. Bangkok. Ministry of Agriculture, Division of Agricultural Economics, Office of the Undersecretary of State.
36. Thailand (a). National Statistical Office. Various Years. Statistical Yearbook Thailand. Bangkok. National Statistical Office, Office of the Prime Minister.
37. Thaiutsa, Bunvong. 1983. Socioeconomics of Forest Plantations in Thailand. Chapter 7 in The 3rd Seminar on Silviculture Forestry for Rural Community, 24–25 February 1983. Bangkhen, Bangkok. Faculty of Forestry, Kasetsart University. Pp. 7-1–7-15. [In Thai].
38. Thomlinson, Ralph. 1971. Thailand's Population: Facts, Trends, Problems and Policies. Bangkok. Thai Watana Panich Press Co., Ltd.
39. United Nations. 1984. Energy Statistics Yearbook 1982. New York. United Nations Department of International Economic and Social Affairs, Statistical Office.
40. Urapeepatanapong, Chawalit and Sompetch Mungkorndin. 1983. Demand and supply of charcoal in urban and rural areas of Thailand. Chapter 33 *in* The 3rd Seminar on Silviculture for Rural Community. Bangkhen, Bangkok. Faculty of Forestry, Kasetsart University, and Food and Agriculture Organization.
41. Wacharakitti, Sathit. 1984. Unpublished data from Landsat and ground survey of land use in Northeastern Thailand: 1973, 1978, 1982.
42. Wacharakitti, Sathit, P. Boonnarm, P. Sanguantan, A. Boonsaner, C. Silapatong, and A. Songmai. 1979. The Assessment of Forest Areas from Landsat Imagery. Bangkok. Faculty of Forestry, Kasetsart University. Forest Research Bulletin 60.
43. Zinke, Paul J., Sanga Sabhasri and Peter Kunstadter. 1978. Soil fertility aspects of the Lua' forest fallow system of shifting cultivation. Chapter 7 *in* Farmers in the Forest, P. Kunstadter, E. C. Chapman and S. Sabhasri, eds., pp. 134–159. Honolulu. The University Press of Hawaii.

Resource Management and Optimization
1990, Volume 7(1–4), pp. 191–208
Reprints available directly from the publisher.
Photocopying permitted by license only.
© 1990 Harwood Academic Publishers GmbH
Printed in the United States of America

PROTECTED AREAS, DEVELOPMENT, AND LAND USE IN THE TROPICS

JEFFREY A. McNEELY
Chief Conservation Officer, International Union for Conservation of Nature and Natural Resources, Av. du Mont-Blanc, 1196 Gland, Switzerland

and

JOHN R. MacKINNON
Consultant, International Union for Conservation of Nature and Natural Resources, Av. du Mont-Blanc, 1196 Gland, Switzerland

CONTENTS

Virtually all tropical countries have found that it is in the national interest to establish a system of protected natural areas, as part of balanced land use in rural areas. Although protected area systems often began in the colonial era, their greatest expansion has been in post-colonial times. However, opportunities for additional major expansion of national parks and other strictly protected natural areas are rapidly closing. The future will therefore witness an increasing concentration on improved management of existing protected areas and on the expansion of conservation-oriented land-use practices in areas which receive less strict forms of legal protection. A full range of protected areas established for different objectives will both provide better coverage and bring more benefits for local people. It is important that these benefits be quantified as part of the effort to integrate protected areas into regional development plans; these benefits include watershed protection, contributions to agriculture and forestry, tourism, wildlife management and conservation of indigenous cultures.

1. INTRODUCTION

As other papers in this volume have shown, the tropics hold tremendous natural resources and a bewildering complexity of natural and man-made ecosystems. Drawing on these resources, human populations have increased greatly in modern times, from about 800 million in 1750 A.D. at the dawn of the industrial age to some 5 billion at present. This population growth has been particularly remarkable over the past 30 years, when the number of people has doubled in the tropics, and the growth of agricultural production has more or less kept pace (it might even be said that it has fueled the population growth (Dumond, 1975)).

The expanding human population is part of an expanding human ecological niche whose growth is at the expense of other species and natural ecosystems (McNeely, 1984a). Recognizing that under current conditions nature is likely to prosper best when it serves the human interest, IUCN (1980) prepared the *World Conservation Strategy* (WCS), which defined "conservation" as "the management of human use of the biosphere so that it may yield the greatest sustainable benefit to present generations while maintaining its potential to meet the needs and aspirations of future generations." It considered the major objectives of conservation to be:

—to maintain life support systems and essential ecological processes;
—to preserve genetic diversity; and
—to ensure that any utilization of species and ecosystems is sustainable.

The concerns of the WCS have been widely accepted in principle by governments, development agencies, and conservation organizations. The question is no longer whether conservation is a necessary part of social and economic development, but rather *how* conservation can be achieved. This question is more important than ever, as much action is still required to ensure that the means employed to meet human needs from renewable natural resources are sustainable and to allow species and ecosystems to be conserved.

Natural habitats and living species should be maintained wherever they occur, but at least in the short term national parks and other protected areas provide the most secure means of conserving samples of natural ecosystems. Protected areas will never be more than a small proportion of what formerly were natural areas, yet they must satisfy the habitat requirements of threatened species and meet certain basic human needs. They will be able to play this role effectively only if habitat management, both within protected areas and outside, compensates for loss of habitat elsewhere and enhances the carrying capacity for species of conservation concern.

This chapter outlines an expanded rationale for protected areas and shows how they help to sustain human society, thereby contributing to the conservation and development of living natural resources in the tropics.

2. PROTECTED AREAS IN THE TROPICS

2.1 An Expanded Rationale for Protected Areas

Does the tremendous increase in human population in the tropics inevitably mean that "nature" is doomed, that the human requirement for more land will be insatiable until everything is turned to human ends? Ironically, hidden within this troubling question is the best hope for natural ecosystems: The better that protecting natural areas serves basic human needs, the better are the chances of survival for natural areas. Since demands on tropical resources can only be expected to continue to grow, it is necessary both to justify existing protected areas ever more convincingly and to establish new areas under a range of management regimes which can adapt to varying local conditions and human requirements. Linking protected areas together with human needs can support ecologically-sound development which takes on practical meaning for governments and local people (see Section 3). In order to demonstrate how protected areas can contribute to sustaining society, Miller (1980) devised a set of 12 broad objectives which can guide management decisions (Table I).

However, the uncontrolled implementation of all of these expanded objectives within any one area could lead to over-exploitation or even destruction of the natural values the protected area was established to protect. Clearly, some objectives are more compatible with others in areas with different natural values. Logging in a national park, for example, is clearly inappropriate, whereas wildlife management and certain types of education, training and research may be compatible and even help support sustained-yield forestry. On the other hand, a national park established to conserve sample ecosystems and ecological diversity can often also support tourism and conserve watersheds.

To accommodate this wider range of management objectives without giving up any of the important gains made by national parks and other strictly-protected categories, IUCN (1978, 1984a) devised a system of categories of conservation units (Table II). These categories show, for example, that while nature reserves (category I) and national parks (II) must be strictly protected against resource extraction, their objectives can be reinforced by adjacent protected landscapes (V), resource reserves (VI), natural biotic areas (VII) and multiple-use management areas (VIII) where some carefully-controlled resource exploitation is permitted.

A series of case studies illustrating how a range of complementary protected area categories can enable governments in the tropics to meet their responsibilities for protecting nature while providing for human development on a sustainable basis is presented in McNeely and Miller (1984).

TABLE I
An Expanded Set of Conservation Objectives for Protected Areas in the Tropics

1. *Sample ecosystems.* To maintain large areas as representative samples of each major biological region of the nation in its natural unaltered state for ensuring the continuity of evolutionary and ecological processes, including animal migration and gene flow.
2. *Ecological diversity.* To maintain examples of the different characteristics of each type of natural community, landscape and land form for protecting the representative as well as the unique diversity of the nation, particularly for ensuring the role of natural diversity in the regulation of the environment.
3. *Genetic resources.* To maintain all genetic materials as elements of natural communities, and avoid the loss of plant and animal species.
4. *Education and research.* To provide facilities and opportunities in natural areas for purposes of formal and informal education and research, and the study and monitoring of the environment.
5. *Water and soil conservation.* To maintain and manage watersheds to ensure an adequate quality and flow of fresh water, and to control and avoid erosion and sedimentation, especially where these processes are directly related to downstream investments which depend on water for transport, irrigation, agriculture, fisheries, and recreation, and for the protection of natural areas.
6. *Wildlife management.* To maintain and manage fishery and wildlife resources for their vital role in environmental regulation, for the production of protein, and as the base for industrial, sport, and recreational resources.
7. *Recreation and tourism.* To provide opportunities for healthy and constructive outdoor recreation for local residents and foreign visitors, and to serve as poles for tourism development based on the outstanding natural and cultural characteristics of the nation.
8. *Timber.* To manage and improve timber resources for their role in environmental regulation and to provide a sustainable production of wood products for the construction of housing and other uses of high national priority.
9. *Cultural heritage.* To protect and make available all cultural, historic and archeological objects, structures and sites for public visitation and research purposes as elements of the cultural heritage of the nation.
10. *Scenic beauty.* To protect and manage scenic resources which ensure the quality of the environment near towns and cities, highways and rivers, and surrounding recreation and tourism areas.
11. *Options for the future.* To maintain and manage large areas of land under flexible land-use methods which conserve natural processes and ensure open options for future changes in land use, incorporate new technologies, meet new human requirements, and initiate new conservation practices as research makes them available.
12. *Integrated development.* To focus and organize conservation activities to support the integrated development of rural lands, giving particular attention to the conservation and utilization of "marginal areas" and to the provision of stable rural employment opportunities.

(after Miller 1980)

Linking the 12 objectives and 10 management categories provides a decision matrix for protected area planners and development agencies (Table III). The matrix suggests that, given an objective, several alternative approaches to management may be fol-

TABLE II
Categories and Management Objectives of Protected Areas

I.	*Scientific Reserve/Strict Nature Reserve.* To protect nature and maintain natural processes in an undisturbed state in order to have ecologically representative examples of the natural environment available for scientific study, environmental monitoring, education, and for the maintenance of genetic resources in a dynamic and evolutionary state.
II.	*National Park.* To protect natural and scenic areas of national or international significance for scientific, educational, and recreational use.
III.	*Natural Monument/Natural Landmark.* To protect and preserve nationally significant natural features because of their special interest or unique characteristics.
IV.	*Managed Nature Reserve/Wildlife Sanctuary.* To assure the natural conditions necessary to protect nationally significant species, groups of species, biotic communities, or physical features of the environment where these require specific human manipulation for their perpetuation.
V.	*Protected Landscapes.* To maintain nationally significant natural landscapes which are characteristic of the harmonious interaction of man and land while providing opportunities for public enjoyment through recreation and tourism within the normal life style and economic activity of these areas.
VI.	*Resource Reserve.* To protect the natural resources of the area for future use and prevent or contain development activities that could affect the resource pending the establishment of objectives which are based upon appropriate knowledge and planning.
VII.	*Natural Biotic Area/Anthropological Reserve.* To allow the way of life of societies living in harmony with the environment to continue undisturbed by modern technology.
VIII.	*Multiple-Use Management Area/Managed Resource Area.* To provide for the sustained production of water, timber, wildlife, pasture, and outdoor recreation, with the conservation of nature primarily oriented to the support of the economic activities (although specific zones may also be designed within these areas to achieve specific conservation objectives).
IX.	*Biosphere Reserve.* To conserve for present and future use the diversity and integrity of representative biotic communities of plants and animals within natural ecosystems, and to safeguard the genetic diversity of species on which their continuing evolution depends.
X.	*World Heritage Site.* To protect the natural features for which the area was considered to be of World Heritage quality, and to provide information for world-wide public enlightenment.

(after IUCN, 1978)

lowed, and shows which objectives may be compatible with other objectives within a management category.

2.2 The Current Status of Protected Areas in the Tropics

While destructive trends are only too obvious, it is no coincidence that protected areas have also expanded greatly in recent years. In the Indomalayan Realm, for example, there were 85 areas protecting 4.75 million ha in 1958, but by 1984, 481 areas protected 26.5 million ha. Similar growth rates could be cited for other parts of the tropics as

TABLE III
Decision-Making Guide to Alternative Protected Area Categories to Attain Conservation Objectives

Objectives for Conservation	Alternative Management Categories									
	I Nature Reserve	II National Park	III Natural Monument	IV Game Reserve	V Protected Landscape	VI Resource Reserve	VII Anthro-pological Reserve	VIII Multiple-Use Mgmt Area	IX Biosphere Reserve	X World Heritage Site
1. Maintain sample ecosystems	2	(1)	(1)	(1)	2	3	3	3	(2)	(1)
2. Maintain ecological diversity	(3)	(1)	(1)	(1)	(3)	(3)	(3)	(3)	(1)	(1)
3. Conserve genetic resources	(1)	(1)	(1)	(1)	(3)	(1)	(3)	(3)	(1)	(1)
4. Education, research, monitoring	(1)	(2)	(2)	(2)	4	4	4	4	(1)	(1)
5. Watershed and soil conservation	3	3	3	3	3	4	4	4	3	3
6. Fisheries and game management	—	—	—	(1)	4	4	(3)	(2)	(2)	—
7. Recreation and tourism	—	(2)	4	1	3	4	4	4	2	(1)
8. Forest production	—	—	—	—	4	—	4	2	4	—
9. Protect cultural treasures	—	(1)	4	—	4	—	(1)	4	4	(3)
10. Maintain Scenic resources	3	(1)	(1)	3	3	4	4	4	4	(1)
11. Maintain Options	—	—	—	—	4	(1)	3	(1)	2	—
12. Support rural development	(3)	(3)	(3)	(3)	(1)	(4)	(1)	(1)	(2)	(3)

( ) Major purposes for employing management system
1 Objective dominates management of entire area
2 Objective dominates management of portions of area through zoning
3 Objective is accomplished throughout portions or all of area in association with other management objectives
4 Objective may or may not be applicable depending upon treatment of other management objectives and the characteristics of the resources.
— Not applicable.

TABLE IV
Coverage of Protected Areas in the Tropics

Realm	Number of Units	Area Protected (km^2)
Afrotropical	444	827,100
Indomalayan	682	323,500
Oceanian	54	49,800
Neotropical	461	771,224
Totals	1,641 units	1,971,624 km^2

well, with most rapid growth coming in the post-colonial years (see Harrison, *et al.* 1982 for a further discussion of the world coverage of protected areas, with more complete details in IUCN 1984b).

Most tropical countries have established protected areas. The exceptions include several small island nations: Barbados, Comoros, Grenada, Kiribati, Maldives, Nauru, Niue, Samoa, Tuvalu, and Vanuatu; and a few larger mainland countries: Bahrain, Burundi, Iraq, Lao People's Democratic Republic, Qatar, and Yemen (IUCN, 1984a). Even the most densely populated parts of the tropics have significant areas under protection. Java, for example, an island the size of Greece or the state of New York inhabited by 90 million people, has 30 protected areas which cover over 530,000 ha. Large mammals such as the Javan rhinoceros (*Rhinoceros sondaicus*) and leopard (*Panthera pardus*) still survive, and the Javan tiger (*P. tigris javanicus*) has become extinct only in the past several years.

A brief summary of protected areas in the major tropical realms, of categories I, II, III, IV, and V is presented in Table IV.

The total of 1.97 million km^2—the size of Iran and nearly the twice the size of Venezuela, Pakistan, or Tanzania—seems quite respectable, especially since virtually all major countries are covered. But is this area really adequate to conserve the species, ecosystems, and ecosystem functions that the areas are established to protect? Leaving aside the question of how effectively the existing areas are managed, biogeography and population genetics suggest that the answer is still a clear "no."

Studies of island biogeography (Wilcox, 1980) have shown that any time the total area of an ecosystem is reduced, species diversity will decline until it reaches a new equilibrium for the size of the ecosystem. No reduced area can ultimately retain all its original species when it becomes an "island" of natural habitat surrounded by other radically different ecosystems (such as plantation forestry or intensive agriculture). Larger reserves lose fewer species at a slower rate, but *any* loss of natural habitat will lead to a loss in species. A rough generalization is that a single reserve containing just 10 percent of the original habitat will support just 50 percent of the original species present.

Recent advances in population genetics has led to new considerations of what protected areas must do to ensure the long-term survival of key species. Short-term fitness versus long-term adaptation, relations between genetic heterozygosity and reproductive behaviour, and the continued opportunity for evolution of species have clearly indicated that large population sizes are required by the species which are the least dense in the tropics, such as the large herbivores, carnivores, scavengers and large trees. For some of these, it may no longer be possible to have reserves of sufficient size to ensure their long-term survival, unless active genetic management measures are instituted.

2.3 Ecological Principles of Protected Area Establishment

Tropical ecosystems are very complex and poorly understood, so it is challenging to attempt to use ecological principles to guide the establishment of protected areas designed to conserve samples of tropical nature. However, options for selecting new areas are being closed rapidly, so ecologists and development planners must do the best they can with the science available, while continuing to promote additional research and quickly applying new findings to management problems.

Soulé and Wilcox (1980) and Frankel and Soulé (1981) summarize the state of the art. The relevant conclusions are:

1. Multiple reserves are necessary for the maintenance of biological diversity, including a range of natural situations, plant successional stages, and several populations of key species (thereby reducing the chances of accidental extinction);
2. Various types and amounts of manipulation of habitats will generally be required to hold or foster succession;
3. The size of each individual protected area will depend on the species to be included, the area required to meet the species' requirements, and the ecological processes which need to be provided special management to ensure their permanent contribution to the area. Ideally, whole ecosystems should be included in the area, but this is seldom possible today; nevertheless, areas should be as large as possible;
4. The shape of protected areas is critical in avoiding insularization effects. The ideal reserve is circular.

2.4 Conclusion: Protected Areas Will Not Work by Themselves

From the above it is clear that while an admirable number of protected areas have been established in the tropics, many more are needed. Further, it is evident that by themselves protected areas will never be able to conserve the species, genetic resources, and ecological processes they were established to protect. The best answer to this dilemma seems to be to select and manage protected areas in the tropics to support the overall fabric of social and economic development; not as islands of anti-development, but rather as critical elements of regionally envisioned harmonious landscapes. Through a planned mix of national parks and other types of reserves, amidst productive forests, agriculture or fisheries, protected areas can serve people today and safeguard the well-being of future generations (McNeely, 1984b).

Garratt (1984) points out that Nature does not recognize Man's laws and boundaries, and that laws by themselves do not change human habits and traditions. Protected area management must therefore consider the physical and social environment of the broader region if it is to be effective. It must account for the need for other means of integrating conservation and development and for examining the impacts of protected areas on local people.

3. PROTECTED AREAS AND REGIONAL LAND USE

The above discussion suggests that protected areas are subject to ecological, physical, cultural, social and economic influences from outside the area, and that, in turn, such areas also influence neighbouring lands. These influences can be identified and integrated into the development process, for the benefit of both protected areas and surrounding land. This section suggests some of the ways and means for promoting linkages between protected areas, regional land use authorities and economic development.

The first requirement is to ensure close cooperation between the various agencies involved in regional land use. This will usually require an institutional structure, such as a Regional Planning Board which oversees a Regional Land Use Plan. Ideally the various agencies will be so convinced by the wisdom and self-interest of cooperation that various informal linkages will spring up as well.

Strong ties between protected areas and other agencies and programmes involved in regional development are important for several reasons:

- —Developing mutually beneficial inter-agency cooperation and dependencies, while settling inter-agency land-use conflicts and overlaps in planning;
- —Modifying plans of other agencies so threats to protected areas are reduced or that the integrity of such areas is enhanced; and
- —Promoting wider acceptance for the role of protected areas in regional development and including protected area management in wider multiple land use packages (including funding for vital protective functions as part of those projects and programmes which derive significant benefits from the protected areas).

3.1 The Benefits of Protected Areas for Regional Development

The objectives defined in Table I suggest many of the benefits which protected areas provide to regional development. MacKinnon (1985) expanded on these to derive 16 main benefits of protected areas for regional communities.

1. Natural balance of environment
2. Stabilisation of hydrological functions
3. Protection of soils
4. Stability of climate
5. Protection of genetic resources
6. Preservation of breeding stocks and population reservoirs
7. Conservation of renewable harvestable resources
8. Promotion of tourism
9. Creation of employment opportunities
10. Provision of research facilities
11. Provision of educational facilities
12. Provision of recreational facilities
13. Maintenance of a high quality living environment
14. Advantages of special treatment
15. Preservation of traditional and cultural values
16. Regional pride and heritage value

MacKinnon also provides details of each of these benefits, pointing out that they are of different scales of magnitude, accrue over different timespans and fall to different groups in the local community. However, he also demonstrates that they are additive and can provide considerable total value to the region as a whole. Some of the benefits will occur automatically with the establishment of the reserve, whereas others will require some management effort to reach their full potential. Ideally, the sum of these protected area benefits compared with the potential values or benefits attainable if the area was designated for alternative use will determine the best land use for a particular area. But such an analysis is seldom made in practice.

Experience has shown that determining socio-economic justification for establishing reserves in inhospitable marginal lands is almost always easier than for reserves in areas of high agricultural or urban potential. Unfortunately, areas of high agricultural potential are often biologically the richest and most valuable for conservation also.

As human pressure on land increases it becomes more important for the protected area management authority to put an economic value on both the tangible and the intangible benefits provided by the national system of reserves and to predict the likely immediate and future costs to the community if the land is designated for alternative uses. Speaking the language of development economists can earn a seat at the decision-making table for protected area managers.

A few of the key benefits of protected areas for development are expanded on in the following sections.

3.2 Protected Areas and Water Resources

Water resources are so vital to human welfare that their proper management is a fundamental concern to society. In the tropics, where there is usually too much or too little water on the land, developing irrigation systems and improving drainage are the most common means of achieving better control of the water flow. Although colossal sums are invested hydraulic development to enhance water supplies these investments can be easily jeopardised by poor protection of the water catchments on which they depend.

Watershed protection has therefore been used to justify many valuable reserves which otherwise might not have been established, and irrigation agencies can make powerful potential allies for protected areas which protect watersheds. For instance, the Guatopo National Park in Venezuela is justified by providing the water for Caracas, and the Tijuca National Park provides water for Rio de Janiero. Garcia (1983) estimated that Venezuela's Canaima National Park safeguards a catchment feeding hydroelectric developments which save the nation an estimated US$4.3 billion per year in fossil fuel. The watershed protection function of Canaima is so important that the government recently tripled the size of the park to 3 million ha, to further enhance its effectiveness.

In many parts of the world it is being discovered that the total costs of establishing and managing protected areas which protect catchment areas can often be met and justified as part of the hydrological investment. In a case study in Indonesia, MacKinnon (1982) examined the condition of the water catchments of 11 irrigation projects for which development loans were being requested from the World Bank. Conditions of the catchments varied from almost pristine to areas of heavy disturbance, deforestation, logging and casual settlements. Even where hydrological protection forests existed they were poorly protected by the Forestry Department, which considered them of low priority and provided them inadequate budgets. By using standard costings for the development of proper boundaries, establishment of guardposts, recruitment of guards and purchase of basic equipment, plus the costs of reforestation where necessary and even resettlement of families in some cases, it was found that the costs of providing adequate protection for the catchments ranged from less than 1% of the development costs of the respective irrigation project—in cases where the catchment was more or less intact—to a maximum of about 10% where resettlement and reforestation were needed. Such costs are trivial compared to the estimated 30% to 40% drop in efficiency of the irrigation systems were catchments not properly safeguarded.

It is evident that the costs for protecting watersheds should be an automatic component of irrigation loan requests and that the protected area authority provide the necessary management paid from the irrigation budgets. In the case of the Dumoga-Bone National Park, in Indonesia, this has already been done in collaboration with the World Bank, and provides one of the country's model protected areas (Sumardja *et al.*, 1984).

Protected areas can also be threatened by development projects which cause changes in hydrological regimes, even if the projects are outside the area itself. Upstream catchments may need to be protected to prevent flooding, siltation or pollution of a protected area. Abnormal depositions of sediments may also influence key ecosystems in reserves, especially when they affect coastal systems such as coral reefs, which are sensitive to the quality of effluent streams.

Hydrological projects may also cause changes in water regime which may threaten the integrity of natural ecosystems in protected areas. For example, a planned dam and hydroelectric plant in Silent Valley, India would have flooded a large area of unique habitat in that reserve; the Manu National Park in Peru is threatened by the planned construction of a canal which will cause major changes in the water regime of the area; and modification of river flow in the Zambesi river below the Kariba dam has resulted in accelerated bank erosion and the river has become wider and shallower in the Mana Pools National Park of Zimbabwe (Thorsell, 1984). Clearly, the protected area management authority must have close working relations with the water resources agencies to avert such threats where possible, or include in such major projects safeguards that will protect the hydrological regime of the affected protected areas.

3.3 Protected Areas and Agriculture

Protected areas often perform a useful service to neighbouring agricultural areas in safeguarding against floods, and in providing water through dry periods and fertile silt in the rainy season. Moreover, many of the wild species from protected areas perform valuable functions for agriculture. Birds help control levels of insect and rodent pests, bees perform vital fertilisation functions, and bats control insects and pollinate many tropical fruits.

A good illustration of the importance of bats is provided by the durian (*Durio zibethinus*), a valuable fruit in Southeast Asia which is highly seasonal in its fruiting and depends on wild nectar-feeding bats for its pollination. When the durian is not in flower the bats (especially *Eonycteris spelaea* and *Macroglossus minimus*) depend on other wild tree species for nectar, so loss of protected areas can lead to loss of the bats and result in failure of the socially and economically important durian crop (Start and Marshall, 1976).

In addition, protected areas can serve as *in situ* gene banks for plants which are of direct benefit to agriculture through their contribution to crop improvement via disease resistence, enhanced yield, and greater hardiness. Prescott-Allen and Prescott-Allen (1983, 1984) have detailed the many values of wild plants to agriculture, and how protected areas are essential to maintaining the value of wild plants. They conclude, "*In situ* genebanks could be set up quite easily in many protected areas of all kinds. The need is great and urgent, because wild genetic resources are enormously important economically and socially. Many are disappearing due to habitat loss and other pressures, and *in situ* protection is the prime means of maintaining them" (Prescott-Allen and Prescott-Allen 1984:638).

Where such benefits are clear the agricultural authorities and neighbouring agricultural communities can become good allies for the protected area. But there are also many less beneficial relationships. Many species of wildlife—parrots, monkeys, deer, rodents, antelope, pigs, elephants, and large carnivores—can be pests and the protected area management authority needs to be involved in any necessary control measures to ensure that appropriate methods are used. In particular, any use of poisons must be carefully monitored to prevent the spread of poisons through often-elaborate food chains or waterways into protected area ecosystems.

Reserves may also harbour reservoirs of certain diseases, weeds, or insect pests which are a danger to man and/or his domestic species (e.g. malaria and trypanosomiasis). These, too, may need control measures to reduce health risks. But such measures can also threaten the integrity of the reserves if inappropriate. Accidental spraying of selective weed killers, fungicides or insecticides onto protected areas, for example, could be disastrous to the natural ecosystem and such chemicals introduced into upstream water sources can also have serious impacts.

As a general principle, intervening buffer zones or some intermediate land use is needed between intensive agriculture and protected areas. Ideally, a natural barrier such as a river, sea, ridge crest or swamp should form a deterrent to human incursion and wildlife excursion. But in other cases the same function can be served by a less strictly controlled category of protected area, production forest, plantations that are unattractive to wildlife, an airport, golfcourse, reservoir or other guarded self-contained area. In many cases, however, it is necessary to plan and develop a specific buffer zone to reduce direct interference between protected areas and conflicting land uses.

3.4 Protected Areas and Forestry

Certain forms of forestry can take place in protected areas of categories V and VIII, and forestry lands are often adjacent to protected areas of all categories. Forestry practices can have effects on wildlife species that range from almost total elimination of wildlife when natural forests are clearfelled and replaced by exotic monocultures, to slight reductions of arboreal species and sometimes increases in terrestrial species when forests are gently thinned by some selective logging systems (Wilson and Wilson, 1975). In areas where selective logging is combined with leaving untouched patches of natural forest as reseeding stock (as in the Malaysian virgin forest system), forestry can yield a sustainable profit while also serving an important nature conservation function.

On the other hand modern methods such as heavy tractors, wide unmettled roads and chain saws, can be very destructive to wildlife and any subsequent timber crop, yielding only short-term benefits. But with more effort given to using more sustainable logging techniques it is often possible to improve forestry yield and reduce damage to the environment and wildlife. Where forestry can be aimed more at providing forest products for local people rather than for international or domestic timber markets, benefits will also flow to nearby protected areas by reducing the need for local villagers to seek forest products inside protected areas.

For these and other reasons, the protected area management authority should work closely with the forestry authority in both formal and informal ways. The clear-cut separation of function between the two departments is not in accordance with current thought, since significant protection and conservation can be achieved in forestry lands and some forestry can be performed in protected areas. The "minor" forest products—wildlife, medicinal plants, fibres, animal fodder, thatch, edible mushrooms, fruits, honey—harvested in a sustained way can often exceed the value of timber, so tropical forestry would benefit by expanding its purview to include all types of valued production on forest lands. In that way the linkage between conservation and development would be strengthened.

3.5 Protected Areas and Indigenous People

Some protected areas, particularly protected landscapes (category V), anthropological reserves (category VII) and biosphere reserves (category IX), may be inhabited by indigenous people. In other categories of protected areas the presence of indigenous peoples may be acceptable where they are living in close and balanced harmony with their environment and can be said to have become a part of the natural ecosystem. In other cases, where no people live in a reserve, traditional harvesting of various resources may be permitted on a seasonal basis along with the use of traditional cultural sites for religious or spiritual purposes.

Protection of indigenous cultures is highly sensitive and requires a judicious balance between the continued practice of traditional rights in national parks or other protected areas, and the pursuit of the advantages of modern development. There must be no question of trying to establish "human zoos" as scientific curios or tourist objects.

There are many areas in which native populations, following their traditional cultures on their own land, protect large areas of essentially natural ecosystems and harvest the renewable resources of their environment on a sustained yield basis (see McNeely and Pitt, 1985 for a number of case studies). These people and protected area managers can be appropriate allies; managers can learn much about resource conservation and use, while the conservation of natural areas can provide the opportunity for traditional cultures to survive. The social and behavioural patterns of these allegedly "primitive" peoples have become so integrated with their natural environment that usually, though not always, they achieve ecologically sound long term-use of an area. Both are easily disturbed by insensitive forces from outside.

Outright conflict between conservationist and indigenous objectives has occurred. Tribes have been expelled from national parks or denied the use of resources within the Park. For example, the Rendille were driven from the Sibiloi National Park in Kenya and the Ik expelled from Kidepo National Park in Uganda, with disastrous results for the tribes (see especially Turnbull, 1973).

Indigenous groups can often link restrictive land use policies to conservation objectives. Brownrigg (1985) offers four options to consider:

1. *Reserves,* where a protected natural area corresponds with the territory of a particular native population;

2. *Native-owned lands,* where the protection of the area is by native peoples;
3. *Buffer zones,* where a protected area serves as a physical or ecological barrier between native lands and the lands of others; and
4. *Research stations,* where certain areas under native management are organized as agricultural or ecological research stations.

The option which is most appropriate will depend on the cultures of the native peoples and the specific objectives of the protected area. In general the best indigene/conservationist relations occur when indigenous peoples see the protected area as helping to maintain their culture (and to provide employment); when indigenous organisations have strong bargaining positions (related to unambiguous title to their lands); and when permitted land-use in the protected area is well-defined.

An outstanding case is that of Kunas of Panama, where the encroachment of slash-and-burn cultivation from outsiders was countered by the Kuna themselves turning part of their traditional territory into a protected area that includes research facilities for foreign scientists and appropriate tourist facilities for visitors. The Kuna have maintained control of their traditional land and culture, served conservation objectives, brought in foreign exchange, and fostered economic development by establishing a protected area (Breslin and Chapin, 1984).

3.6 Other Contributions of Protected Areas to Regional Development

Two additional contributions of protected areas to regional development are worth mentioning: tourism; and harvesting of wildlife and other natural products.

Protected areas are major tourist attractions for many countries, bringing significant economic benefits to the country and, when properly carried out, to the local communities. In Kenya, for example, where the tourist industry is the largest earner of foreign exchange, 1977 foreign earnings from tourism totalled US$ 125 million, of which a third was provided by seven national parks; much of the other revenues from international travel, hotels, and souvenir sales, expended outside of reserves, were also largely dependent on the protected area attraction.

Since protected areas which are developed for tourism became showpiece areas of a country, local government is often willing to promote development in surrounding areas. It pays to make clear to the local people the fact that they are getting preferential treatment, and that this is due to their privileged location close to the protected area. These benefits should be emphasised by the extension and information programme of the reserve. The example of Amboseli National park in Kenya shows how successful this approach can be (Western, 1984).

In short, tourism development in and around protected areas can be one of the best ways to bring economic benefits to remote areas, providing local employment, stimulation of local markets, improvement of transportation and communication infrastructure, and many others. But careful planning is necessary to avoid some of the negative side-effects of tourism, particularly the tendency for local people to view protected areas as being established for the benefit of foreigners rather than themselves.

National parks which draw tourists can also provide other direct benefits to local people. In Nepal's Chitwan National Park, for example, the local people are allowed into the park during a specific two-week period each year to harvest excess thatch grass, worth some US$ 600,000 per year. Since virtually the entire area around Chitwan has been denuded of natural vegetation, the park now provides almost the only source of thatch, the most important traditional roofing material in the region (Mishra, 1984).

In some categories of protected areas, hunting and other forms of direct harvesting of wildlife are appropriate and economically important. In Zimbabwe, for example, controlled culling of excess elephant populations in the Chirisa Safari Area brought US$ 556,230 in profits to the local District Council at a critical stage in the history of the area (Child, 1984). Where wildlife production is conducted with proper controls and the application of sensible quotas it can be regarded as sound conservation—sound use of resources—and can help justify the maintenance of wild populations in protected areas.

Sometimes species are harvested inside protected areas, as in hunting reserves. In other cases, protected areas protect a proportion of the population or a vulnerable life stage of the harvested population, e.g. fish spawning areas or bird nesting colonies, a function which can be of very high economic importance. In India, for example, the prawn production from a partially protected mangrove swamp was estimated at 110 kg/ha/year, while in a nearby estuary where the mangroves were damaged or removed, prawn production was just 20 kg/ha/year (Krishnamurthy and Jeyaseelan, 1980).

4. THE ROLE OF PROTECTED AREAS IN IMPROVING TROPICAL LAND USE PRACTICES

The poverty of many tropical countries is caused or aggravated by poor land use, where the recent history of population growth has been at the expense of land fertility. Original tropical ecosystems were highly productive but produced little human food per hectare, so many have been converted to agroecosystems. While this has been very productive on the best lands, especially those which have been converted to irrigated rice, cutting of trees and opening of the land to the effects of rain, sun, and wind and the added loss of organic material due to regular burning, has resulted in speedy deterioration of agricultural potential in many of the hilly or less fertile areas. Only where man is able to apply expensive irrigation, terracing, fertilizers and insecticides does tropical agriculture approach the productivity of temperate farmlands.

The research needed to improve tropical land use practices involves preservation of a wide range of natural species to select from and provision of opportunities to study wild species to determine their properties. National scientific institutes in tropical countries should give high priority to research programmes aimed at determining potential uses of wild species, improving qualities of wild genetic resources and making better use of tropical soils and ecosystems.

Protected area management authorities should recognise the importance of such research and encourage and accommodate as much research as is compatible with the objectives of particular protected areas. To do this they must foster good relationships with the scientific authorities in their respective countries.

More important, protected areas must continue to make their significant contributions to regional land use in the sorts of fields outlined in this paper. Because of their proximity to environmentally critical areas and places of showcase status, well-managed protected areas can serve as the focus of regional development, thereby helping to maintain a more natural balance to the ecosystem over a much wider area.

REFERENCES

1. Breslin, P. and M. Chapin. 1984. Conservation Kuna-style. *Grassroots Development* 8(2):26–35.
2. Brownrigg, Leslie. 1985. Native cultures and protected areas: Management options. pp. 33–44 in McNeely, Jeffrey A. and David Pitt (eds). 1985. *Culture and Conservation: The human dimension in environmental planning*. Croom Helm, London. 308 pp.
3. Child, Graham. 1984. Managing wildlife for people in Zimbabwe. pp. 118–123 in McNeely, Jeffrey A. and Kenton R. Miller. 1984 (eds). *National Parks, Conservation, and Development: The role of protected areas in sustaining society*. Smithsonian Institution Press, Washington D.C. 838 pp.
4. Dumond, D.E. 1975. The limitation of human population: A natural history. *Science* 187:713–721.
5. Frankel, O.M. and Michael E. Soulé. 1981. *Conservation and Evolution*. Cambridge University Press, New York. 327 pp.
6. Garcia, José Rafael. 1984. Waterfalls, hydropower, and water for industry: Contributions from Canaima National Park. pp. 588–591 in McNeely, Jeffrey A. and Kenton R. Miller (eds). 1984. *National Parks, Conservation, and Development: The role of protected areas in sustaining society*. Smithsonian Institution Press, Washington D.C. 838 pp.
7. Garratt, Keith. 1984. The relationship between adjacent lands and protected areas: Issues of concern for the protected area manager. pp. 65–71 in McNeely, Jeffrey A. and Kenton R. Miller. 1984 (eds). *National Parks, Conservation, and Development: The role of protected areas in sustaining society*. Smithsonian Institution Press, Washington D.C. 838 pp.
8. Harrison, Jeremy, Kenton Miller and Jeffrey McNeely. 1982. The World Coverage of Protected Areas: Development Goals and Environmental Needs. *Ambio* 9(5):238–245.
9. IUCN, 1978. *Categories, objectives and criteria for protected areas*. IUCN, Morges, Switzerland. 26 pp.
10. IUCN, 1980. *The World Conservation Strategy*. IUCN/WWF/UNEP, Gland, Switzerland, and Nairobi, Kenya. 44 pp.
11. IUCN, 1984a. Categories, objectives and criteria for protected areas. pp. 47–55 in McNeely, Jeffrey A. and Kenton R. Miller. 1984 (eds). *National Parks, Conservation, and Development: The role of protected areas in sustaining society*. Smithsonian Institution Press, Washington D.C. 838 pp.
12. IUCN, 1984b. *1984 United Nations List of National Parks and Protected Areas*. IUCN, Gland, Switzerland. 167 pp.
13. Krishnamurthy, K. and M.J.P. Jeyaseelan. 1980. The impact of the Pichavaram managrove ecosystem upon coastal natural resources: a case study from southern India. Asian Symposium on Mangrove Environments: Research and Management. Kuala Lumpur.
14. MacKinnon, John R. 1982. Irrigation and watershed protection in Indonesia. Report to the World Bank.
15. MacKinnon, John R. 1985. *Protected Areas in the Tropics: A manager's sourcebook*. IUCN, Gland, Switzerland. 250 pp.
16. McNeely, Jeffrey A. 1984a. Biosphere reserves and human ecosystems. pp. 492–498 in *Conservation, Science and Society*. Unesco, Paris.

17. McNeely, Jeffrey A. 1984b. Introduction: Protected areas are adapted to new realities. pp. 1-7 in McNeely, Jeffrey A. and Kenton R. Miller. 1984 (eds). *National Parks, Conservation, and Development: The role of protected areas in sustaining society*. Smithsonian Institution Press, Washington D.C. 838 pp.
18. McNeely, Jeffrey A. and Kenton R. Miller. 1984 (eds). *National Parks, Conservation, and Development: The role of protected areas in sustaining society*. Smithsonian Institution Press, Washington D.C. 838 pp.
19. McNeely, Jeffrey A. and David Pitt (eds). 1985. *Culture and Conservation: The human dimension in environmental planning*. Croom Helm, London. 308 pp.
20. Miller, Kenton R. 1980. *Planning National Parks for Ecodevelopment*. University of Michigan, Ann Arbor. 500 pp.
21. Mishra, Hemanta. 1984. A delicate balance: Tigers, rhinoceros, tourists and park development vs. the needs of the local people in Royal Chitwan National Park, Nepal. pp. 197–205 in McNeely, Jeffrey A. and Kenton R. Miller. (eds). 1984. *National Parks, Conservation, and Development: The role of protected areas in sustaining society*. Smithsonian Institution Press, Washington D.C. 838 pp.
22. Prescott-Allen, Robert and Christine Prescott-Allen. 1983. *Genes from the Wild*. Earthscan, London. 101 pp.
23. Prescott-Allen, Robert and Christine Prescott-Allen. 1984. Park your genes: Managing protected areas for genetic conservation. Pp. 634–638 in McNeely, Jeffrey A. and Kenton R. Miller. (eds). 1984. *National Parks, Conservation, and Development: The role of protected areas in sustaining society*. Smithsonian Institution Press, Washington D.C. 838 pp.
24. Soulé, M.E. and Bruce A. Wilcox (eds). 1980. *Conservation Biology*. Sinauer, Sunderland, Massachusetts. 395 pp.
25. Start, A.N. and A.G. Marshall. 1976. Nectarivorous bats as pollinators of trees in West Malaysia. pp. 141–150 in J. Burley and B.T. Styles (eds). *Tropical Trees: Variation, breeding and conservation*. Academic Press, London.
26. Sumardja, Effendi, Tarmudji, and Jan Wind. 1982. Nature conservation and rice production in Dumoga, North Sulawesi. pp. 224–227 in McNeely, Jeffrey A. and Kenton R. Miller (eds). 1984. *National Parks, Conservation, and Development: The role of protected areas in sustaining society*. Smithsonian Institution Press, Washington D.C. 838 pp.
27. Thorsell, James W. 1984. Protected areas in danger. IUCN, Gland, Switzerland. 35 pp.
28. Turnbull, Colin. 1973. *The Mountain People*. Jonathan Cape Ltd, London. 253 pp.
29. Western, David. 1984. Amboseli National Park: Human values and the conservation of a savanna ecosystem. pp. 93–100 in McNeely, Jeffrey A. and Kenton R. Miller. 1984 (eds). *National Parks, Conservation, and Development: The role of protected areas in sustaining society*. Smithsonian Institution Press, Washington D.C. 838 pp.
30. Wilcox, Bruce. 1980. Insular ecology and conservation. pp. 95–118 in Soulé, M.E. and Bruce A. Wilcox (eds). 1980. *Conservation Biology*. Sinauer, Sunderland, Massachusetts. 395 pp.
31. Wilson, Caroyln C. and W.L. Wilson. 1975. The influence of selective logging on primates and some other animals in East Kalimantan. *Folia Primatologica* 23:245–274.

Resource Management and Optimization
1990, Volume 7(1–4), pp. 209–226
Reprints available directly from the publisher.
Photocopying permitted by license only.
© 1990 Harwood Academic Publishers GmbH
Printed in the United States of America

ECOLOGICAL ASPECTS OF TROPICAL PASTURE RESOURCES

M. NUMATA
Dean, Faculty of Science, Chiba University, Yayoi-cho, Chiba 260, Japan

CONTENTS

Temperate and tropical grasslands (pastures and meadows) have been studied particularly from the standpoint of vegetation dynamics and condition diagnosis. However, in this chapter, phytosociological studies of tropical and subtropical grasslands, various types of grasslands in relationship to vegetation dynamics and the influences of burning and grazing were reviewed.

Following that Numata's methodological viewpoints, such as criteria for grassland condition diagnosis and pasture management, phase dynamics in plant succession, and diversity, production and grassland conditions were summarized based on his own data. The methods using DS and IGC are applicable not only to temperate grasslands, but also tropical grasslands as a general methodology of dynamic viewpoints.

1. INTRODUCTION

The humid tropics is, in general, a world of forests, and there are few large-scale pastures. Scattered, large-scale grasslands dominated by *Imperata cylindrica* occur, but these are not pastures for animal husbandry. Rather, under the influence of high rainfall and soil erosion, they are established on natural forests sites clearly for shifting cultivation and then abandoned. In Eastern Nepal, grasslands dominated by *Cynodon dactylon* and *Imperata cylindrica* were used for grazing (Numata, 1965). In that case, *Imperata cylindrica* was used as the state of a short grass under continuous grazing.

Imperata cylindrica usually makes up tallgrass type grasslands with *Heteropogon contortus, Botriochloa glabra, Rottboellia exaltata* and *Themeda triandra*, among others, in Sulawesi, but they shift to shortgrass type grasslands dominated by *Ischaemum timorense, Zoysia tenuifolia, Cynodon dactylon* and *Panicum indicum* (Yoshida, 1957). In temperate Japan, *Imperata cylindrica* var. *koenigii* is a mid-grass, and its communities are distributed in coastal and riverside sandy habitats. With a few exceptions, they are not used for pastures (Numata and Ohga 1976).

Dicranopteris linearis fernlands and bamboo thickets are, like *Imperata* grasslands, also found on sites cleared by felling and burning in the tropics. Such fernlands are unfertile, but bamboolands are fertile and suitable for teak plantations according to a study conducted in Burma (Stamp, 1926). In general, grasslands maintained for a long time in the humid tropics, such as of *Imperata* and *Dicranopteris*, may cover unfertile eroded lands. Although some projects in Southeast Asia are attempting to grow maize and mullberry after cultivating the soil of such unfertile grasslands by tractors, agricultural development is being inhibited by problems of soil nutrients, monoculture and disease.

On the other hand, there exist large-scale pastures in arid and semi-arid lands in the tropics that are used for cattle-raising. Typical is the *caatinga* area of northeast Brazil, where many large stock farms exist. As a tropical pasturage, grasses in a semi-arid areas like the *caatinga* are more important than those in humid areas.

A research project on Tropical Grassland Ecology was undertaken. The report of that, published with Suganuma, Hayashi, Naito, Nishimura, Kayama and Ohga (Numata, 1983), forms the basis of this chapter refers.

2. PHYTOSOCIOLOGICAL STUDIES OF TROPICAL AND SUBTROPICAL GRASSLANDS

There have been few phytosociological studies of tropical grasslands. Among them, Werger (1977) explained the relationship of phytosociological studies of tropical and subtropical pastures in Africa using the Zürich-Montpellier method for management and conservation. Natural and semi-natural pastures in Africa cover woodlands, savannas, grasslands, grass steppes, and dwarf shrub steppes. Although animal husbandry in such vegetation types is the main primary industry in Africa, there have been few trials to establish an ecologically-based management plan for it.

The application of the Zürich-Montpellier method to grassland vegetation with a rich flora was discussed by Mullenders (1954) and Werger et al. (1972). The area of sample quadrats should be greater than that of temperate grasslands, for example, 5 × 10 m quadrats. One example is the grassland vegetation of the Kalahari Gemsbock National Park (Leistner and Werger, 1973), that is *Stipagrostietum amabilis, Hirpicio echini-Asthenatheretum, Monechma ineanum-Stipagrostis ciliata* community, *Peliostomo-Stipagrostietum obtusae, Aizoo-Indigoferetum anricomae, Sporobolo lampranthi-Zygophylletum tenuis, Sporoboletum coromandeliani, Sporoboletum vangei, Lycium tenue* community, *Panicetum colorati* etc. Their distribution is explained on the basis of soil types, pH and geomorphology.

Coetzee (1976) classified six community types using presence data in relation to geology and geomorphology. These included the *Cynodon dactylon-Themeda triandra* community and *C. dactylon-Conyza podocephala* community on abandoned fields. The dry-grass communities on Oahu, Hawaii were classified using the Braun-Blanquet method and the ordination technique (Kartawinata and Mueller-Dombois 1972), into *Eragrostis variabilis, Heteropogon contortus-Rhynchelytrum repens, Trichachne insularis, Chloris barbata, Dicanthium aristatum* and *Panicum maximum* communities on the windward side, and *Rhynchelytrum repens, Melinis minutiflora* and *Andropogon virginicus* communities on the leeward side. Most were dominated by exotic grasses.

Subtropical grasslands in Japan, particularly those of the Ryukyu Islands, have been phytosociologically classified as *Miscanthion sinensis* (*Lygodio-Miscanthion sinensis, Thelyplero-Miscanthetum sinensis*), *Imperation cylindricae* (*Oxalido-Imperatetum cylindricae, Paspalo-Imperatum cylindricae*) and *Zoysion tenuifoliae* (*Hedyotio- Zoysietum tenuifoliae, Euphorbio-Zoysietum tenuifoliae*), among others. These grassland associations were classified in relation to habitat conditions, such as soils, water, wind, geology and geomorphology, as well as such biotic factors as burning, grazing and mowing. The latter is closely related to succession (Suganuma 1983). There are many descriptions of tropical grasslands based on physiognomy and dominants.

3. VARIOUS TYPES OF GRASSLANDS IN RELATIONSHIP TO VEGETATION DYNAMICS

Various types of tropical and subtropical grassland are situated on a dynamic status along a sere. In the Congo, the *Hyparrhenia cymbaria-Echinops amplexicaulis* community is maintained by the burning of scrub, but it advances to *Albizzia* forests without burning and is occupied by *Cymbopogon afronardus, Digitaria vestita* and *Sporobolus pyramidalis* under the pressure of heavy grazing (McLiroy 1972). In Uganda, *Pennisetum purpureum* grasslands shift to *Imperata cylindrica* and *Digitaria scalarum* pastures when subjected to overgrazing. Many shortgrasses, such as *Hypar-*

rhenia spp., *Brachiaria* spp., *Setaria sphacelata, Chloris gayana, Panicum maximum, Sporobolus* spp. and *Andropogon* spp., among others, are distributed in lowlands. *Cynodon dactylon* and *Eragrostis* spp. are widely distributed and suitable for grazing. *Pennisetum clandestinum* and *Trifolium johnstonii* are dominant in the highlands, and they are good forages under strong grazing (McLiroy 1972).

In secondary succession in the South African highveld the sequence of stages in abandoned fields is observed, *viz.*, the annual grass stage of *Erigeron canadensis* and *Eleusine indica*, the first perennial grass stage of *Eragrostis curvale* and *Cynodon dactylon*, the second perennial grass stage of *Hyparrhenia hirta* (after ten years), and the subclimax grassland of *Trachypogon spicatus* and *Tristachya hispida*. The first perennial grass stage is that most suitable for grazing (Jones 1968). In the humid savanna with the climax grassland of *Hyparrhenia, Imperata cylindrica* var. *africana* dominated on the fallow (Knapp 1973). This type of pioneer grassland is used for grazing.

In tropical deciduous forests of Mexico *Bouteloua filiformis* grasslands are distributed with *B. hirsuta* and *B. curtipendula* after felling. They change to shortgrass pastures of *Opizia stolonifera, Sporobolus capensis, Andropogon bicornis, Panicum trichoides* with Leguminosae, such as *Sesbania, Polichos, Phaseolus, Crotalaria, Calopogonium, Teramnus* and *Indigofera*. After the felling of tropical evergreen forests, pastures of *Paspalum notatum, P. conjugatum, P. minus, Axonopus affinis* and *A. compressus* occur. In the savannas of Mexico *Imperata brasiliensis, Trichachne insularis, Paspalum virgatum, Andropogon bicornis, A. glomeratus, Homolepis atrensis* and other grasses occur. In flood plains are good forages like *Leersia hexandra* and *Paspalum fasciculatum*, together with sedges (McLiroy 1972).

In semi-arid subtropical India a retrogressive succession of grassland on a loose soil is *Dichanthium-Cenchrus-Lasiurus* grasslands to *Cenchrus-Lasiurus, Cynodon-Eleusine, Aristida*, and *Cenchrus biflorus* grasslands under the influence of grazing (Whyte 1968).

The *Phragmites-Saccharum-Imperata* grassland in riverside plains in India changes to *Saccharum-Imperata-Sclerostachya* grasslands under burning and mowing, and to *Desmostachya-Imperata* grasslands and to *Sporobolus-Paspalum-Chrysopogon* grasslands after burning and grazing. The *Themeda-Arundinella* grasslands in the northern highlands of India change to *Arundinella-Chrysopogon, Heteropogon-Bothriochloa* and *Cynodon* pastures under grazing (Whyte 1968).

In the retrogressive succession of grasslands in the tropics and subtropics of the Western Himalayas of India, *Cenchrus ciliaris* → *Cynodon dactylon* → *Sporobolus marginatus* → *Chloris* spp. pastures in lowlands (400–600 m in alt.), *Heteropogon contortus* → *Chrysopogon fulvus* → *Bothriochloa* pastures in midlands (500–1,000 m in alt.), and *Arundinella* → *Arundinella-Heteropogon contortus* → *Arundinella-Chrysopogon fulvus* → *Bothrochloa-Cynodon* pastures, or *Themeda anathera* → *Themeda-Arundinella* → *Themeda anathera-Chrysopogon fulvus* → *Bothriochloa-Cynodon* pastures in the highlands (Gupta 1978).

The primary productivity of savanna grasslands in Africa attains a maximum of

1750g/m^3/yr in *Imperata*-dominant grasslands and a minimum of 37g/m^{3}yr in *Aristida pappasa*-dominant grasslands (Bouliere and Hadley 1970).

As the basis of vegetation dynamics in pasture management, the environmental adaptation of tropical pasture plants (Humphreys 1981) is very important as is the avoidance or tolerance of environmental stresses such as: 1) those associated with climate (drought, chilling, frost, etc.), 2) edaphic and physiographic situations (flooding, impeded drainage, soil acidity, mineral toxities, salinity and soil texture), and 3) biotic hazard (grazing and cutting, burning, pests and diseases). In relation to this, the adaptation of tropical pasture plants is shown in origin, perenniality, habit, mode of reproduction, control flowering, climate, drought, cold, shade, waterlogging, acidity, Al excess, Mn excess, salinity, fire defoliation, and high fertility response on grasses and adding rhizobium on legumes. There is another table on latitudinal and climatic characteristics of pasture species, such as *Cynodon dactylon*: 31,4±7,5 in latitude, 1156±557 mm in annual rainfall, 17,8±3,5 in annual mean temperature, and 4,6±5,8 in mean minimum temperature.

4. THE INFLUENCES OF BURNING AND GRAZING

The grasslands of the tropics and subtropics are mostly called savannas. In savannas, deciduous trees like *Acacia* and *Adansonia* are scattered, and grasses such as *Stipa, Andropogon* and *Panicum, Paspalum,* among others occur. One opinion holds that grasslands and savannas in tropical lowlands are biotic climaxes resulting from burning, edaphic climaxes not suitable for tree growth, or a stage of hydrosere, and that a tropical grassland climate does not exist (Beard 1946). In northeastern South America rain forests change to savanna forests owing to the leaching of soils, then to savannas as a consequence of repeated burning. In that case the savanna is a fire climax closely related to soil conditions (Richards 1976).

The burning of vegetation is used for slash and burn agriculture and to obtain tender grasses for grazing. In particular, grazing and burning are closely related to grassland conditions.

The *caatinga* of northeast Brazil is strongly influenced by grazing and burning, and branched trees and shrubs are beneficial growth forms under such conditions (Hayashi and Numata 1976). When the basal area per ha is an indicator of human impact, particularly burning, *cerradão* < *cerrado* < *campo cerrado* < *campo sujo* is the order of influence of burning to tree growth (Goodland 1971).

In the chaparral of south Texas, burning decreases trees and herbs, but increases some grasses, particularly fire-resistant grasses such as *Buchloe dactyloides, Panicum filipes* and *Sporobolus asper*. Then, chaparral gradually changes to grassland dominated by those fire-resistant grasses (Box et al. 1967). Burning, in general, stimulates grasses, herbs and trees to produce new buds, and dry weight and grass height increases remarkably. Further, growth is accelerated by higher rainfall after burning. The number of species per unit area increases after burning in Nigerian savanna. In particular the constituents, hemicryptophytes, geophytes and therophytes stimulate their growth (Hopkins 1965,1968).

In Kenya, creeping perennials such as *Cynodon dactylon* and *Digitaria scalarum,* good tufted perennials, such as *Panicum maximum* and *Themeda triandra,* and poorly tufted perennials, such as *Cymbopogon pospischilii, Harpachne schimperi* and *Sporobolus pyramidalis* increased their frequency. However, the frequency of annuals such as *Aristida keniensis, Brachiaria semiundulata, Dactyloctenium aegyptium, Digitaria ternata, Eragrostis tenuifoila, Panicum atrosanguineum, Setaria pallide-fusca* and *Sporobolus panicoides* did not increase. Tufted perennials increased their coverage after burning (Pratt and Kinight 1971).

Fire is a great ecological factor in Australian grasslands, and plant traits adaptive to fire occur, such as the subterranean protection of buds, resprouting after fire, fire-stimulated flowering, the storage of seeds in the soil and fire-stimulated flowering, the storage of seeds in the soil and fire-stimulated germination in relationship to hardseededness (Gill 1978).

On ridges and well-drained slopes in Papua New Guinea, a subclimax grassland dominated by *Gleichenia* spp. and *Machaerina rubiginosa* develops. Fire-susceptible species, such as *Deyeuxia sclerophylla, Hierochloe redolens* (except tussocks), *Eriocaulon hookerianum, Hypericum mutilum* and *Lycopodium* spp., among others, are strongly affected by fire. The post-fire community includes such fire-resistant species as *Rynchospora rugosa, Hierochloe redolens* (tussocks), *Miscanthus floridulus, Isachne globosa,* and *Blumea lacera.* Some species are later extinguished, and the community gradually proceeds to *Gleichenia* subclimax.

After the felling of lowland *Dipterocarpus* forests in Malaysia a grassland vegetation composed of *Paspalum conjugatum, Cynodon dactylon, Digitaria* spp., *Echinochloa colonum* and *Eleusine indica,* develops. It then proceeds to *Imperata cylindrica* grasslands under the influence of burning (Verboom 1968).

The *Paspalum conjugatum* community established after the felling of *Dipterocarpus* forests mentioned above changes to a *Chrysopogon aciculatus-Axonopus compressus-Desmodium triflorum* community under the influence of grazing and cutting (Verboom 1968).

The directions of vegetation dynamics are different with burning and grazing. Similar to the Malaysian case, a ruderal grassland composed of *Echinochloa colonum, Eriochloa punctata, Digitaria horizontalis, Eleusine indica, Cenchlus echinatus* and *Paspalum fimbriatum,* among others, in the coastal grassland of Puerto Rico changes to a grassland of *Sporobolus indicus* and *Paspalum conjugatum.* It then shifts to a grassland composed of *Axonopus compressus, Paspalum notatum, Stenotaphrum secundatum,* and *Sporobolus indicus.* Under heavy grazing, *Paspalum conjugatum* and *Sporobolus indicus* rapidly invade the grassland (Molinari 1949).

A prairie vegetation dominated by *Andropogon scoparius* changes to a grassland dominated by *Bouteloua rigidiseta* and *Stipa leucotricha* under the influence of grazing. In the case of various stocking rates, the decreasers include *Andropogon scoparius, Bouteloua curtipendula, Andropogon gerardi,* and the increasers include *Boute-*

loua rigidiseta, Stipa leucotricha, Sporobolus asper, Andropogon saccharoides and *Paspalum plicatum*. Decreasers and increasers under the influence of mowing have a similar tendency (Launhbaugh 1955). A similar experiment was made with various grades of mowing on a temperate grassland (Numata 1976).

According to data on the influence of drought on grasses in Texas (Box 1967), the percentage of mortality of grasses, such as *Sorghastrum mutans, Panicum virgatum, Eragrostis trichodes, Bouteloua curtipendula* and *Andropogon saccharoides* under the influence of grazing is remarkably higher than that under ungrazed conditions. The mortatlity of those grasses is closely related to clump size. Larger clumps can survive longer under drought conditions. The clumps become smaller at higher stocking rates.

On the other hand, in California *Bromus carinatus, Chorazanthe douglasii, Lotus micranthus, Plagiobothrys californicus,* and *Styloeline gnaphloides* invade *Stipa pulehra* grassland under the influence of grazing. The increasers in that grassland are *Bromus arenarius, Torilis nodosa, Plantago hookeriana* var. *california* and *Aira caryophyllea* as well as *Stipa pulehra*. Those grasses increased the frequency of clumps and the density of culms stimulated by the impact of cattle raising. The mortality percentage also decreased (White 1967).

In Africa, there are tallgrass types grasslands of *Themeda triandra, Hyparrhenia hirta, Chrysopogon aucheri,* and *Bothriochloa insculpta,* among others (Box 1967), which shift to shortgrass type pastures of *Pennisetum ramosum, Sporobolus pyramidalis, Cenchrus ciliaris,* and *Brachiaria brizantha* under heavy grazing. With increased stocking rates *Themeda triandra, Bothriochloa insculpta, Hyparrhenia filipendula* and *H. dissoluta* decrease remarkably, whereas *Sporobolus festivus, Microchloa kunthii, Tragus* and *Sida* grow well under the heaviest grazing pressure (Heady 1966). Retrogressive succession under grazing stress in East Africa is: tallgrass stage (*Themeda triandra, Bothriochloa* spp., *Hyparrhenia filipendula, Setaria sphacelata* and *Chloris gayana*) → intermediate grass stage (*Cynodon dactylon, Pennisetum mezianum, Heteropogon contortus, Sporobolus angustifolius* and *S. marginatus*) → shortgrass stage (*Microchloa kunthii, Eragrostis tenuifolia, Sporobolus festivus, Aristida adscensionsis, Digitaria macroblephara, Tragus berteronianus* and *Harpachne schimperi*) → weed stage (*Sida ovata, Aster hyssopifolius, Dyschoriste radicans, Solanum campylacanthum* and *Tribulus terrestris*) (Heady 1966).

In Australia *Themeda australis* grasslands shift to *Heteropogon contorta/ Chrysopogon fallax* grasslands under grazing pressure and regular burning. The tendency of the replacement of dominants is similar to that in Africa (Tothill 1974). In India several retrogressive seres occur under grazing pressure, such as *Sehima-Dicanthium* → *Bothriochloa* → *Eremopogon* → *Aristida-Eragrostis-Melanocenchris, Themeda-Arundinella* → *Arundinella-Chrysopogon* → *Heteropogon-Bothriochloa* → *Cynodon* (Whyte 1977).

In the grasslands of dry areas such as Kashmir and Australia two types of grazing occur, one with nomadism (Kashmir) and the other without nomadism (Australia). In the former there is "Ziegenweide" by "Ziegenleute". In "Ziegenländer" in Turkey, Yugoslavia and the Karakorum, prolonged overgrazing thoroughly destroyed the vegetation and soil. The maintenance of soil conditions is most important in the tropical and subtropical dry area. Overgrazing in sparse dry vegetation is most dangerous since it results in soil erosion (Uhlig 1965).

Grazing is a complementary occupation on the lower hills, but at higher elevations, as barreness slopes increases, grazing becomes the main source of livelihood (Chatterjee 1980). In addition limitations on agriculture, particularly rice production at higher elevations increases the importance of animal husbandry.

In the dry area of the western Himalayas, there occurs the *Themeda-Arundinella* type, the *Heteropogon* type associated with *Bothriochloa pertusa, Arundinella setosa, Chrysopogon fulvus* and *Eulaliopsis binata,* and the *Cenchrus ciliaris* type associated with *Cynodon dactylon, Bothriochloa pertusa, Saccharum spontaneum, Imperata cylindrica* and *Sporobolus marginatus* (Gupta 1980). These varieties of pasture types correspond to the *Cynodon-Imperata* type in the eastern Himalayas. These pastures are under pressure from overgrazing caused by over-population and rapid population growth in tropical and subtropical highlands (Ruddle and Manshard 1981).

5. CRITERIA FOR GRASSLAND CONDITION DIAGNOSIS AND PASTURE MANAGEMENT

A survey of Japanese grasslands used for grazing and mowing has been conducted since 1957. The data derived were summarized and analyzed from the viewpoint of plant succession and to judge grassland conditions and trends. A grassland on a subsere was estimated by the degree of succession (DS), $DS = \frac{\Sigma d \cdot l}{n} \cdot v$ a quantitative measure of successional progression and retrogression (Numata 1961, 1969), where d is the relative importance (i.e. the summed dominance ratio [SDR] expressed as a percentage), l the lifespan of the constituent species, and is assumed according to the life-form (i.e. Th = 1, H,G and Ch = 10,N = 50, and M and MM = 100), n the number of species and v the ground cover rate (taking 100 percent as 1). The major grassland types of Japan from north to south are ordinated against the axis of DS (Numata 1969,1974). This same method was applied to the data of Eastern Nepal. The DS-frequency curve of Nepalese pastures is very similar to shortgrass pastures dominated by *Zoysia japonica* in Japan (Numata 1980, Fig. 1).

Because the temperature response of grasses is evolutionarily well-diversified, the altitudinal grassland zones in Nepal Himalaya are classified into two types: upper, cool and lower, warm. Compared with the altitudinal grassland zones, the altitudinal climax forest zones are more diversified (Numata 1981, Table I). Semi-natural pastures in eastern Nepal have two vertical zones, *Imperata-Cynodon* type (lower) and *Festuca-Poa-Agrostis* type (higher). The montane pastures in the tropics are divided into two, mainly depending on the difference in the mean temperatures in the coldest month, i.e., lower and higher than 10°C (Hartley 1954, Numata 1965). Hartley proposed to estimate the type of plant climate from the standpoint of grasses. He made a normal

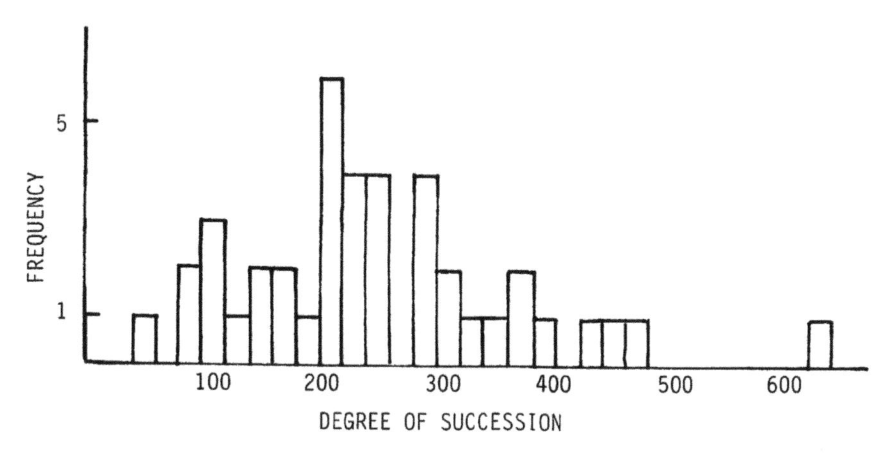

FIGURE 1.

spectrum of the main tribes of Gramineae and designated an area with an iso-agrostological index. This is a useful guideline for assessing the introduction of some fodder species to an area.

Successional diagnosis using DS shows the exact dynamic status of grasslands, corresponding to the diagnosis by floristic composition, life-form spectra, spatial pattern, productive structure and bryophytic indicator (Numata 1970, 1979). The major grassland types of Japan from north to south were ordinated against the axis of DS (Numata 1974).

However, such a diagnosis does not show the utilization value of grasslands, for example, for cattle-raising. An index of grassland condition (IGC),

$$\mathrm{IGC} = \left[\frac{\sum d \cdot l \cdot g}{n}\right] \cdot v$$

was devised to develop the formula of DS (Numata 1962).

Here, g is the grazing rate with a maximum of 1. The grazing rate is a practical expression of the forage value and palatability of plants. The DS and IGC methods have been applied to Korean (Park 1965) and to Himalayan grasslands (Numata 1965). This method has proved useful as a criteria for comparing grassland conditions and trends in general. There is a great need for a comparative study of the productivity of the world's grasslands. The maximum biomass as well as the productivity of a grassland depend on the dynamic status as well as phytosociological unit of the grassland. However, research has not yet approached this methodology (Numata 1970).

An attempt has been made to bridge the gap between plant sociology and production ecology using DS during the IBP (1964~1972). To do that, a model of the biomass-DS

TABLE I
Life Zones in Eastern Nepal (Numata 1966)

Climatic Zone	Climax Species	Secondary Species	Grassland Species and Weeds
Arctic 3,900–5,000m	Androsace, Cassiope, Saussurea, Sedum, Myricaria, Arenaria, Juniperus wallichiana, Rhododendron campanulatum, R. barvatum, Betula utilis		
Subartic 3,000–3,900m	Abies spectabilis, Tsuga dumosa, Betula utilis, R. arboreum, R. barvatum, Arundinaria maling, Viburnum		Poa, Festuca, Agropyron, Agrostis, Arundinella, Brachiaria, Calamagrostis, Eragrostis, Hordeum, Trisetum, Carex, Kobresia
Cool–temperate 2,500–3,000m	Tsuga dumosa, Quercus semicarpifolia, Acer campbellii, Sorbus cuspidata, Magnolia campbellii, Pentapanax leschenaultii	Pinus excelsa	
Temperate 1,900–2,500m	Quercus lamellosa, Q. lineata, Cinnamomum sp. Machilus sp.		Cynodon, Imperata,
Warm–temperate 1,200–1,900m	Castanopsis indicus, C. tribuloides, Schima wallichii, Alnus nepalensis	Pinus griffithii	Alopecurus, Arundinella, Avena, Chrysopogon, Dactyloctenium, Dichanthium,
Subtropical < 1,200m	Shorea robusta, Lagerstroemia, Terminalia, Bombax, Bauhinia, Ficus, Engelhardtia spicata, Wendelandia tinctoria, Albizzia sp.		Digitaria, Echinochloa, Eleusine, Eragrostis, Eragrostiella, Lolium, Panicum, Paspalidum, Paspalum, Pennisetum, Phalaris, Polypogon, Saccharum, Setaria, Tripogon, Cyperus

relationship (Numata 1966,1970) was proposed. The biomass dynamics of meadow communities and dominant species are closely related in their degree of succession (Okuda and Numata 1957, Tsuchida and Numata 1978-Fig. 2). The biomass curves of meadow communities (c′), and the dominants $D_1 \sim D_n$ correspond to DS. $M_1, \ldots,$ M_n are the modes of the dominant biomass curves. The location of each individual meadow is shown against the DS axis, $M_1, \ldots$ M_n are the modes of dominants of the biomass-DS curves. The location of the individual meadow is shown in the DS as the ecological distance from the modes, or as the typical situation of the biomass-DS curve of the dominant (Numata 1976).

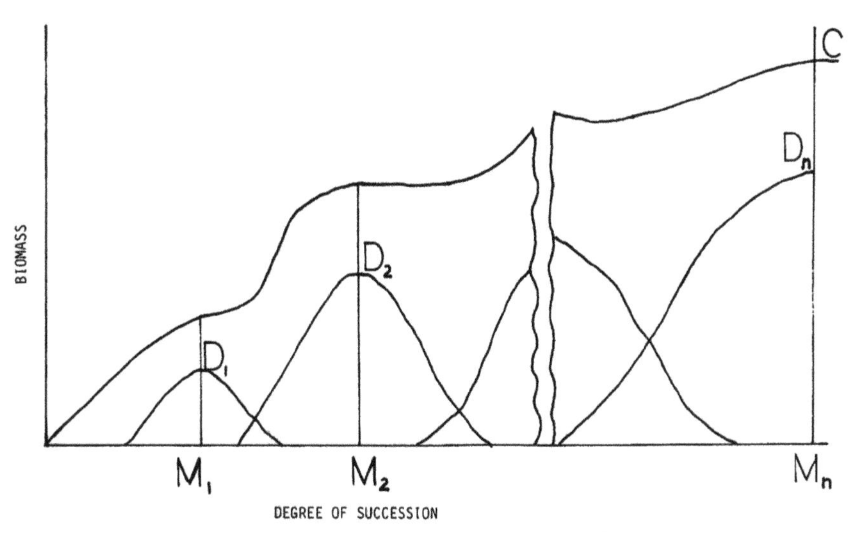

FIGURE 2.

The range of the value of DS for the *Miscanthus sinensis* community at the IBP site in northeastern Japan is between 140 and 740. The biomass of *M. sinensis* increases to the peak biomass corresponding to the mode of the DS frequency, and then falls off after the mode (about 500). The total biomass of the community dominated by *M. sinensis* changes up and down corresponding to the biomass curve of the dominant contrary, as a rule, to early expectation on the gradually ascending curve (Numata 1966). This may be because the increase of the total biomass drops at the border of the different seral stages. Such related changes of biomass against DS are different for each plant association or subassocation (Okuda and Numata 1975; Tsuchida and Numata 1979). Therefore the relationships between the phase within a successional stage and its production are identified as the relationship between the production and DS within the framework of each phytosociological vegetation unit. These observations fit the proposed model.

6. PHASE DYNAMICS IN PLANT SUCCESSION

The course of plant succession was formalized by Clements (1916) and others. Although descriptive and qualitative data on succession in special areas have been accumulated, no attempt at quantification has yet appeared. The degree of succession (DS) has been devised, and the relationship between the biomass of a community and its dominant, and DS is shown, as mentioned above. Due to such an idea, inter- and intra-stage, and inter- and intra-phase discriminations are done for seral grasslands as well as secondary and climax forests.

The situation of a plant community on a successional course is quantitatively indicated using DS. DS was originally proposed as a basis for judging the quantitative condition and trend of a grassland community.

For example, the *Miscanthus sinensis* stage is an ordinary perennial grass stage in Japan, however it covers a range of about 600 (from 200 to 800) in DS with a peak of 500. The peak is of DS and biomass. The former half, 200–500 in DS has an ascending trend, and the latter half, 500–800 in DS has a declining trend in the biomass of the dominant species as well as in that of the whole community. However, in the latter half, the biomass curve of the whole community ascends in an undulating way with the influence of the biomass of the following dominant.

The distinction of a successional stage from others, for example, discrimination of a *Miscanthus sinensis* stage from a *Zoysia japonica* stage, is an interstage diagnosis of grasslands. Compared with this, distinction of a phasic community from others within the framework of a *Miscanthus sinensis* stage, for example discrimination of a phasic community with 300 in DS and another with 500, is an intrastage diagnosis of grasslands. The interstage diagnosis is not usually difficult due to the recognition of the dominant species of successional stages. However, the intrastage (phasic) discrimination and diagnosis including grassland conditions and trends under the same dominant is rather difficult. For that purpose, DS was devised as an effective tool.

The idea of phase was proposed by Watt (1947) as a regeneration complex including pioneer, building, mature and degenerating phases within a successional stage, such as brecklands and beechwoods. In relation to this he proposed the idea of cyclic succession, which was proposed by Thoreau (1860) (Whitford and Whitford 1951).

There are many studies on the natural regeneration of a beech forest starting from gaps (Nakashizuka and Numata 1982). In our field survey of a natural beech forest at Kiyadaira, Nagano Prefecture in Central Japan, phases 1~5 were distinguished. Phase 1 is a grassy *Sasa* phase with beech seedlings, 2 is a relatively dense young tree phase, 3 is an intermediate tree phase with vigorous growth, 4 is a mature phase with big trees 300 to 400 years old, and 5 is an overmature, declining phase with the oldest trees. This series of phases is sometimes considered a sere from the pioneer to the climax stage. Certainly, the phasic development is somewhat similar to progressive succession. However, a beech forest as a climax has all phases 1~5. A beech forest is a phase-complex. A climax forest is not only phase 4, but also has phases 1~3 and 5 which form a sustaining and regenerating phase-complex. Such a mosaic of various phases under the influence of a dominant, beech (*Fagus crenata*) is just a climax. The state as such is not exactly expressed by a climax being a static stable stage, but is in dynamic equilibrium.

This kind of phasic change including retrogression as well as progression is found not only in a climax but also in every seral stage. For example, a cyclic change of phases is found even in *Miscanthus sinensis* meadows.

In our observation, seedlings of *Miscanthus* grass are frequently found on the ground

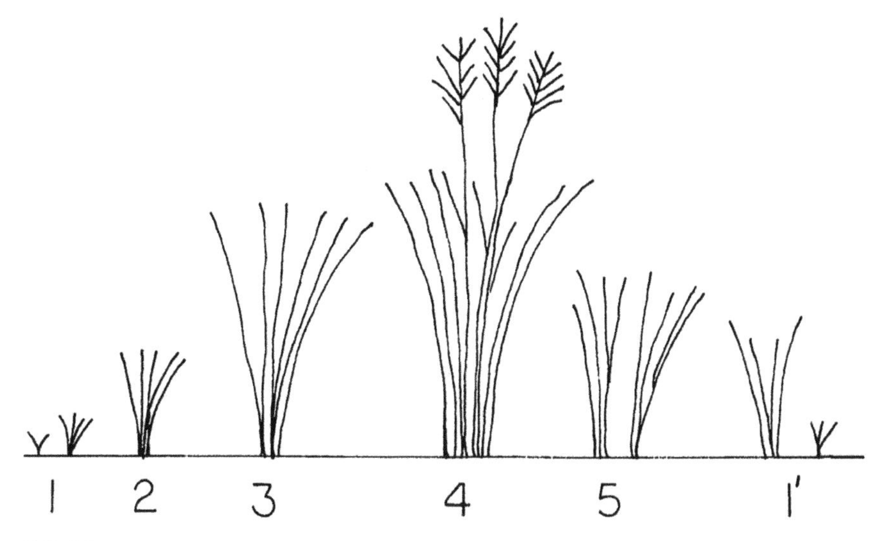

FIGURE 3.

after burning (if not burned, there are very few seedlings under a thick layer of litter) or in upland fields near the meadow (due to the wind dispersal of the seeds). Some of those small seedlings of a monocotyledonous type grow into tufted forms or tussocks. In central and southern Japan, tussocks grow 2 m or so in shoot height during vegetative growth and higher if their heads and stalks are included. One clump of tufted aerial shoots consists of one hundred to several hundred shoots which are sometimes densely and sometimes loosely gathered. After vigorous vegetative growth for several years, clumps have dead centers. A clump is broken into small pieces of clonal growth which are the starting points of the following generation, phase 1. That is to say, one generation of *Miscanthus* grass starting from seed dispersal is Phase 1: seedlings; Phases 2-3: tufted growth developing clumps; Phase 4: vigorous vegetative growth and production of seeds; Phase 5: clumps coming to pieces with dead centers; Phase 1′: seedlings in the vacant niche or the regeneration of small pieces of clones. Such a sequence of 1-2-3-4-5-1′-2-3. . . . is certainly a phasic change in a semi-natural meadow (Figs. 3). Therefore, a *Miscanthus* meadow is a regeneration complex of various phases which is very similar to the cyclical process seen in the European heathland (Gimingham 1972). Thus, cyclical phasic change is observed in seral stages as well as in climax stages. The *Miscanthus sinensis* stage meadow is an orthoseral course of succession is a transitional stage between a perennial herb and a shrubby stage which proceeds from phase 1 to 5 without a phasic cycle returning to 1′. However, when interference from outside, such as burning and/or mowing, stops the progressive succession from going to the following stage, a cyclical phasic change will be seen. Biotic factors in a seral stage and climatic factors in a climax cause a cyclical phasic change with dynamic equilibrium.

TABLE II
The Index of Diversity, of Some Grassland Types in Eastern Nepal

Dominant	α
Pteridium aquilinum	5.61
Potentilla kleiniana	4.98
Senecio chrysanthemoides var. *spectabilis*	4.57
Artemisia dubia	3.95
Plantago erosa	4.38
Plantago erosa	5.19
Plantago erosa	4.77
Juncus effusus	3.08
Juncus leschenaultii	1.66
Imperata cylindrica	3.74
Cynodon Dactylon	4.57
Paspalum distichum	4.98

Nevertheless, the mature phase of a meadow as a quasi-stable state is not always homogeneous. It includes patches of clumps of various phases (1~5) under the dominant state of mature phase (4) clumps. This kind of heterogeneity is shown using the DS isogram (Numata 1980).

7. DIVERSITY, PRODUCTIVITY AND GRASSLAND CONDITIONS

In the grassland survey in eastern Nepal, two zones at the upper limit of rice cultivation (about 2500 m) were studied; several vertical zones based on climax forests were recognized. This finding (Numata 1976, 1978) is closely related to the findings of Hartley (1954). According to Hartley (1954), the isotherm of 10°C in monthly mean temperature at mid-winter is the border between cool-grasses (Festuceae and Agrosteae) and warm-grasses (Paniceae, Eragrosteae and Andropogoneae). Species-area curves have been drawn to determine the minimum sampling areas (Numata 1965) and the indices of diversity are calculated (Table II) (Fisher et al. 1943). The data from Nepalese pastures are comparable to that from Japan.

For a renewable natural resource, productivity is important from the standpoint of utilization. The productivity (annual net production) is calculated based on the SDR. The value obtained is the biomass of the present edible and palatable plants among the many pasture species. The IGC from 14 to 225, indicates the potential amount of forage plants in relationship to the pasture condition and the trend to be used for grazing (Table III). The relationship of the forage value to the IGC is not linear (Fig. 4), and the asymptote of the forage value, IGC curve, will be about 60% more than the pasture communities under grazing. On the other hand the IGC is an expression of not only the present palatable biomass based on the SDR, but also of the durability of

TABLE III
DS, IGC and Forage % of Semi-natural Pastures in Nepal

No.	Dominant	Site	Alt. (m)	dl	n	v	DS	IGC	Forage (%)
1	*Cynodon dactylon*	Dalaghat	580	2490	8	100	311	177	41
2	*Chrysopogon aciculatus*	Dalaghat	580	1800	8	100	225	72	32
3	*Chrysopogon aciculatus*	Bustikhola	840	5502	18	100	289	26	24
4	*Cynodon dactylon*	Pachkhal	860	3711	7	80	424	87	42
5	*Chrysopogon aciculatus*	Kirantechap	1100	6983	9	80	622	69	26
6	*Cynodon dactylon*	Namdu	1350	2085	6	100	347	172	41
7	*Oxalis corymbosa*	Kathmandu	1290	1903	16	85	118	23	21
8	*Cynodon dactylon*	Kathmandu	1290	2047	11	90	167	126	51
9	*Imperata cylindrica*	Kathmandu	1290	3266	15	100	217	117	40
10	*Cynodon dactylon*	Kathmandu	1290	1860	8	75	174	126	44
11	*Imperata cylindrica*	Katakote	1420	5334	12	100	444	121	45
12	*Cynodon dactylon*	Peta	1440	5861	16	100	366	68	41
13	*C. dactylon*	Banepa	1480	4554	18	100	253	121	47
14	*C. dactylon*	Kenza	1540	3785	14	90	243	98	40
15	*Imperata cylindrica*	Namdu	1550	4088	11	60	223	58	34
16	*Brachyaria villosa*	Risingo	1600	4405	15	100	293	88	36
17	*Cynodon dactylon*	Kapure	1720	1472	10	100	147	74	56
18	*Eragrostis tenella*	Chaubas	1850	5615	10	85	478	142	36
19	*Cynodon dactylon*	Yarusa	1900	4836	15	80	258	84	31
20	*Inperata cylindrica*	Manga Daurari	1920	2727	9	50	152	117	33
21	*I. cylindrica*	Chagma	1920	1776	14	80	101	48	27
22	*I. cylindrica*	Chagma	1920	7230	18	80	321	175	38
23	*I. cylindrica*	Jiri	1900	2420	18	100	133	61	41
24	*Paspalum distichum*	Jiri	1900	3620	11	100	329	225	57
25	*Juncus leschnaultii*	Jiri	1900	3820	16	85	202	66	17
26	*Cynodon dactylon*	Jiri	1900	6223	18	85	293	84	41
27	*Imperata cylindrica*	Jiri	1900	3613	15	85	204	128	49
28	*Erigeron canadensis*	Jiri	1900	2446	9	15	41	48	0.1
29	*Cynodon dactylon*	Manga Daurari	2040	4392	15	100	292	57	32
30	*Imperata cylindrica*	Chitre	2200	4658	12	100	388	91	39
31	*I. cylindrica*	Chaubas	2200	3055	15	100	205	83	38
32	*Cynodon dactylon*	Seta	2430	2318	11	90	189	82	41
33	*Artemisia*	Seilo	2500	2965	23	85	109	23	19
34	*Senecio chrysanchemoides* var. *spectabilis*	Junbesi	2700	2414	20	80	97	14	11
35	*Fragaria nubicola*	Junbesi	2700	5658	21	80	215	27	26
36	*Epilobium* sp.	Junbesi	2700	3251	11	80	236	79	37
37	*Plantago erosa*	Junbesi	2700	1874	11	55	94	24	23
38	*Fragaria nubicola*	Junbesi	2700	4544	16	75	213	26	17
39	*Festuca rubra*	Junbesi	2700	4164	12	100	347	80	24
40	*Juncus effusus*	Junbesi	2700	2571	10	100	257	70	34
41	*Senecio chrysanthemoides* var. *spectabilis*	Junbesi	2700	1843	8	100	230	52	18

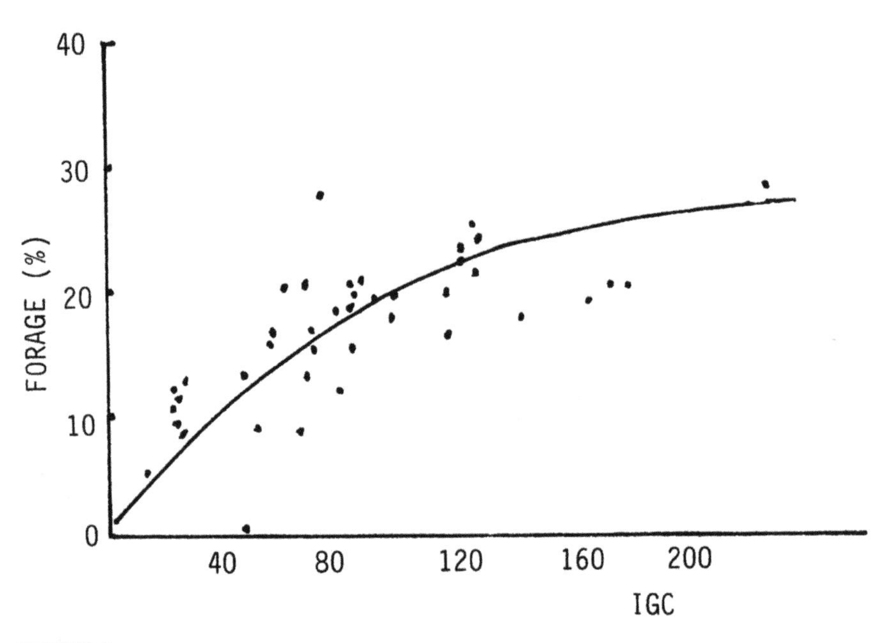

FIGURE 4.

potential palatable biomass based on the life span depending on the life-forms. Thus, a high value of the IGC accompanying a high value of forage percentage will be most desirable for pastures. From the forage value and IGC relationships (Fig. 4), the first group with the IGC under 50 is a deteriorated pasture. The second group, between 50 and 120, is an intermediate pasture, and the third group, over 120, is an excellent pasture land.

As mentioned above, initially the pasture vegetation is measured by the summed dominance ration (SDR), then its dynamic states on the course of succession is estimated by the degree of succession (DS). Following that, the condition and trend of vegetation, as a pasture for grazing, are evaluated by the forage value of the present palatable biomass and by the index of grassland condition (IGC) as the durable potential palatable biomass. Such measurements and evaluations will be useful measures for evaluating and managing pastures.

REFERENCES

1. Beard, J.S. (1964). The natural vegetation of Trinidad. *Oxford For. Mem.* No. 20.
2. Bouliere, F. and Hadley, M. (1970). The ecology of tropical savannas. *Annual Review of Ecology and Systematics*, 1, 125–152.

3. Box, T.W., Powell, J. and Drawe, D.L. (1967). Influence of fire on South Texas chaparral communities. *Ecol.* 48, 955–961.
4. Box, T.W. (1967). Influence of drought and grazing on mortality of five West Texas grasses. *Ecol.* 48, 654–656.
5. Chatterjee, P.C. (1980). Nomadic graziers of Garhwal. Singh, T. and Kaur, J. ed.: *Studies in Himalayan Ecology and Development Strategies*, 121–133. New Delhi.
6. Clements, F.E. (1916). Plant Succession. Carnegie Inst. Wash. Pub. No. 242.
7. Coetzee, B.J. (1974). Improvement of association analysis classification by Braun-Blanquet technique. *Bothalia* 11, 356–367.
8. Gill, A.M. (1978). Evolution of vascular plant traits adaptive to fires. A paper for Workshop on "*The Role of Past and Present Fire Frequency and Intensity of Ecosystem Development and Management.*" The East-West Center, Hawaii.
9. Goodland, R. (1971). A physiognomic analysis of the cerrado vegetation of Central Brazil. *J. Ecol.* 59, 411–419.
10. Gupta, R.K. (1978). Impact of human influences on the vegetation of the Western Himalaya. *Vegetatio* 37, 111–118.
11. Gupta, P.K. (1980). Development problems and potential for increased fodder production in the submontane regions of western Himalaya, Singh, T. and Kaur, J. ed.: *Studies in Himalayan Ecology and Development Strategies*, 134–151. New Delhi.
12. Hartley, W. (1954). The global distribution of tribes of the Gramineae in relation to historical and environmental factors. *Austr. J. Agr. Res.* 1, 355–373.
13. Heady, H.F. (1966). Influence of grazing on the composition of *Themeda triandra* grassland, East Africa. *J. Ecol.* 54, 705–727.
14. Hopkins, B. (1965). *Forest and Savanna.* Ibadan and London.
15. Hopkins, B. (1968). Vegetation of the Olokemeji Forest Reserve, Nigeria, V. The vegetation on the savanna site with special reference to its seasonal changes. *J. Ecol.* 56, 97–115.
16. Humphreys, L.R. (1981). *Environmental Adaptation of Tropical Pasture Plants.* McMillan, London.
17. Jones, R.M. (1968). Seed production of species in highveld secondary succession. *J. Ecol.* 56, 389–419.
18. Kartawinata, K. and Mueller-Dombois, D. (1972). Phytosociology and ecology of the natural dry-grass communities on Oahu, Hawaii. *Reinwardtia* 8, Part 3, 369–494.
19. Knapp, R. (1973). *Die Vegetation von Africa.* VEB Gustav Fischer Verlag, Jena.
20. Launghbaugh, J.L. (1955). Vegetational changes in the San Antonio Prairie associated with grazing, retirement from grazing and abandonment from cultivation. *Ecol. Monog.* 25, 39–57.
21. Leistner, O.A. and Werger, M.J.A. (1973). Southern Kalahari phytosociology. *Vegetatio* 28, 353–399.
22. McLiroy, R.J. (1972). *An Introduction to Tropical Grassland Husbandry.* 2nd ed. Oxford Univ. Press.
23. Molinari, O.G. (1949). Succession of grasses in Puerto Rico. *Rev. Agr. Puerto Rico* 39, 199–217.
24. Mullenders, W. (1954). La Végétation de Kaniama. *Publ. I.N.E.A.C. Sér. Scient.* 61, 1–500.
25. Nakashizuka, T. and Numata, M. (1982). Regeneration process of climax beech forest, 1. Structure of a beech forest with the undergrowth of *Sasa. Jap. J. Ecol.* 32, 57–67.
26. Numata, M. (1961). Problems in ecological succession, particularly on secondary succession and successional diagnosis. *Biol Sci.* 13, 147–152 (In Japanese).
27. Numata, M. (1962). Grassland diagnosis by the degree of succession (DS) and the index of grassland condition (IGC). *Kagaku (Science)* 32, 658–659 (In Japanese).
28. Numata, M. (1965). Grassland vegetation in eastern Nepal. Numata, M. ed.: *Ecological Study and Mountaineering of Mt. Numbur in Eastern Nepal*, 1963, 74–94.
29. Numata, M. (1966). Some remarks on the method of measuring vegetation. *Bull. Mar. Lab., Chiba Univ.* No. 8, 71–77.
30. Numata, M. (1969). Progressive and retrogressive gradients of grassland vegetation measured by degree of succession. *Vegetatio* 19, 96–127.
31. Numata, M. (1970). Primary productivity of semi-natural grasslands and related problems in Japan. Coupland, R.T. and Van Dyne, G.M. eds.: *Grassland Ecosystems: Review of Research*, 52–57.
32. Numata, M. ed. (1974). *The Flora and Vegetation of Japan.* Kodansha and Elsevier, Tokyo and Amsterdam.
33. Numata, M. (1976). Ecological studies of temperate semi-natural meadows of the world- particularly on primary production. *J. Japan. Grassl. Sci.* 22, 17–37.

34. Numata, M. (1980a). Facts, causal analysis and theoretical considerations on plant succession. Miyawaki, A. and Okuda, S. eds.: *Vegetation und Landschaft Japans*, 71–92. Yokohama.
35. Numata, M. (1980b). Semi-natural pastures and their management in the Himalayas. Furtado, J.I. ed.: *Trop. Ecol. and Develop.*, 1980, 399–409.
36. Numata, M. (1981). The altitudinal vegetation and climate zones of the humid Himalayas. *Geological and Ecological Studies of Qinghai-Xizang Plateau.* Vol. 2, 1963–1970, Peking.
37. Numata, M. ed. (1983). *Ecology and Utilization of Tropical Grasslands.* 189 pp. Tokyo (In Japanese).
38. Numata, M. and Ohga, N. (1976). Grasslands in the Oki Island. *Grassland Ecology* No. 15, 45–54 (In Japanese).
39. Okuda, S. and Numata, M. (1975). Relationship between succession of *Miscanthus sinensis* community and its productivity on phytosociological analysis. Numata, M. ed.: *Ecological Studies in Japanese Grasslands.* JIBP Synthesis Vol. 13, 45–50. Univ. Tokyo Press.
40. Park, B.K. (1965). Ecological studies on native grassland vegetation in Korea. *J. Korea Cult. Res.* Inst. 5, 177–183.
41. Pratt, D.J. and Kinight, J. (1971). Bush control studies in the drier area of Kenya, V. Effects of controlled burning and grazing management on *Tarchonathus-Acacia* thicket. *J. Appl. Ecol.* 8, 217–237.
42. Richards, P.W. (1976). *The Tropical Rain Forest.* Cambridge Univ. Press.
43. Ruddle, K. and Manshard, W. (1981). *Renewable Natural Resources and the Environment.* Tycooly Intern. Publ. Dublin.
44. Suganuma, T. (1983). Phytosociological classification and distribution of tropical grasslands. Numata, M. ed.: *Ecology and Utilization of Tropical Grasslands*, 1–16 (In Japanese).
45. Stamp, D. (1926). Some special aspects of vegetation survey in the tropics. Tansley, A.G. and Chipp, T.F. ed.: *Aims and Methods in the Study of Vegetation.* London, 238–258.
46. Thoreau, H.D. (1860). The succession of forest trees in *The Writings of H.D. Thoreau*, Vol. 9, Excursions (1893).
47. Tsuchida, K. and Numata, M. (1978). Relationship between successional situation and production of *Miscanthus sinensis* communities at Kirigamine Heights, Central Japan. *Vegetatio* 39, 15–23.
48. Uhlig, H. (1965). Die geographischen Grundlagen der Weidewirtschaft in den Trockengebieten der Tropen und Subtropen. Knapp, R. ed.: *Weide-Wirtschaft in Trockengebieten*, 1–28. Gustav Fischer, Stuttgart.
49. Verboom, W.C. (1968). Grassland succession and associations in Pahang, Central Malaya. *Trop. Agr. Trin.* 45, 47–59.
50. Watt, A.S. (1947). Pattern and process in the plant community. *J. Ecol.* 43, 490–506.
51. Werger, M.J.A. (1977). Applicability of Zürich-Montpellier methods in African tropical and subtropical range lands. Kause, W. ed.: *Application of Vegetation Science to Grassland Husbandry*, 123–145. Dr. W. Junk. The Hague.
52. Werger, M.J.A., Kruger, F.J. and Taylor, H.C. (1972). Pflanzensoziologische Studie der Fynbos-Vegetation am Kap der Huten Hoffnung. *Vegetatio* 24, 71–89.
53. Whitford, P. and Whitford, K. (1951). Thoreau. Pioneer ecologist and conservationist. *Sci. Month.* 73, 291–296.
54. Whyte, R.O. (1968). *Grasslands of the Monsoon.* Faber, London.
55. Whyte, R.O. (1977). Analysis and ecological management of tropical grazing lands. Krause, W. ed.: *Application of Vegetation Science to Grassland Husbandry.* Handbook of Vegetation Science 13, 3–121.
56. Yoshida, S. (1957). Studies on the grassland in Makassar Peninsula of Selebes Island. *Sci. Rep. Res. Inst. Tohoku Univ.* D-8 (2), 173–213.

Resource Management and Optimization
1990, Volume 7(1–4), pp. 227–250
Reprints available directly from the publisher.
Photocopying permitted by license only.
© 1990 Harwood Academic Publishers GmbH
Printed in the United States of America

TROPICAL AGROFORESTRY SYSTEMS AND PRACTICES

P.K.R. NAIR
Department of Forestry, IFAS, University of Florida Gainesville, Florida 32611, USA.

CONTENTS

Agroforestry is a new field of organized scientific pursuit although the practice encompasses some age-old land use activities. It involves elements of agriculture and forestry, wherein woody perennials are deliberately mixed or retained with crop or animal production units.

A global overview of the current agroforestry situation indicates that there are several examples of agroforestry systems and practices in different ecological and geographic regions of the world, especially the tropics. Depending on the dominant components, these systems can broadly be classified into agrosilvicultural, silvopastoral and agrosilvopastoral. Prominent examples of each are given from different parts of the tropics.

The role of woody perennials in agroforestry systems can be both productive (such as producing food, fodder, fuel and wood) and protective (like soil conservation, windbreaks and shelterbelts).

Although agroforestry has the most apparent potential in "marginal lands", it can equally be effective in high-potential areas. In both types of area it can have a special role where socio-economic or physical constraints force farmers to produce most of their basic needs from their own land. However, there are several scientific, institutional, developmental and extension constraints and impediments facing the development of agroforestry.

While developing management approaches in agroforestry, special emphasis has to be given to the overall performance of the system, and components should be viewed from such a perspective. Some fundamental aspects relating to the two major disciplinary components of land use systems—plant and soil—are also examined in the light of these considerations.

1. INTRODUCTION

In the recent history of developments in tropical land use, agroforestry as a term as well as a concept is perhaps second only to multiple cropping in terms of the rate and magnitude of the enthusiasm it has generated. It was not until the report of Béné et al. (1977) that the term became internationally current. Since then, and especially in the last five years, there has been a veritable explosion of interest and activities relating to agroforestry. The word has now become so firmly implanted that despite its alleged linguistic inadequacy (Stewart, 1981) and the likelihood of it being erroneously portrayed as a branch of forestry, it would now probably cause more confusion were another term introduced and popularised to encompass the same concept.

The concepts and principles of agroforestry have recently been elucidated in several publications from the International Council for Research in Agroforestry (ICRAF) as well as other organizations. However, in view of the newness of the topic, some of these fundamental aspects need to be recapitulated, even at the risk of repetition.

2. AGROFORESTRY APPROACH TO LAND USE

It is certainly not (only) the currently prevalent "fad" for new terminologies that has activated and accelerated agroforestry. As pointed out by Lundgren and Raintree (1983), agroforestry is the first concrete concept that builds on a synthesis of much of the practical experience and scientific knowledge acquired over the past decades in tropical agriculture, forestry, ecology, soil management and rural socioeconomics. Increasing dependence of modern agricultural technology on high-value inputs on the one hand, and the deteriorating economic situation of most of the developing countries on the other, have caused a renewed awareness of the potentials of age-old conservation farming technologies. At the same time, the seriousness of forest destruction and its alarming consequences are also being increasingly understood. The major cause of deforestation is now recognized to be man's search for additional areas to produce food to meet the ever-increasing demand for this basic need. Thus, in the wake of the mounting pressures of food and fuel shortage, and the serious environmental problems associated with deforestation, it is no longer prudent to ignore the conservational benefits and the potential for sustained yields provided by agroforestry farming systems based on or involving trees and other woody perennials, some forms of which have long been in existence in various parts of the world.

What would be a suitable definition for agroforestry that would embody all these concepts and encompass all the complexities? Certainly there is no concensus of opinion. Many definitions have been proposed. Some have even forwarded exaggerated and presumptuous claims that agroforestry, by definition, is a superior and more successful approach to land use than others. The definition adopted by ICRAF is as follows:

Agroforestry is a collective name for land-use systems and technologies where woody perennials (trees, shrubs, palms, bamboos, etc.) are deliberately used on the same land-management unit as agricultural crops and/or animals, either on the same form of spatial arrangement or temporal sequence. In agroforestry systems there are both ecological and economical interactions between the different components.

This definition implies that:

i) agroforestry normally involves two or more species of plants (or plant and animals), at least one of which is a woody perennial;
ii) an agroforestry system always has two or more outputs;
iii) the cycle of an agroforestry system is always more than one year; and
ix) even the simplest of agroforestry systems is more complex, ecologically (structurally and functionally) and economically, than a monocropping system.

3. VARIETY OF AGROFORESTRY SYSTEMS AND PRACTICES

3.1 Classification of Agroforestry Systems

Whatever the definition of agroforestry, it is now generally agreed that it represents an approach to land use involving deliberate retention of trees and other woody perennials in the crop/animal production fields (Lundgren and Raintree, 1983; Nair, 1983 a & b). When the existing land use systems are examined from such a broad concept as agroforestry, several of them can be considered to encompass the principles of agroforestry.

Various attempts have been made authors to classify different agroforestry systems. Obviously, a classification scheme will depend on its intended use. On a global basis, there can be geographical considerations, and within each geographical region there can be ecological factors that determine the types of systems in a locality. Social factors, especially demographic, coupled with economic background of the population can add another dimension. However, the basic structure of a system is decided primarily by the type and arrangements of its components. Therefore, one of the primary criteria in classifying agroforestry systems is the components that constitute the systems.

Following the definition mentioned in section 2, the basic groups of components in an agroforestry system can be two or three: woody perennials, herbaceous crops and/or animals. Since the woody perennial forms the common denominator in all agroforestry systems, a component-based classification scheme will logically have to be based on this predominant form. Here again there can be several the criteria, as pointed out by Torres (1983 a): the *type* of woody perennial, its *role and function* in the system, the nature of *interactions* between the woody and other components, and so on. All component-based classification schemes of agroforestry systems have so far considered the type of woody perennials as the first step in the exercise. Based on that, three broad subdivisions have been proposed by Nair (1983 d): agrosilvicultural, silvopastoral and agrosilvopastoral.

The agrosilvicultural system combines the production of tree crops (forest-, horticultural-, or agricultural plantation-) with herbaceous crops, in space or time, to fulfill productive or protective roles within the land management systems. Examples can be hedgerow intercropping (alley cropping), improved "fallow" species in shifting cultivation, multistorey multipurpose crop combinations, multipurpose trees and shrubs on farm lands, shade trees for commercial plantation crops, integrated crop combinations with plantation crops, agroforestry fuelwood production systems, shelterbelts and windbreaks and so on. The silvopastoral systems integrate woody perennials with pasture and/or livestock. Examples include animal production systems in which multipurpose woody perennials provide the fodder (protein bank), or function as living fences around grazing land, or are retained as commercial shade/browse/fruit trees in pasture lands. The agrosilvopastoral systems, as the name implies, combine trees and herbaceous crops with animals and/or pastures. The use of woody hedgerows for browse, mulch and green manures and for soil conservation, the crop/tree/livestock mix around homesteads, and the like, are good examples of this system. It is also common in some places to have sequential patterns (integration in time) of a agrosilvicultural phase followed by a silvopastoral one, so that initially trees and crops are established, and later on the crops are replaced with pasture and animals introduced.

3.2 Field Examples of Agroforestry Systems in the Tropics

ICRAF has undertaken a global inventory of existing agroforestry systems and practices in the tropics and subtropics (Nair, 1989). The basic document that was prepared for the project included a preliminary overview of the situation as a "Systems Overview Table" indicating the most prominent examples found in different geographical regions. An up-dated version of that is presented here as Table I.

The ICRAF survey collected information pertaining to the functioning of these systems, as well as analyzed their merits, weaknesses and constraints, with a view to identifying research needs and extending the system to other areas. A summary account of some of these extensively practised agroforestry systems and practices is given in Table II. (For a more detailed account of the woody species involved, see Nair, 1983 d and Nair et al., 1984). Without going into detail, suffice it to say that there are several widely practised land use systems which though not known by the name agroforestry, do encompass the agroforestry approach to land use.

4. PRODUCTIVE AND PROTECTIVE ROLES OF AGROFORESTRY

The field examples of agroforestry systems and practices presented in Table II show not only that they are widespread in different ecological regions, but are also important in terms of the role of woody perennials in producing the basic needs and/or protecting and prolonging the sustainability of the system. These primary roles (productive/protective) of the woody perennials, the type of their interactions (temporal/spatial) with the other major components and the spatial arrangement of the components (mixed/zonal) in the major agroforestry systems are summarized in Table III.

4.1 Productive Role

The productive role of the woody perennials in agroforestry systems includes the production of food, fodder, fuelwood and various other products. One of the most promising technologies of this kind that is applicable in a wide range of situations is the hedgerow intercropping (alley cropping) in crop production fields. Promising results have been obtained from this type in trials conducted at the International Institute of Tropical Agriculture (IITA), Ibadan, Nigeria (Wilson and Kang, 1981), where the practice is called alley cropping. The most promising system based on those trials is *Leucaena leucocephala*/maize alley cropping. IITA studies showed that leucaena tops maintained maize grain yield at a reasonable level even without nitrogen input on a low-fertility sandy Inceptisol, the nitrogen contribution by leucaena mulch on maize grain yield being equivalent to about 100 kg ha^{-1} for every 10 t ha^{-1} of fresh prunings (Kang *et al.*, 1981). The hedgerow intercropping system offers the advantage of incorporating a woody species with arable farming system without impairing soil productivity and crop yields. The potential of nutrient (N) contribution by several candidate species of woody legumes suggests that a wide range of such species could be integrated into crop production systems (Nair, 1984; Nair *et al.*, 1984).

Integration of trees in crop production fields is an essential part of traditional farming systems in the dry regions also. Two typical examples are the extensive use of *Acacia albida* in the groundnut and millet production areas of sub-Saharan Africa (Felker, 1978) and the dominant role of *Prosopis cineraria* in the arid North-Western parts of India (Mann and Saxena, 1980).

The role of woody perennials for producing fuelwood on farmlands is another example of the productive role of species in agroforestry. The seriousness of the fuelwood situation has been well-recognized all over the world, so that several initiatives and studies on this aspect are being undertaken at present. Several fast-growing fuelwood crops, most of them legumes, suitable for different environmental conditions, have been identified (NAS, 1980); most of them combine well with conventional agricultural crops (Nair, 1980).

In the "animal agroforestry" systems, the woody components could be used either as a source of fodder to improve livestock productivity or to obtain another commodity such as fuel, fruit, or timber. Based on this "productivity objective", silvopastoral systems can be either browse grazing or forest/plantation grazing systems. The role of woody perennials in these systems has been reviewed excellently by Torres (1983 b).

TABLE I.

Some Examples of Prominent Agroforestry Systems and Practices in the Developing Countries

Systems		Examples from Different Geographic Regions*						
Major Systems	**Sub-Systems/ Practices**	**Pacific**	**Southeast Asia**	**South Africa**	**Middle East and Mediterranean**	**East and Central Africa**	**West Africa**	**American Tropics**
Agrosilvicultural systems	Improved "Fallow" (in shifting cultivation areas)		Forest villages of Thailand. Various fruit trees & plantation crops used as "fallow" species in Indonesia	Improvements to Shifting Cultivation; several approaches e.g. in the north-eastern parts of India		Improvements to Shifting Cultivation e.g. gum gardens of the Sudan	*Acioa barteri*, *Anthonontha macrophylla*, *Gliriricidia sepium*, etc. tried as "fallow" species	Several forms
	The Taungya System	(e.g. Taro with *Cedrella*. and *Anthocephalus* trees)	Widely practised; Forest villages of Thailand is an inproved form	Several forms, several names		The 'Shamba' System	Several Forms	Several forms
	Tree Gardens	Involving fruit trees	Dominated by fruit trees	In all ecological regions	The Dehesa system; "Parc arboreé"			e.g. 'Paraiso Woodlot' of Paraguay
	Hedgerow Intercropping (Alley Cropping)		Extensive use of *Sesbania grandiflora*, *Leucaena leucocephala* and *Calliandra callothyrsus*	Several experimental approaches e.g. Conservation Farming in Sri Lanka		The Corridor System of Zaire	Experimental systems on alley cropping with *Leucaena* and others	

Multipurpose Trees and Shrubs on Farmlands	Mainly fruit or nut trees (e.g. *Canarium*, *Pometia*, *Barringtonia*, *Pandanus*, *Artocarpus altilis*)	Dominated by fruit trees; also *Acacia mearnsii*—cropping system, Indonesia	Several forms both in lowlands and highlands e.g. Hill Farming in Nepal; 'Khejri'-based system in the dry parts of India	The Oasis system; Crop combinations with the Carob tree; The Dehesa system; Irrigated systems; Olive trees + cereals	Various forms; The Chagga system of Tanzanian highlands; The Nyabisindu system of Rwanda	*Acacia albida*-based food production systems in dry areas; *Butyrospermum* + *Parkia* systems; "Parc arboreé"	Various forms in all ecological regions
Crop Combination with Plantation Crops	Plant. crops and other multipurpose trees: e.g. *Casuarina* and coffee in the highlands of PNG; also *Gliricidia* and *Leucaena* with cacao	Plant. crops + fruit trees; smallholder systems of crop combinations with plantation crops; plantation crops with spice trees	Integrated production systems in smallholdings; shade trees in plantations; other crop mixtures including various spices	Irrigated systems; Olive trees + cereals	Integrated production; shade trees in commercial plantations; mixed systems in the highlands	Plantation crop mixtures; smallholder production systems	Plantation crop mixtures; shade trees in commercial plantations; mixed systems in small-holdings; spice trees
AF Fuelwood Production	Multipurpose fuelwood trees around settlements	Several examples in different ecological regions	Various forms		Various forms	Common in the dry regions	Several forms in the dry regions
Shelterbelts, Windbreaks, Soil Conservation Hedges	*Casuarina oligodon* in the highlands as shelterbelts and soil improver	Terrace stabilization in steep slopes	Use of *Casuarina* spp. as shelterbelts; several windbreaks	Tree spices for erosion control	The Nyabisindu system of Rwanda	Various forms	Live fences, windbreaks especially in highlands

(continued)

TABLE I.

Some Examples of Prominent Agroforestry Systems and Practices in the Developing Countries (continued)

Systems		Examples from Different Geographic Regions*						
Major Systems	Sub-Systems/ Practices	Pacific	Southeast Asia	South Africa	Middle East and Mediterranean	East and Central Africa	West Africa	American Tropics
Sylvopastoral Systems	Protein Bank (Cut-and-carry Fodder Production)		Very common, especially in highlands	Multipurpose fodder trees on or around farmlands especially in highlands		Very common	Very common	Very common
	Living Fence of Fodder Trees and Hedges		*Leucaena*, *Calliandra*, etc. used extensively	*Sesbania*, *Euphorbia*, *Syzigium* etc. common		Very common in all ecological regions		Very common in the highlands
	Trees and Shrubs on Pastures	Cattle under coconuts, pines and *Eucalyptus deglupta*	Grazing under coconuts and other plantations	Several tree species being used very widely	Very common in the dry regions; the Dehesa system	The *Acacia* dominated system in the arid parts of Kenya, Somalia and Ethiopia	Cattle under oil palm Cattle and sheep under coconut	Common in humid as well as dry regions e.g. Grazing under plantation crops in Brazil
Agrosilvopastoral Systems	Woody Hedges for Browse, Mulch, Green Manure, Soil Conservation, etc.	Various forms: *Casuarina oligodon* widely used to provide mulch and compost	Various forms	Various forms especially in lowlands			Very common	

	Home Gardens (involving a large number of herbaceous and woody plants with or without animals)		Very common; Java Home Gardens often quoted as good examples	Common in all ecological regions	The Oasis system	Various forms (The Chagga homegardens; the Nyabisindu system)	Compound farms of humid lowlands	Very common in the thickly populated areas
Other Systems	Agro-Silvo-Fishery ('Aquaforestry')		Silviculture in mangrove areas; trees on the bunds of fish-breeding ponds					
	Various forms of Shifting Cultivation	Common	e.g. Swidden Farming	Very common; various names		Very common	Very common in the lowlands	Very common in all ecological regions
	Apiculture with Trees	Common	Common		Common	Common	Common	

TABLE II
Field Examples of Some Common Agroforestry Systems and Practices in the Tropics

Sub-system/ Practice	Country/ Region	Some examples of the woody species involved	Remarks/ Major references
I.A. AGROSILVICULTURAL SYSTEMS—Humid/Sub-humid Lowlands			
Improved "Fallow" (in shifting cultivation areas)	Indonesia	*Aleurites molucana* *Erythrina* spp. *Styrax* spp.	Kunstadter et al. (1978)
Woody species planted and left to grow during the "Fallow phase"	Nigeria	*Acioa barteri* *Anthonotha macrophylla*	Getahun et al. (1982)
Tree Gardens	Nigeria	*Daniellia oliveri* *Gliricidia sepium* *Parkia clappertoniana* *Pterocarpus africana*	Getahun et al. (1982)
Multilayer, multi-species plant associations with no organized planting arrangement	Pacific Islands	*Inocarpus edulis* *Morus nigra* *Spondias dulce*	Richardson (1982)
	India, Sri Lanka	*Areca catechu* *Artocarpus* spp. *Cocos nucifera* *Mangifera indica*	Coconut intercropping: Nair (1979; 1983); Liyanage et al. (1984)
	Paraguay	*Melia azedarach*	The Paraiso woodlot (Evans and Rombold, 1984)
	SE Asia	*Albizia falcataria* *Artocarpus* spp. *Bambusa* spp. *Durio zebethinus* *Nephelium lapaceum*	Ambar (1982) Forest Villages of Thailand (Boonkird et al., 1984)
Hedgerow intercropping (Alley cropping)	SE Asia	*Calliandra callothyrsus*	
Woody species in hedges; agri. species in between hedges (alleys)	Nigeria	*Leucaena leucocephala*	Wilson and Kang (1981)
Multipurpose trees and shrubs on farmlands	Brazil	*Cassia excelsa* *L. leucocephala* *Mimosa scabrella*	
Trees scattered haphazardly or according to some systematic patterns	India	*Derris indica* *Emblica officinalis* *Moringa oleifera* *Tamarindus indica*	NAS (1980)
	Kenya	*Anacardium occidentale* *Ceiba petandra* *Mangifera indica* *Manilkara achras*	

TABLE II
Field Examples of Some Common Agroforestry Systems and Practices in the Tropics (continued)

Sub-system/ Practice	Country/ Region	Some examples of the woody species involved	Remarks/ Major references
	SE Asia	*Acacia mangium* *Artocarpus* spp. *Durio zibethinus* *Gliricidia sepium* *Sesbania grandiflora*	
Crop combinations with plantation crops		Plantation crops	
1) Integrated production of plantation crops and other crops in intimate plant associations		*Anacardium occidentale* *Camellia sinensis* *Cocos nucifera* *Coffea arabica* *Elaeis guineensis* *Hevea brasiliensis* *Piper nigrum* *Theobroma cacao*	
2) Mixtures of plantation crops, e.g. coconut and cacao			
3) Shade trees for commercial plantation crops	Brazil	*Bertholletia excelsa* *Copernicia prunifera* *Cordia alliodora* *Inga* spp. *Orbignya* spp. *Samanea saman*	Hecht (1982) Alvim and Nair (1986)
	Costa Rica	*Cordia alliodora* *Erythrina poeppigiana* *Gliricidia sepium* *Inga* spp.	Budowski (1983) Heuveldop and Lagemann (1981)
	India	*Albizia* spp. *Cassia* spp. *Erythrina* spp. *Grevillea robusta*	Coconut intercropping (Nair, 1979; 1983; Liyanage et al., 1984)
	SE Asia	Various fruit trees	
	West Indies	*Inga vera*	
	Western Samoa	*Erythrina variegata* *Gliricidia sepium* *Leucaena leucocephala*	Richardson (1982)
AF for fuelwood Production	India	*Albizia* spp. *Cassia siamea* *Derris indica* *Emblica officinalis*	ICAR (1979) NAS (1980)

(continued)

TABLE II
Field Examples of Some Common Agroforestry Systems and Practices in the Tropics (continued)

Sub-system/ Practice	Country/ Region	Some examples of the woody species involved	Remarks/ Major references
Interplanting firewood species on or around agricultural lands	Indonesia	*Albizia falcataria* *Calliandra callothyrsus* *Sesbania grandiflora* *Trema orientalis*	NAS (1980)
Shelterbelts, windbreaks, soil conservation hedges	India	*Casuarina equisetifolia* *Syzygium cuminii*	NAS (1980)
Planting around agricultural lands as windbreaks and shelterbelts; planting along contours for terrace stabilization and soil conservation	Indonesia (and other parts of SE Asia)	*Gliricidia sepium* *Leucaena leucocephala* *Sesbania grandiflora*	NAS (1980)
I. B. AGROSILVICULTURAL SYSTEMS—Tropical Highlands			
Multipurpose trees and shrubs on farmlands	India	*Albizia* spp. *Bauhinia variegata* *Dalbergia sissoo*	NAS (1980)
	Kenya	*Ceiba petandra* *Eriobotrya japonica* *Grevillea robusta*	
	Nepal	*Bauhinia* spp. *Erythrina* spp. *Ficus* spp. *Litsea polyntha*	Hill farming in Nepal (Fonzen and Oberholzer, 1984)
	Paraguay	*Melia azedarach*	The Paraiso woodlot (Evans and Rombold, 1984)
	Tanzania	*Albizia* spp. *Cordia africana* *Croton macrostachys* *Trema guineensis*	The Chagga system (Fernandes et al., 1984)
Crop combinations with plantation crops	Brazil	*Alnus acuminata* *Enterolobium contorsiliquum* *Erythrina velutina*	
	Costa Rica	*Alnus acuminata* *Erythrina poeppigiana* *Inga* spp.	Budowski (1983)
	India, Sri Lanka	*Albizia* spp. *Grevillea robusta*	
	Kenya	*Grevillea robusta*	
	Papua New Guinea	*Casuarina olygodon*	Bourke (1984)
	Philippines	*Trema orientalis*	
	Rwanda	*Albizia* spp.	Fernandes et al. (1984)

TABLE II
Field Examples of Some Common Agroforestry Systems and Practices in the Tropics (continued)

Sub-system/ Practice	Country/ Region	Some examples of the woody species involved	Remarks/ Major references
	Tanzania	*Cordia africana* *Grevillea robusta* *Trema guineenis*	Neumann (1983)
AF Fuelwood Production	India, Nepal	*Albizia stipulata* *Bauhinia* spp. *Grewia* spp.	ICAR (1979) NAS (1980)
Shelterbelts, Windbreaks, Soil Conservation Hedges		(same as in lowlands)	
I.C. AGROSILVICULTURAL SYSTEMS—Arid and Semi Arid Regions			
Multipurpose Trees and Shrubs on Farmlands	Brazil	*Caesalpinia ferrea* *Prosopis juliflora* *Zizyphus joazeiro*	Johnson (1983)
	Central African Republic	*Adansonia digitata* *Balanites aegyptiaca* *Borassus aethiopium*	Yandji (1982)
	India	*Cajanus cajan* *Derris indica* *Prosopis cineraria* *Tamarindus indica*	NAS (1980)
	Kenya	*Acacia* spp. *Balanites aegyptiaca* *Cajanus cajan*	
	Tanzania	*Acacia* sp. *Combretum* spp.	
AF Fuelwood Production	Chile India	*Prosopis tamarugo* *Albizia lebbek* *Cassia siamea* *Prosopis* spp.	NAS (1980) Little (1983) ICAR (1979)
	Sahel	*Acacia albida* *A. senegal* *A. tortilis*	von Maydell (1984)
Shelterbelts and Windbreaks	India, Pakistan	*Azadirachta indica* *Cajanus cajan* *Cassia siamea* *Eucalyptus* spp. *Pithecellobium dulce* *Populus* spp.	Sheikh and Chima (1976), Sheikh and Khalique (1982)
II. SILVOPASTORAL SYSTEMS—Humid/Sub-humid Lowlands			
Protein Bank (Multi-purpose Fodder Trees on or around Farmlands)	India, Nepal, Sri Lanka	*Artocarpus* spp. *Anogeissus latifolia* *Bombax malabaricum* *Cordia dichotoma*	ICAR (1979) Pandey (1982) Singh (1982)

(continued)

TABLE II
Field Examples of Some Common Agroforestry Systems and Practices in the Tropics (continued)

Sub-system/ Practice	Country/ Region	Some examples of the woody species involved	Remarks/ Major references
		Dalbergia sissoo Eugenia jambolana Samanea saman Zizyphus jujuba	
Living Fences of Fodder Trees and Hedges	Costa Rica	*Diphysa robinoides* *Gliricida sepium*	
	Ethiopia SE Asia	*Erythrina abyssinica* *Sesbania grandiflora*	
Trees and Shrubs on Pastures (similar to multipurpose trees on farmlands)	Brazil	*Acacia* spp. *Anacardium occidentale* *Cedrela odorata* *Cordia alliodora*	Hecht (1982) Johnson and Nair (1984)
	Costa Rica	*Enterolobium cyclocarpum* *Erythrina poeppigiana* *Samanea saman*	De las Salas (1979)
	India	*Derris indica* *Emblica officinalis* *Psidium guajava* *Tamarindus indica*	Singh (1982)
II. SILVOPASTORAL SYSTEMS—Tropical Highlands			
Protein Bank	Indian subcontinent	*Albizia stipulata* *Bauhinia* spp. *Ficus* spp. *Grewia oppositifolia* *Morus alba*	
Living Fences	Costa Rica Ethiopia East Africa	*Gliricidia sepium* *Erythrina abyssinica* *Dovyalis caffra* *Euphorbia tirucalli* *Iboza multiflora*	
Trees and Shrubs on Pastures	Brazil	*Desmanthus varigatus* *Desmodium discolor*	
	Costa Rica Indian subcontinent	*Alnus acuminata* *Albizia stipulata* *Alnus napalensis* *Grewia* spp.	
II. C. SILVOPASTORAL SYSTEMS—Arid and Semi-arid Regions			
Protein Bank	India	*Acacia nilotica* *Ailanthus excelsa*	Singh (1982)

TABLE II
Field Examples of Some Common Agroforestry Systems and Practices in the Tropics (continued)

Sub-system/ Practice	Country/ Region	Some examples of the woody species involved	Remarks/ Major references
		Opuntia ficus indica *Prosopis* spp. *Rhus sinuata*	
Living Fences	East Africa	*Acacia* spp. *Commiphora africana* *Euphorbia tirucalli* *Zizyphus mucronata*	
Trees and Shrubs on Pastures	India	*Acacia* spp. *Prosopis* spp. *Tamarindus indica*	
	Middle East and Mediterranean	*Acacia* spp. *Ceratonia siliqua* *Haloxylon* spp. *Prosopis cineraria* *Tamarix aphylla*	
III. AGROSILVOPASTORAL SYSTEMS			
Woody Hedgerows for Browse, Mulch, Green Manure and Soil Conservation	Indian sub-continent (Humid low-lands), SE Asia	*Erythrina* spp. *Leucaena leucocephala* *Sesbania* spp.	
Tree-Crop-Livestock Mix around Homestead (known as Home Gardens, these associations are found in almost all ecological regions and several countries; only some examples are given)	South and SE Asia (Humid lowlands)	Fruit trees and some plantation crops mentioned under Agro-silvicultural systems	e.g. Home Gardens of Java (Wiersum, 1982)
	Nigeria (Humid lowlands)	*Cola acuminata* *Garcinia kola* *Irvingia gabonensis* *Pterocarpus soyauxii* *Treculia africana*	
	Latin American countries	Several species mentioned under Agrosilvicultural systems	Wilken (1978)
	Tanzania (Highlands)	*Albizia* spp. *Cordia africana* *Morus alba* *Trema guineensis*	Chagga Homegardens (Fernandes et al., 1984)

4.2 Protective Role

The protective role of woody perennials in agroforestry stems from their soil improving and soil conserving functions. There are various avenues through which the leguminous woody perennials could improve and enrich soil conditions; these include fixation of atmospheric nitrogen, addition of organic matter through litterfall and dead and

TABLE III
The Role of Woody Perennials, Their Arrangement and Interaction with Other Components in Some Common Agroforestry Systems

Systems	Sub-systems/ Practices	Primary role of woody perennials	Arrangement of components	Nature of interaction between major components
Agrosilvicultural	Hedgerow intercropping (Alley cropping)	Protective (soil productivity)	Zonal	Spatial
	Improved fallow	Protective (soil productivity and productive)	Zonal	Temporal
	Multistorey crop combination	Productive	Mixed	Spatial and temporal
	Multipurpose trees on farmlands	Productive	Mixed	Spatial
	Shade trees for commercial plantation crops	Protective and productive	Mixed or zonal	Spatial and temporal
	AF fuelwood production	Productive	Zonal	Temporal and spatial
	Shelterbelts and windbreaks	Protective	Zonal	Spatial
	Protein bank	Productive (and protective)	Zonal	Temporal
Silvopastoral	Living fence	Protective	Zonal	Spatial
	Trees over pastures	Productive (and protective)	Mixed and zonal	Spatial
Agrosilvopastoral	Woody hedgerows for browse, mulch, green manure and soil conservation	Productive and protective	Mixed	Temporal and spatial
	Tree-crop-livestock mix around homesteads	Protective and protective	Mixed	Spatial and temporal
	Agrosilvicultural to silvopastoral	Productive	Mixed	Temporal and spatial

decaying roots, modification of soil porosity and infiltration rates leading to reduced erodibility of soil and improving the efficiency of nutrient cycling within the soil-plant system (Nair, 1984). However, the main protective function of woody perennials is in physical conservation of the soil.

The long tradition of planting *Leucaena leucocephala* in contour hedges for erosion control and soil improvement in Southeast Asia, especially Indonesia, is a typical example. Indirect terraces are also formed when the washed-off soil is collected behind the hedges. Loppings and prunings from such hedgerow species could also provide mulch to aid in preventing sheet erosion between trees (Zeuner, 1981; Neumann, 1983). The presence of more plant cover on the soil, either alive or as mulch, also reduces the impact of raindrops on the soil and thus minimizes splash and sheet erosion. Therefore, as pointed out by Lundgren and Nair (1985), the potential role of agroforestry in soil conservation lies not only in woody perennials acting as a physical barrier against erosive forces, but also in providing mulch and/or fodder and fuelwood at the same time.

Other protective functions of woody perennials in agroforestry include their role as live fences, shelterbelts and windbreaks. Use of trees and other woody perennials to protect agricultural fields from trespass or against the adverse effects of wind is a wide-spread practice in many agricultural systems. For example, a large number of multi-purpose woody perennials are being used as effective live fences at CATIE (Centro Agronómico Tropical de Investigación y Enseñanza), Turrialba, Costa Rica (Budowski, 1983). Similarly, very encouraging results on shelterbelts and windbreaks have been obtained at the Pakistan Forestry Research Institute, Peshawar (Sheikh and Chima, 1976; Sheikh and Khalique, 1982), as indicated in Table II.

5. CONSTRAINTS AND POTENTIALS

5.1 Constraints

There are several scientific, institutional, developmental and management constraints and impediments to be overcome before scientifically sound agroforestry technologies can be developed and adopted in areas where other land use systems are breaking down.

Scientifically, agroforestry has as yet no distinct identity or separate existence. By its very nature, it is an integrated and multidisciplinary approach encompassing complex systems. Existing land use research institutions, both national and international, are mostly oriented to specific commodities, disciplines or ecological regions, so that they are poorly equipped to handle complex topics such as agroforestry. The scientists themselves are mostly too discipline-oriented so that it is no easy task to persuade them to relegate and reorient their disciplinary pursuits to the interdisciplinary needs of a multidisciplinary team. Moreover, the experimental methods and procedures that have been developed over the decades for specific disciplines and components must be modified to make them applicable and relevant to integrated and complex systems, which also is not easy.

Institutional constraints to agroforestry development are also equally complex. As mentioned above, rigid boundaries often separate departments dealing with different aspects of land use. This leads to increasing competition for scarce developmental resources at governmental and administrative levels. As pointed out by Lundgren and Raintree (1983), even where formal agroforestry programmes exist, they fall under the forestry departments. These generally have little knowledge of, let alone interaction with, the agriculture departments (which, usually, are more 'prestigious' and 'powerful' than the respective forestry departments). The situation is much the same in the international sphere. Thus there exists a vicious circle: on the one hand agroforestry has not yet developed enough to earn a separate identity in terms of resources allocation, and, such a respectability and identity can, on the other, be achieved only by research investments for development.

The transfer of technology to the masses is another major step involved in the adoption of such land use practices. Most farmers in the tropics are preoccupied by their efforts to satisfy the day-to-day basic needs of food, fuel and shelter so that they cannot easily evaluate the merits of long-term approaches and investments. Thus, whereas it may be relatively easy to introduce short-term technologies, such as new species or better varieties of agricultural crops, or to make marginal improvements in the management of existing tree components, it will be considerably more difficult and challenging to convince farmers to incorporate tree components over existing crop or animal enterprises, especially if land tenure is uncertain and success of the system is not guaranteed. The problem is compounded by poor and inadequate extension services that can seldom handle such complex problems as those of agroforestry.

Management constraints of agroforestry are also several and of varied nature. Special skills and sustained efforts are needed for undertaking the various management aspects of trees, about which many crop or livestock farmers may not be aware. Interaction between components, especially the hypothetical adverse effects of trees on crops, is an area about which farmers who are not experienced with such systems are apprehensive. Further, researchers are not yet equipped enough to allay such apprehension. It is interesting to note in this context that a survey on the extent of intercropping in coconut lands in Sri Lanka identified seven important problems/constraints faced by farmers in expanding their intercropping activity (Liyanage *et al.*, 1984). In order of relative importance these were: drought, lack of funds, price instability, lack of technical know-how, problems of timely availability of labour, availability of planting materials and theft. On an average, each intercropper faced at least three of these problems, their nature and extent depending on the size of holding and type of intercrop.

5.2 Potentials

Notwithstanding those constraints, agroforestry has a great potential for application over vast areas of land. As indicated earlier, the most apparent potential for agroforestry exists where soil fertility is low and depends mainly on soil organic matter fraction, and where erosion hazards are high. Such "marginal lands" cover a majority of land areas in the tropics. Proper integration of appropriate woody species in the land use systems in these areas can enhance both land productivity and sustainability.

However, the potential of agroforestry is by no means confined to such "marginal" lands; it is equally applicable to high-potential areas. Indeed, indigenous agroforestry systems can be found wherever there has been a history of population pressure and a long-standing need for efficient management of scarce resources (Lundgren, 1982). Some of the most successful smallholder systems mentioned in Table II are, in fact, found on high-potential, fertile soils, where such integrated systems are often superior and preferred to other forms of land use. In both low- and high-potential areas agroforestry can have a special role in situations where land tenure system or infrastructural limitations (road, transport, markets and the like) make it imperative that farmers satisfy most of their basic needs from their own land resources.

The potential role of agroforestry in production systems that aim to satisfy such basic needs as well as and in protecting the environment has already been indicated in section 4.2. Agroforestry approaches have also been suggested as alternatives to resource-depleting shifting cultivation (Nair, 1983 d) as well as in other specific environments (Nair, 1983 c). It has also a special role in combating desertification and deforestation because the primary reason for forest destruction is man's ever-increasing demand for more land for producing food and agroforestry offers possibilities for producing food and wood at the same time from the same piece of land (King, 1979; Nair, 1982).

6. MANAGEMENT APPROACHES FOR DEVELOPMENT OF AGROFORESTRY SYSTEMS

Basically, there are two approaches in the study of an entity. First, to consider the different components and study them individually, paying particular attention to their cause-effect relationships. Most agricultural research conducted in the past has been of this nature, and these studies have advanced knowledge considerably. However, problems often arise when attempts are made to put the pieces together and predict the behaviour of the system, which often consists of something more than the individual components. The second approach is to study the system in its totality—a system will, of course, be considered to consist of different sub-systems.

In agroforestry, there are numerous individual components and interactions to be studied. Moreover, the studies are normally long-term. Inadequate planning and uncoordinated data-gathering without a central theme, as is likely to occur in individual disciplinary experiments, might lead to the drawing of incorrect conclusions with respect to the system as a whole. In addition, the extrapolation of results obtained from such piece-meal research might be extremely dangerous. Therefore, technological assessment in agroforestry research should concentrate less on types of component analyses, in which the factors and organisms are treated as if they were independent entities, but focus more on approaches in which the interactive, integrative, and emergent properties are also included.

This is, however, not to suggest that approaches aiming at gathering basic information on the components of the system are not required. In fact, when the land use system is examined in its totality certain aspects of the components that need to be studied in detail to produce the expected technologies will come to the fore. In most cases such technologies and management approaches that require immediate attention will be related to the plant and soil components.

6.1 Plant Aspects of Agroforestry

Because of the newness of agroforestry there are no conventional plant species that can be categorized as ''agroforestry species''. All species that can grow well in combined production systems fall under the domains of ''agricultural'', ''forestry'', ''horticultural'', or other established classes. It is therefore important to examine the ''suitability'' of an economic plant species for agroforestry, regardless of whether or not it is known to belong to any of the conventional disciplinary groups. Nair (1980) examined the ''agroforestry potential'' of several of the better-known as well as some lesser-known ''agricultural'' and ''horticultural'' species, and found that most of them can grow and produce reasonable yields under combined production systems.

When considering integration of trees on farmlands where some agricultural species are already being grown, it is assumed that there will be little or no change in the type of such herbaceous species: they will continue to be limited to what the local population or established markets require. On the other hand, the compatability and complementarity of the woody perennial with such herbaceous species will be the important consideration. In addition to the genotype of the woody species as such, its resource-sharing capabilities, potential microsite enrichment capability, and environmental amelioration are also or prime importance. Thus appropriate management measures (pruning, lopping, pollarding, browsing, time of sowing in the case of herbaceous species) have to be practised in order to optimize the benefits in combined production systems. Peculiar phenological characteristics of economically useful species may become very convenient in some contexts. A typical example is the tree *Acacia albida,* which produces leaves prior to the onset of rains and shed them during the rainy season. Thus millet and groundnut can be grown close to the tree in the rainy season without being shaded, and at the same time they benefit from the microsite enrichment by the tree (Felker, 1978).

Plants, especially woody species, that have hitherto been little studied and understood may prove valuable for agroforestry. Prime candidates will be species that can grow well with other, that thrive in environments too harsh for most other species, that simultaneously yield several products (food, fuel, fodder), that provide environmental amelioration (e.g. soil conservation) and that enrich the micro-site, such as by nitrogen fixation or nutrient cycling. Luckily, a few species have been identified that possess some, if not all, of these attributes (NAS, 1975; Ritchie, 1979), and they are now receiving scientific attention.

Arrangement of component plant species in space and time is also an important but difficult factor in agroforestry because of the many variations in the types of agroforestry practices and the conditions under which they are practised. The motivation for most of the various kinds of smallholder agroforestry systems prevalent throughout the world (Tables I and II) has been to find plants that provide multiple products and that can be grown on the available land. When attempting to improve such systems or to devise new ones it is therefore necessary to know about both the short-term productivity of the plants and the long-term sustainability of the system. Thus, depending on whether the tree-crop interaction is favourable or not, plant arrangements must be devised to maximize the beneficial interactions and minimize the undesirable ones. There are also several other factors to be taken into account, including growth habits and requirements of the component species when grown near other species, simplicity of management procedures for the whole system, and the realization of additional benefits like soil conservation (Huxley, 1983).

6.2. Soil Aspects of Agroforestry

State-of-the-art of soils aspects of agroforestry was brought out in an ICRAF Consultation in 1979 (Mongi and Huxley, 1979). Since agroforestry is particularly suitable for farmers with limited resources in "marginal areas", where sedentary agriculture or forestry systems may not be the most feasible and desirable, the system must be self-maintaining. This means that the system should attain maximum efficiency in inputs and maintain productivity of soil, with a strong emphasis on resource conservation. In view of the importance of the self-sustaining and resource-conserving attributes of agroforestry, the likely effects of agroforestry on the long-term productivity of soil have been examined using existing knowledge derived from similar land use systems (Nair, 1984). This involved an evaluation of soil productivity changes under shifting cultivation, *taungya,* plantation forestry, integrated systems involving plantation crops and multiple cropping. It also entailed assessment of the role of trees in soil productivity and protection.

The analysis revealed that several advantages in terms of soil productivity and protection could be anticipated by the proper incorporation of appropriate woody species into land use systems. Some expedient soil management technologies of a general nature were also suggested, based on these considerations.

In conclusion, agroforestry has generated considerable enthusiasm among various groups of people. There are several types of agroforestry systems and all of them are very complex in nature. The scientific approach to the study of these complex systems is difficult, time-consuming and requires a multidisciplinary approach. Most of the hypothesis concerning the potential as well as management approaches of agroforestry are still in the hypothetical and speculative stage. To validate these hypotheses and devise sound management technologies, research must be undertaken on various aspects in a systematic manner and in different agro-ecological situations. While interpreting results from such research and trying to extrapolate them to other situations, the overall systems perspective of agroforestry must be given adequate attention.

ACKNOWLEDGMENTS

ICRAF's Agroforestry Systems Inventory project, some results of which are included in this paper, was funded partially by the United States Agency for International Development (USAID). Mr. Erick C.M. Fernandes, my colleague in the project, has contributed substantially to this paper, and especially to Tables I and II.

REFERENCES

1. Alvim, R. and Nair, P.K.R. (1986). Agroforestry practices involving agricultural plantation crops in southeast Bahia, Brazil, No. 9 of AF System Description Series, *Agroforestry Systems*. 4: 3–15.
2. Ambar, S. 1982. Overview of the results of traditional agroforestry study in Citarum river basin, West Java. Paper presented to *The Regional Seminar-Workshop in Agroforestry*. 18–22 October, 1982. SEARCA, College, Laguna, The Philippines.
3. Béné, J.G., Beall, H.W. and Côté, A. (1977). *Trees, Food and People—Land Management in the Tropics*, IDRC 084e, International Development Research Centre, Ottawa.
4. Boonkird, S.-A., Fernandes, E.C.M. and Nair, P.K.R. (1984). Forest villages—an agroforestry approach to rehabilitating forest lands degraded by shifting cultivation in Thailand, No. 2 of AF System Description Series, *Agroforestry Systems*. 2: 87–102.
5. Bourke, M. (1984). Food, coffee and Casuarina: an agroforestry system from the Papua New Guinea highlands, No. 6 of AF System Description Series, *Agroforestry Systems*. 2: 273–279.
6. Budowski, G. (1983). An attempt to quantify some current agroforestry practices in Costa Rica, in *Plant Research and Agroforestry* (Ed. Huxley, P.A.), pp. 43–62, ICRAF, Nairobi.
7. De las Salas, G. (Ed). (1979). *Proceedings of the Workshop on Agroforestry Systems in Latin America*. CATIE, Turrialba, Costa Rica.
8. Evans, P.T. and Rombold, J.S. (1984). Paraiso (*Melia azedarach* var. "Gigante") woodlots: an agroforestry alternative for the small farmer in Paraguay, No. 5 of AF System Description Series, *Agroforestry Systems* 2: 199–214.
9. Fernandes, E.C.M., O'King'ati, A. and Maghembe, J. (1984). The Chagga home-gardens: a multistoried agroforestry cropping system in Mt. Kilimanjaro, N. Tanzania, No. 1 of AF System Description Series, *Agroforestry Systems* 2: 73–86.
10. Felker, P. (1978). *State of the Art: Acacia albida as a complementary intercrop with annual crops*, Univ. California, Berkeley, California (Grant No. AID/afr C-1361; mimeographed report).
11. Fonzen, P. and Oberholzer, E. (1984). Use of multipurpose trees in hill farming systems in western Nepal, No. 4 of AF System Description Series, *Agroforestry Systems* 2: 187–197.
12. Getahun, A., Wilson, G.F. and Kang, B.T. (1982). The role of trees in the farming systems in the humid tropics, in *Agroforestry in the African Humid Tropics*, (Ed. MacDonald, L.H.), pp. 28–35, UNU, Tokyo.
13. Hecht, S.B. (Ed). (1982). *Amazonia: Agriculture and Land Use Research*, CIAT, Cali, Colombia.
14. Heuveldop, J. and Lagemann, J. (Eds). (1981). *Agroforestry: Proc. of a seminar held at CATIE*, 23 February—3 March, 1981, CATIE, Turrialba, Costa Rica.
15. Huxley, P.A. (Ed). (1983). *Plant Research and Agroforestry, Proceedings of an Expert Consultation*, ICRAF, Nairobi.

16. Indian Council of Agricultural Research. 1979. *Proceedings of the National Seminar on Agroforestry, May 1979.* ICAR, New Delhi.
17. Johnson, D.V. (1983). *Agroforestry Systems in Northeast Brazil,* Report of the Special Consultant, ICRAF, Nairobi (unpublished).
18. Johnson, D.V. and Nair, P.K.R. (1984). Perennial crop-based agroforestry systems in northeast Brazil, No. 8 of AF System Description Series, *Agroforestry Systems* 2: 281–292.
19. Kang, B.T., Wilson, G.F. and Sipkens, L. (1981). Alley cropping maize and leucaena in southern Nigeria, *Plant and Soil,* 63: 165–179.
20. King (1979) Agroforestry and the utilization of fragile ecosystems. *Forest Ecology and Management* 2: 161–168.
21. Kunstadter, P., Chapman, E.C. and Sabhasri, S. (Eds). (1978). *Farmers in the Forest: Economic Development and Marginal Agriculture in Northern Thailand,* East-West Center, Honolulu, Hawaii.
22. Little, E.L. (1983). *Common Fuelwood Crops: A Handbook for their Identification,* McClain Printing Co., Parsons, West Virginia.
23. Liyanage, M. de S., Tejwani, K.G. and Nair, P.K.R. (1984). Intercropping under coconuts in Sri Lanka, No. 7 of AF System Description Series, *Agroforestry Systems.* 2: 215–228.
24. Lundgren, B.O. (1982). Agroforestry approaches to land use in the tropics, *Fourth International Congress of INTERFORST,* Munich.
25. Lundgren, B.O. and Nair, P.K.R. (1985). Agroforestry for soil conservation, *Soil Erosion and Conservation,* (Ed. El-Swaify, S.A., Moldenhauer, W.C., and Lo, A.), pp. 703–717. Soil Cons. Soc. Am., Ankeny, Iowa.
26. Lundgren, B.O. and Raintree, J.B. (1983). Sustained agroforestry, in *Agricultural Research for Development: Potentials and Challenges in Asia* (Ed. Nestel, B.), pp. 37–49, ISNAR, The Hague.
27. Mann, H.S. and Saxena, S.K. (Eds). (1980). *Khejri (Prosopis cineraria) in the Indian Desert,* CAZRI Monograph No. 11, Central Arid Zone Research Institute, Jodhpur, India.
28. Mongi, H.O. and Huxley, P.A. (Eds). (1979). *Soils Research in Agroforestry—Proceedings of an Expert Consultation,* ICRAF, Nairobi.
29. Nair, P.K.R. (1979). *Intensive Multiple Cropping with Coconuts in India: Principles, Programmes and Prospects,* Verlag Paul Parey, Berlin (West).
30. Nair, P.K.R. (1980). *Agroforestry Species: A Crop Sheets Manual,* ICRAF, Nairobi.
31. Nair, P.K.R. (1982). Agroforestry: a sustainable land-use system for the fragile ecosystems in the tropics, *Malay. Nat. J., 35, 109–123.*
32. Nair, P.K.R. (1983 a). Multiple land-use and agroforestry, in *Better Crops for Food, CIBA Foundation Symposium 97,* pp. 101–115, Pitman Books, London.
33. Nair, P.K.R. (1983 b). Tree integration on farmlands for sustained productivity of smallholdings, in *Environmentally Sound Agricultural Alternatives,* (Ed. Lockeretz, W.), pp. 333–350, Praeger, New York.
34. Nair, P.K.R. (1983 c). Some promising agroforestry technologies for hilly and semi-arid regions of Rwanda, in *Report of a Seminar on Agricultural Research in Rwanda: Assessment and Perspectives.* (Ed. Chang, J.), pp. 93–99, ISNAR, The Hague.
35. Nair, P.K.R. (1983 d). Alternative and improved land use systems to replace resource-depleting shifting cultivation, *Expert Consultation on Strategies, Approaches and Systems for Integrated Watershed Management,* Forest Resources Division, FAO, Rome.
36. Nair, P.K.R. (1984). *Soil Productivity Aspects of Agroforestry,* ICRAF, Nairobi.
37. Nair, P.K.R. (Ed.) (1989). *Agroforestry Systems in the Tropics.* Kluwer Academic Publishers, Dordrecht, The Netherlands.
38. Nair, P.K.R., Fernandes, E.C.M. and Wambugu, P.N. (1984). Multipurpose leguminous trees and shrubs for agroforestry, *Agroforestry Systems.* 2: 145–163.
39. NAS (1975). *Underexploited Tropical Plants with Promising Economic Value,* National Academy of Sciences, Washington, D.C.
40. NAS (1980). *Firewood Crops: Shrubs and Trees for Energy Production,* National Academy of Sciences, Washington, D.C.
41. Neumann, I. (1983). Use of trees in smallholder agriculture in tropical highlands, in *Environmentally Sound Agriculture,* (Ed. Lockeretz, W.), pp. 351–374, Praeger, New York.
42. Panday, K. (1982). *Fodder Trees and Tree Fodder in Nepal,* Swiss Devpt. Corp., Berne, and Swiss Federal Inst. of Forestry Research, Birmensdorf, Switzerland.

43. Richardson, S.D. (1982). Agroforestry education in the south Pacific—opportunities and constraints, *International Workshop on Professional Education in Agroforestry,* December 1982, ICRAF, Nairobi.
44. Ritchie, G.A. (Ed). (1979). *New Agricultural Crops,* Selected Symp. No. 38, American Assoc. Adv. Sci., Westview Press, Colorado.
45. Sheikh, M.I. and Chima, A.M. (1976). Effect of windbreaks (tree rows) on the yield of wheat crop, *Pakistan Journal of Forestry, 26(1),* 38–47.
46. Sheikh, M.I. and Khalique, A. (1982). Effect of tree belts on the yield of agricultural crops, *Pakistan Journal of Forestry, 32,* 21–23.
47. Singh, R.V. (1982). *Fodder Trees in India,* Oxford and IBH Pub. Co., New Delhi.
48. Stewart, P.J. (1981). Forestry, agriculture and land husbandry, *Commonw. For. Rev., 60(1),* 29–34.
49. Torres, F. (1983 a). Agroforestry: concepts and practices, in *Agroforestry Systems for Small-scale Farmers,* (Eds. Hoekstra, D.A. and Kuguru, F.M.), pp. 27–42, ICRAF/BAT, Nairobi.
50. Torres, F. (1983 b). Role of woody perennials in animal agroforestry, *Agroforestry Systems 1:* 131–163.
51. von Maydell, H.-J. (1984). *Agroforestry Systems and Practices in the Arid and Semi-Arid Parts of Africa,* Report of the Special Consultant, ICRAF, Nairobi (unpublished).
52. Wiersum, K.F. (1982). Tree gardening and taungya in Java: examples of agroforestry techniques in the humid tropics, *Agroforestry Systems 1,* 53–70.
53. Wilken, G.C. (1977). Integration of forest and small scale farm systems in middle America, *Agro-Ecosystems 3,* 291–302.
54. Wilson, G.F. and Kang, B.T. (1981). Developing stable and productive biological cropping systems for the humid tropics, in *A Scientific Approach to Organic Farming,* (Ed. Stonehouse, B.), pp. 193–203, Butterworth, London.
55. Yandji, E. (182). Traditional agroforestry systems in the Central African Republic, in *Agroforestry in the African Humid Tropics,* (Ed. MacDonald, L.H.), pp. 52–55, UNU, Tokyo.
56. Zeuner, T.H. (1981). An ecological approach to farming: some experiences of the agro-pastoral project, Nyabisindu, Rwanda, in *Kenya National Seminar on Agroforestry* (Ed. Buck, L.), pp. 329–353, ICRAF, Nairobi.

Resource Management and Optimization
1990, Volume 7(1–4), pp. 251–267
Reprints available directly from the publisher.
Photocopying permitted by license only.
© 1990 Harwood Academic Publishers GmbH
Printed in the United States of America

TROPICAL RESERVOIR FISHERIES

T. O. PETR and JAMES M. KAPETSKY
Inland Water Resources and Aquaculture Service, Fisheries Resources and Environment Division; Fisheries Department, UN Food and Agriculture Organisation, Via delle Terme di Caracalla, 00100 Rome, Italy

CONTENTS

Reservoirs as sources of fish for food are a twentieth century phenomenon. These reservoirs can serve as significant food sources in the tropics. It is necessary to manage these resources carefully. Transfer of fishery management and development experience among regions, while improving, is still inadequate.

1. INTRODUCTION

By definition, a reservoir is an artificial water body established by damming a river. In this chapter, only the tropical reservoirs are discussed. Reservoirs vary in size, shape and depth, and they will usually undergo seasonal changes in volume. Reservoirs are established for various purposes, with the stored water in the largest reservoirs being used for water supply, flood control, irrigation, hydro-electric power production and for some other purposes, amongst which fisheries are almost never given a priority. Fisheries have often been considered as a secondary benefit, or as an unanticipated profit of reservoir formation, and thus in the past, fisheries rarely have been adequately planned for in the preparatory stages of dam construction and reservoir formation. This is now changing. For example, Mexico has a programme to alleviate hunger through the development of fish production in small reservoirs and has initiated a very large-scale programme of pond and reservoir construction accompanied by loans and extension services for this purpose (COPESCAL, 1982). In China, small reservoirs are now designed and constructed exclusively for fishery purposes, including grading and smoothing of bottoms to facilitate harvest (Song, 1980) and in Southeast Asia increasing effort is being spent on optimizing fish production from these water bodies (Baluyut, 1983).

The oldest reservoirs are found in Asia, where reservoir building commenced about 4000 years ago (Fernando, 1977). The greatest majority of the large tropical reservoirs have been constructed in this century when heavy machinery for large-scale earth movement became available, especially after World War II.

The number of dams in the tropics, and that of the resulting reservoirs is steadily increasing. Bernacsek (1984) lists for the tropical belt of west, central, east and southern Africa—without South Africa and Namibia—112 reservoirs ranging in size from 5 to 827,000 ha. Existing African reservoir surface area, up to 1982, was put at 41,000 km2 and new reservoirs projected to be formed by the end of this century for Africa will amount to an additional 56,000 km2 (Clay, 1984). In comparison, southeast Asian reservoirs currently cover 30,000 km2 and are projected to expand to 200,000 km2 by the year 2000 (Fernando, 1984). Some of the world's largest tropical reservoirs exist in Latin America, and the continued construction of reservoirs there, both large and small, can be expected to continue.

This expansion of reservoir surface area has important implications for fisheries. Reservoirs currently provide about 10 percent of the inland fishery yield from Africa (Kapetsky, 1986). The median fishery yield from a sample of 92 tropical reservoirs from Africa, Asia and Latin America is 63 kg ha-1 yr-1 (Fig.1). If this yield rate could be maintained in the tropical reservoirs, then the catch from new African reservoirs could amount to some 350 thousand tons by the end of the century. Likewise, if the tropical Asian reservoir surface area expands by as much as 170,000 km2 by year 2000, then the new Asian reservoir area could then be supplying more than 1 million tons.

Yields among tropical reservoirs vary widely (Table I). In addition to morphometric

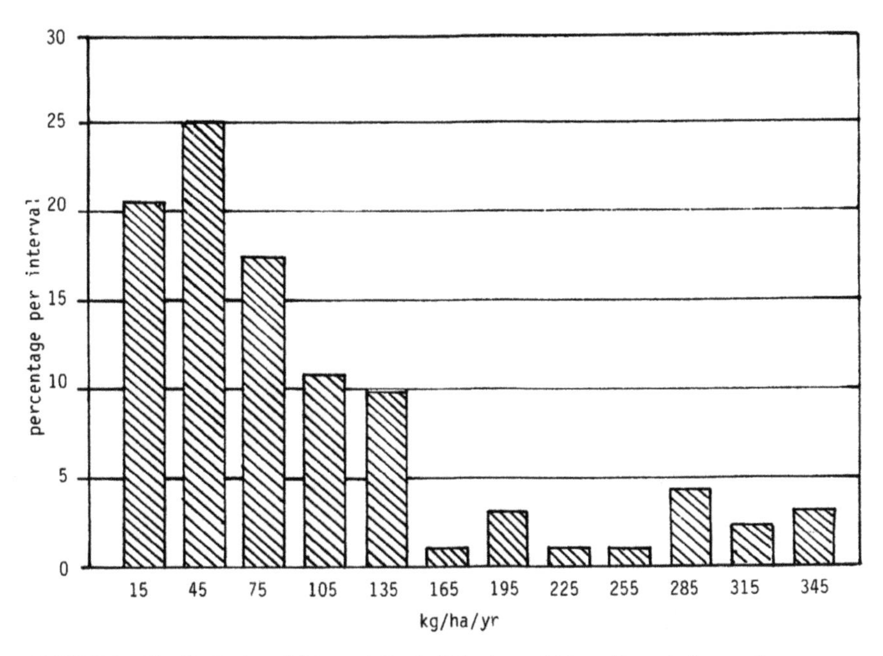

FIGURE 1. The distribution of fishery yields (in 30 kg intervals) from 91 tropical reservoirs.

features and the basic biological productivity of the waters, much also depends on how the reservoirs are exploited and how they are managed. The effect of exploitation pressure on reservoir fishery yield variability has not been taken into account here. Based on the available statistics, Africa's reservoirs have a median yield of 70 kg ha-1 yr-1. In Asian countries (Bangladesh, India, Indonesia, Sri Lanka and Thailand), in spite of introductions and stocking, the median yield is lower, only 57 kg ha-1 yr-1. Even in Sri Lanka where prolific tilapias are stocked, the median yield remains low, only 55.5 kg ha-1 yr-1. Introductions and stocking programmes have perhaps provided the best results in Brazil where the yields from 16 reservoirs in the northeast tabulated by Paiva (1976) have a median of 90 kg ha-1 yr-1, with up to 28 species stocked (Gurgel, 1984).

Little information is available for other countries of the tropical belt of Asia, and even less so on the Pacific and Indian ocean island countries with the exception of the above (Petr, 1984).

In tropical Latin America the first large reservoir built, the Brokopondo in Suriname of 1,800 km2, was neglected for many years from the fisheries point of view; later on, interest in developing fisheries on this lake led to an evaluation of its fishery potential (Kapetsky, 1978); however, because of the poverty of the waters, isolation, and low human population as well as competition from estuarine and sea fisheries, no devel-

TABLE I
Yields from 92 Tropical Reservoirs on Three Continents (Numbers of Reservoirs)

Yield Interval (kg/ha/yr)	Africa	Indonesia	Thailand	Sri Lanka	Bangladesh & India	Brazil	Totals	Percent
0–29	2	1	4	6	5	1	19	20.7
30–59	6	1	4	6	3	3	23	25.0
60–89	6	1	2	3	0	4	16	17.4
90–119	2	1	2	0	1	4	10	10.9
120–149	2	0	1	2	2	2	9	9.8
150–179	1	0	0	0	0	0	1	1.1
180–209	0	1	0	1	0	1	3	3.3
210–239	0	0	0	1	0	0	1	1.1
240–269	1	0	0	0	0	0	1	1.1
270–299	1	1	0	1	1	0	4	4.3
300–329	0	1	0	1	0	0	2	2.2
330–359	0	1	0	1	0	1	3	3.3
Totals	21	8	13	22	12	16	92	

opment has taken place. The best summary of the fishery potential of Brazilian reservoirs is that by Paiva (1976) who listed 46 reservoirs covering an area of 1,898,662 ha and estimated to have a 123,091 tonnes yr-1 potential fish production. Actual yields from 100 reservoirs comprising 1,500 km2 in Northeast Brazil have averaged 120 kg ha-1 in recent years (Gurgel, 1984).

2. PHYSICO-CHEMICAL AND BIOLOGICAL FACTORS OF SIGNIFICANCE TO FISHERIES

The number of physico-chemical parameters required for fish production prediction purposes is smaller than that usually collected for water quality assessment. Marshall (1984a) has reviewed empirical models for the prediction of fishery yields based on simple, easily obtained data in an attempt to improve the fishery predictive capabilities of pre-impoundment reservoir studies. Morphometric features well-related to fish yield were surface area and shoreline development, the latter especially so for large African reservoirs. The Morpho-Edaphic Index (MEI), much used for fish production estimates in African reservoirs and lakes is based only on two basic parameters: the mean water depth and water conductivity (Henderson and Welcomme, 1974). Deep, V-shaped reservoirs are considered to be less productive than wide shallow ones. Thus, when predicting fish production from reservoirs, in deep reservoirs the MEI should be calculated separately for the littoral ''tablefish'' production zone, and separately for the pelagic fish. A comparison is then possible between the littoral zone of such reservoirs, and total production of shallow lakes in which the total lake area functions as littoral.

With only morphological and edaphic data, however, the range of confidence in the yield prediction is from about 0.5 to 2 times the point estimate, and Toews and Griffith (1979) have added surface area to the morpho-edaphic relationship to improve confidence in the yield estimates. MEI has not been applied elsewhere in the tropics, largely due to insufficient data on fish landings needed for the construction of a reliable prediction model. Other parameters of importance for fish production in African reservoirs were reviewed by Petr (1975) and Bernacsek (1984). In addition to reservoir size, these include water level fluctuations, erosion, aquatic plants, and area of submerged terrestrial vegetation, all of which influence the fish stock recruitment and production. Bernacsek (1984) found a direct correlation between the drawdown magnitude and ichthyomass of all fish with the exception of cyprinids where there is a decrease in ichthyomass with an increase in drawdown magnitude.

Although nutrient content of tropical waters may be low when compared with temperate waters, the fast turnover of nutrients may yield a high production (e.g., Ganf, 1974). Eutrophication of inland waters has been known to result in an increase in total fish production. This has been documented for example by Marshall (1982) for McIlwaine reservoir in Zimbabwe where a direct relationship between the total dissolved solids and fish yield was established. Marshall (1982) also illustrates the sensitivity of reservoir fishery yield to annual nutrient inputs using Lake Kariba as an example. Evidence suggests that populations of *Limnothrissa miodon,* a small clupeid providing the bulk of the reservoir's yield, fluctuate from year to year because of variation in river flow and nutrient input. Marshall (pers. comm.) points out that because of drought in southern Africa, Kariba reservoir has been decreasing in surface area since 1981 with unforeseeable consequences for the future fish production.

In reservoirs largely dominated by illiophagous detritivorous fish, the determination of organic matter content could supplement the studies of productivity of reservoirs. In Salto Grande reservoir in Latin America organic matter "quality" has been used to explain the abundance of fish (Quiros and Baigun, pers. comm.).

Turbidity appears of importance especially where pelagic stocks, together with their predators, are concerned. In Cahora Bassa, Mozambique, a high clay load during some months results in a low phytoplankton production, restricting zooplankton mainly to feeding on organic matter and bacteria bound to clay particles. The low food concentration may be a reason for the low zooplankton concentrations which in turn limit the planktivorous sardines (Gliwicz, 1984). On the other hand, the high turbidity may reduce the predation by the tiger fish, *Hydrocynus,* and the catfish, *Eutropius,* due to poor visibility.

Primary productivity was found to be correlated with fish yields in Indian reservoirs, where commercial fish yields increase logarithmically as primary productivity increases arithmetically (Melack, 1976). A regression equation was constructed for the same relationship for eight African lakes, but not yet for African reservoirs.

Ageing of reservoirs is characterized by an initial release of nutrients from flooded soils and from the decomposing submersed terrestrial vegetation. The result is a high primary and secondary production, followed after several years by a decline (e.g.,

Balon, 1974a; Petr, 1975). Fish biomass, as the final link in most aquatic food chains, is a good indicator of the total aquatic production.

The littoral is considered the major production zone of reservoirs. However, where pelagic stocks develop, either from indigenous or introduced species the pelagic fish yield may far exceed that of the littoral. In Kariba reservoir in Zimbabwe/Zambia the Zimbabwe pelagic sector has yielded up to 12,000 t while that of the littoral only 3,000 t (Marshall, 1984b).

Reservoirs are rarely cleared of terrestrial vegetation prior to their formation. The flooded vegetation becomes a source of nutrients, a substrate for aufwuchs, and provides a barrier against waves. Floating aquatic plants usually find such an environment conducive and may form dense cover between partially emersed trees. All the above enhances the lake production, but excessive growth of floating macrophytes will impede fish production and make fishing very difficult, if not impossible. Such explosive growth is usually a temporary phenomenon as floating macrophytes retreat gradually as nutrients become depleted.

The littoral zone in reservoirs is subject to drawdown, and such areas are known to be more productive than the rest of the water body. This is due to the development of terrestrial vegetation during the dry period. Upon its submersion, this vegetation decays, and if it was grazed upon by domestic or wild animals, their dung further enhances aquatic production of the littoral. Balon (1974b) found the littoral fish production in a drawdown zone of Lake Kariba 35 times higher than in the permanently submersed area.

Fish stocks in new reservoirs originate largely from the original riverine fish, although there is almost always a chance for some lacustrine species to enter the reservoirs from submersed backwaters or stagnant water bodies such as ponds and lakes. Invariably, the lacustrine species become dominant, as reservoir conditions are not conducive to the river fish. Thus, cichlids become dominant in the African tropical reservoirs while characids may be second (Petr, 1978; Kapetsky, 1986).

Some of the success of cichlids is probably related to conditions for reproduction. In floodplain situations cichlids migrate laterally to reproduce and in reservoirs find approximately similar conditions in the littoral. Cyprinids, however, generally are river spawners and find favourable conditions for reproduction only in upstream reaches of reservoirs, or in inflowing streams and rivers.

In some areas of Southeast Asia, a large number of fish species is landed as for example in the Ubolratana reservoir in Thailand, where 47 species are frequently encountered in catches and are economically valuable (Bhukaswan, 1983). The three major representatives in this reservoir belong to cyprinids, which in 1970 represented 55 percent of the total catch. Kaptai reservoir in Bangladesh has 58 species in 25 families of which 22 species are of commercial importance and of which 20 percent of the yield by weight is major carps (cyprinids) (Hye, 1983). However, in other areas the number of native species available to reservoir fisheries may be small, as for example, in Malaysia and the Philippines where only 15 species are indigenous (Baluyut, 1983).

In Brazilian reservoirs the most important fish are *Prochilodus* (Prochilodontidae),

Plagioscion (Sciaenidae) and *Hoplias* (Erythrinidae) (Paiva, 1976). These species contributed about 80 percent of the total fish yield. In northeast Brazil *Serrasalmus* dominates the fish (Welcomme, pers. comm.). Mexican reservoir fisheries are based on tilapias. In the few Cuban reservoirs managed for food fish production, introduced *Oreochromis niloticus* and *O. aureus* comprise 80 to 90 percent of the catch. The other, mainly smaller reservoirs, serve sport fisheries which are based on the introduced black bass, *Micropterus salmoides,* (Henderson, pers. comm.).

3. SOCIO-ECONOMIC ASPECTS

Reservoir fisheries usually face the problem of shortage of fishermen during the first years—the most biologically productive—of the reservoir. The dilemma of how to use the peak fish production potential without raising false expectations of a continuity in high fish yields has not been solved with satisfaction. In remote areas and on large reservoirs with insufficient number of people around, the peak fishery production available may never be harvested for shortage of man-power and equipment. In other reservoirs, unregulated inflow of people—not necessarily all being fishermen—will cause heavy pressure on fish stocks, as well as on the riparian environment. This need not be entirely negative. The wealth accumulated by the early fishermen taking advantage of the trophic upsurge may later permit many of them to enter other business enterprises which benefit the local economy. Only good planning beginning with a prediction of "trophic upsurge" and later "stabilized" fishery potential and enforcement of fishery regulations will prevent disappointment. This is easier to say than to do. However, in some situations, there is an inherent self-regulatory mechanism in the fisheries based on tradition. In other places, a high fishing pressure will lead to overfishing of the most sought after fish species, and this may lead to changes in fish stock composition, accompanied by a lowering of demand for the usually smaller and more abundant replacement species.

The socio-economic consequences of impoundment have been summarized, for example by Barrow (1981), who emphasizes the resettlement aspects in African and Asian schemes. Fisheries opportunities created by large impoundments during the peak production phase are indisputable, and may lead to a large influx of fishermen: within two years of the closure of the Nam Pong dam in Thailand about 1,000 fishermen arrived (SCOPE, 1972). Even larger numbers of fishermen reached Volta reservoir during the first years, and six years after dam closure 12,500 fishermen were still present (Barrow, 1981). In 1975, there were 1,479 fishing settlements in the shore area of the reservoir (Coppola and Agadzi, 1977), and an estimated 87,000 persons living primarily off the fishery in the area (Thomi, 1981).

4. HEALTH

Health problems are associated especially with reservoirs in Africa and Latin America, in regions where bilharziasis is endemic. In such areas parasites are transmitted into reservoir water by human hosts, and if suitable snail hosts are present, the cycle is completed. Malaria generally becomes less common as a result of reservoir formation in areas formerly rich in small water bodies. However, outbreaks of malaria have been recorded after dam construction in India and Sri Lanka (Brown and Deom, 1973), and in the Parana-Paraguay basin. Arbovirus diseases may increase, although this may be more closely related to irrigation schemes than to impoundments themselves. The incidence of sleeping sickness and river blindness is greatly reduced due to flooding the vector habitat. Human sleeping sickness has not been recorded as a result of flooding, although there was evidence of decimation of cattle by bovine trypanosomiasis in the Lake Kariba area (Brown and Deom, 1973). Prevention and control of vector-borne diseases in water resources develoment projects have been discussed at the World Health Organization conferences and in WHO technical reports. The general conclusion is that in Asia, as in Africa, clear evidence of the adverse effects of water impoundment schemes on health, with the exception of schistosomiasis, is not often found. However, in small impoundments, which tend to serve more purposes than do large reservoirs, and where therefore the human and animal population contact with water is high, the disease transmission rates are correspondingly increased.

5. PLANNING AND MANAGEMENT

5.1 Constraints on Management

Management in the classical sense of regulation of fisheries—such as limits on fishermen numbers, fishing gears, gear characteristics, the setting aside of closed areas or closed seasons—is little practised on tropical reservoirs at present, mainly because such regulations are politically and economically unacceptable. For example, many of the African countries suffer from food shortages, underemployment and high rates of unemployment; so African reservoir fisheries are important first as producers of food and secondly as providers of employment. Any type of regulation that would limit employment, either directly or indirectly, would not find favour. Furthermore, there are many reservoirs which remain underharvested, for example those of Sri Lanka (Oglesby, 1981), a number of smaller Indo-Pacific island countries (Petr, 1984), and some in Africa as well (Kapetsky, 1986) and they would benefit from development activities that would stimulate fishing rather than from restrictive regulations. Finally, the output of many reservoir fisheries appears to be relatively unaffected by heavy fishing; species compositions of the catch change, but total catch remains little affected, as for example on Kainji Lake in Nigeria. Additionally, where regulation might be beneficial, efforts to translate resource and economic information into management policy, and management policy into the implementation of management programmes, have not always been successful. Thus, the most viable reservoir management activities have been those which take advantage of the available but underutilized biological potential of reservoirs. These mainly have been introductions of exotic species and periodic stocking of native or exotic species. However, reservoir fishery yields have also been enhanced qualitatively and quantitatively by aquaculture, mainly cage culture.

5.2. Information on Fishery Resources for Management

Reservoir fishery management and development activities have to be guided and eventually evaluated on the basis of their effect on the fishery resources and the fishery. For this reason, considerable time and effort have been expended on developing stock and fishery assessment techniques for tropical reservoirs.

Periodic fishing of standard gillnet fleets at fixed stations has been a universal stock assessment tool for African reservoirs. This type of sampling offers the advantages that results can be made available readily, and analyses are straightforward. Moreover, data can be obtained regularly and inexpensively by a small field staff, and the equipment is inexpensive, simple and easily maintained.

Other more sophisticated sampling techniques have been tried. Among them was the use of toxicants in blocked-off areas on Lakes Kariba (Balon, 1974a), Volta (Loiselle, 1972) and Kainji (Ita, 1984), and sampling with an explosive grid on Lake Kariba (Marshall, 1984b). Major drawbacks of these methods are that they are relatively demanding of time, equipment, supplies, manpower and finances, and they cannot be effectively applied in all habitats.

Acoustic techniques have been used along with test fishing to survey pelagic resources on Lake Kariba (Woodward, 1974), Lake Nasser, and most recently on Cahora Bassa reservoir (Lindem, 1983). Although this is a useful tool, the field equipment is complex and fairly delicate and repairs must usually be made abroad.

Catch and effort data along with socio-economic statistics on the fishery, although recognized by fishery managers as fundamental for resource evaluation, have not always been collected in a consistent or timely manner. One reason for this is that it has been difficult to make the operational and administrative shift from full-time, labour-intensive, census-based programmes to the streamlined, stratified sampling programmes which use coordinated aircraft and ground surveys to provide the essential data on a part-time basis.

6. MANAGEMENT TECHNIQUES

6.1 Fish Introductions

The most utilized reservoir management technique throughout tropical regions has been the introduction of non-indigenous species, most frequently tilapias, but also common, Indian and Chinese carps, and predators such as the peacock bass and largemouth black bass. The major purpose of introductions of exotics into tropical waters is to increase fish yield by filling a vacant niche. The pelagic clupeid *Limnothrissa miodon* introduced into Lake Kariba in Africa is now fished by Zimbabwe fishermen at a rate of approximately 10,000 t yr-1 (Marshall, 1984c), and in the Zambian waters of Kariba a fishery is developing for it with the current yield of nearly 5,000 tons. There are several notable points about this introduction. First, the introduction permitted the establishment of a fishery where none had existed previously; secondly, the pelagic fishery has increased by more than three-fold the output from the reservoir, the littoral potential of which is only about 2,500–3,000 tons. Also, the sardine fishery has contributed to local employment. Finally, from the economic point of view of a developing country the sardine is an ideal species. Processing is simple resulting in a low cost of the marketed product. From Lake Kariba, *Limnothrissa* entered the downstream Cahora Bassa reservoir in Mozambique where it now has the potential to support a 7 to 8,000 ton yr-1 fishery (Bernacsek and Lopes, 1984; Vostradovsky, 1984).

The most common introductions, however, are those of cichlid tilapias, which have become widespread in numerous tropical reservoirs of Asia and Latin America. The register of international transfers of inland fish species (Welcomme, 1988) lists 10 Asian and 15 Latin American countries into which *O. niloticus* has been introduced. Many of these introductions were made into reservoirs. Tilapias, for example, highly dominate Sri Lankan reservoir fish production (Fernando, 1984). In northeast Brasil, *Oreochromis niloticus* was introduced in many public and private reservoirs in the early 1970's and by the early 1980's it was contributing about 30 percent by weight to the entire reservoir catch (Gurgel, 1984).

On the other hand, in some countries of the Pacific, where tilapias have been accidentally or deliberately stocked, their rapid reproduction has been considered a problem. In some situations, ways are being sought to control them so that other introduced species may thrive. In spite of that tilapias are of considerable importance as protein source and will continue to substantially contribute to fish production of many tropical reservoirs.

Among other introductions, in the best investigated Thai reservoir Ubolratana, the introduced gouramy *Trichogaster pectoralis* became the dominant fish for a three-year period, but other introduced fish have had little influence on the total yield. In the Chulaporn reservoir bighead carp was introduced successfully and in Lam Takong reservoir the introduced *O. niloticus* eventually constituted 15.8 percent of catches (Chukajorn and Pawapootanon, 1977).

Predatory species have also been introduced into some locations, either for sport fishery purposes, such as black bass into a number of reservoirs on tropical islands (Petr, 1984), or incidentally such as *Cichla orientalis* into the Gatun Lake in Panama. Although an excellent fish, it played havoc with the native fish dramatically changing the biotic community of this reservoir (Zaret and Payne, 1973).

In summarizing the results for the southeast Asia countries, Baluyut (1983) pointed out that the introductions were being done frequently without much prior thought and planning and with poor knowledge of the biology of the introduced species and local fauna. As a result, some of these introductions have failed, and others have succeeded but with negative consequences on the existing fishery. Still other introductions were initially successful, but eventually failed for lack of continued stocking. In spite of that, introduction of exotic species remains the only successful means of maximizing fish production in reservoirs in a number of tropical countries.

The most obvious advantages of introductions and follow-up stocking programmes appear for the islands of the Indian Ocean and the Pacific, where native fish stocks suffer from a low species diversity and from being mostly riverine fish of marine origin. Introductions and then stocking of natural lakes and reservoirs has often increased the yield spectacularly, with the introduced exotic fish quickly dominating the fishery. An example of Sri Lankas reservoirs has already been mentioned. In Madagascar, the mountain reservoir Mantasoa, as well as the natural lakes Alaotra and Itasy, have fish stocks dominated by introduced species. In Mantasoa, the native fish disappeared soon after the reservoir was created in 1935, and it is now dominated by *T. rendalli* and *Cyprinus carpio,* with a total fish production of 15.4 kg ha-1yr-1 (Moreau, 1984). In Hawaii, Maciolek (1969) found fish lacking as a native component of the natural lakes. The Hawaii capture fishery is therefore entirely based on exotic fish species. Some 30 species were introduced into the fresh waters since the end of the last century. The most important recreational fishery is now on Waihiawa Reservoir on Oahu Island. An angler's creel census has shown a high predominance of tilapias (80.4 percent), followed by *Cichla ocellaris* and *Lepomis macrochirus* (each 6.1 percent), *Carassius auratus* (2.6 percent), *Ictalurus punctatus* (1.6 percent), *Cyprinus carpio* (1.5 percent) and five other species all less than 1 percent each. Today 11 species are well established, forming the basis of the recreational fishery.

In Papua New Guinea, New Caledonia and Guam, reservoir fisheries are undeveloped. In Yate Reservoir in New Caledonia, blackbass (*Micropterus salmoides*) is being caught by recreational fishermen and *O. mossambicus* is known to exist in the lake. The Sirinumu Reservoir of Papua New Guinea has three exotic species represented by *O. mossambicus, Trichogaster pectoralis* and *Gambusia affinis,* and four native species (Berra *et al.*, 1975). There is no regular fishing on the reservoir and fishery research is still to be carried out. In Guam, *O. mossambicus* and *Cichla ocellaris* are established in Fena Reservoir, but exotic fish introduced into Masso Reservoir have been virtually eliminated by poaching using chemicals. In Fiji, tilapias are the sole fish in Monasavu, and black bass and tilapias in Vaturu—both new reservoirs established recently. Both species are introduced.

There is considerable potential for expansion of inland capture fisheries by introductions and stocking but the purpose of introductions must be clearly identified and their consequences thought through prior to any interference with new reservoirs to avoid difficulties with eradication of unwanted species in the future.

6.2 Stocking

Stocking, in contrast to introduction, is the attempt to maintain or improve fish production by periodically providing a reservoir or other water body with fish from elsewhere. Usually, the fish stocked are juveniles, and usually they are hatchery-reared. The idea is to stock fish sufficiently large so that predation on them is light so that most stocked fish can later be harvested. Where fishing pressure is so heavy that insufficient numbers of spawners are left to replenish populations, or where environmental conditions are seasonally too extreme to support fish life as in the case of the dry-zone reservoirs in Sri Lanka (Oglesby, 1981), stocking is virtually the only management method to maintain fish production. Stocking of very large reservoirs using hatchery methods is usually not economic because the large capital investments and running costs to provide sufficient fish for replenishment would be excessive.

The impact on fish yields in tropical reservoirs of introduced and stocked fish has been evaluated by Sreenivasan (1984). Although he acknowledged that stocking plays a very important role in increasing the fish yield, in a number of situations there is a difficulty in evaluating the increments in yield from such stocking. He has criticized the cost of stocking major carps into some south Indian reservoirs, where frequently the return is less than the cost of the fingerlings stocked. He suggested that there is no relationship between stocking and harvesting especially in reservoirs dominated by either *O. mossambicus* or *Rhinomugil corsula*. Poor understanding of interactions between stocked species may also contribute to poor economics of the resulting harvest. In Indian reservoirs with *Oreochromis* it appears a wasted effort to stock carps as the latter have a poor chance for success. Economics play a role, as for example in south India the fishermen earn 4–5 times less per unit weight from tilapia than from the carp fisheries (Sreenivasan, 1984). The stocking rate cannot be prescribed as a general formula, as is done for ponds. However, Sreenivasan has suggested that in the initial years, 250–500 fingerlings ha-1yr-1 may be stocked into Indian reservoirs. By contrast, Chinese reservoirs are stocked at much heavier rates of 1,400 to 75,000 per ha per year with the highest rates being used in the smaller reservoirs. Economically a 10 percent recapture rate would meet the cost of stocking. Gross income and profits rise with increases in recapture rate.

6.3 Other Management Techniques

On Lake Nasser an ''artificial nurseries'' management scheme has been attempted. The scheme is aimed at overcoming two related problems. The first is heavy fishing on Nile tilapia spawners, with 40 percent of the annual tilapia catch being captured during the 3-month spawning period. The second problem is the increasing proportion of the smaller, slower-growing, Galilaean tilapia in the catch. It now makes up more than 50 percent as compared with less than 20 percent ten years ago. Because nearly all of the tilapia are sold as fillets, the larger 1 kg-size tilapia are desired.

Basically, the management techniques of the nurseries scheme are predator control and regulation of fishing. The first scheme consists of a 1,000 ha arm blocked off from the remainder of the reservoir with a fine mesh net. Within this larger blocked-off area much smaller secondary enclosures (fishpens) were placed along the littoral. After predators had been removed from these small enclosures, they were stocked with Nile tilapia breeders. In this first pilot scheme ducks were being farmed to enhance aquatic production. Early results from the pilot scheme were said to be good in that the desired species, the Nile tilapia, made up 80 percent the catch in the blocked-off arm; however, with the data avilable it has not been possible to determine if overall yield had increased above levels from the open, unmanaged areas of the reservoir.

7. FISHERY DEVELOPMENT ACTIVITIES

Fishery management, as discussed above, has been taken to include only those activities that contribute to increased fishery yield through increasing biological production. Fishery development, in comparison, is used here to define those activities that increase fishery output through more efficient fishing, processing, transporting, and marketing. In most reservoirs fishery development activities, such as fish marketing, have been undertaken in a variety of ways—by the private sector, with government funds, with technical assistance from international organizations, or with financing from religious or other non-governmental agencies. From whatever source, the purpose has usually been to get development started; later, when established, it can be taken over by private industry, state companies, or by fishermen's cooperatives.

Some general fishery development activities on Volta Lake provide a good cross-section of those carried out elsewhere. As described by Vanderpuye (1984), the fishing villages on Volta Lake sprang up as the reservoir formed in the 1964–8 period; they were founded by fishermen who had emigrated from downstream on the Volta River. Government-planned development included the establishment of fishing complexes at each important fish landing point. Each complex was to include landing facilities, a store for fishermen's supplies, a market, fish smoking houses, and training facilities for fishermen, boatbuilding and boat motor repair. Only one such complex has been built thus far.

Less ambitious than the fishery complex, but also contributing to fishery development on Volta Lake have been a mobile fishery school; a fish processing demonstration programme; an extension service to introduce more efficient nets and improved smoking methods; establishment of a revolving fund for the construction of boats and the sale of outboard motors to fishermen; the organization of fishermen into cooperatives; and the placing of floating jetties to facilitate fish landings.

Other more general development efforts that have served to stimulate reservoir fisheries by providing improved living conditions have been rural health schemes, assistance with agriculture, and the establishment of schools and communications. These have been an important impetus to fisheries, especially where living conditions are harsh. For example, one of the major constraints on fishery output from Lake Nubia has been the difficulty in attracting fishermen to live and work in isolated desert fishing camps (Ali, 1984) and this also was a factor affecting the Lake Nasser fishery until recent times. There, however, government and private efforts have been instrumental in providing small pumps so that fishermen could irrigate land for subsistence farming, and houses were designed to be comfortable during hot desert days and cold nights.

8. CONCLUSIONS: FUTURE TROPICAL RESERVOIR FISHERIES DEVELOPMENT

Tropical reservoirs as a source of fish are largely a 20th century phenomenon. Prior to that, virtually all inland fish captured came from natural lakes, swamps, rivers and their floodplains. Although dams have a negative impact on the downstream aquatic production, especially that on floodplains, reservoir fish production is considered to counterbalance or exceed the loses. However, a detailed evaluation of losses against gains is still to be done. The growing awareness of government planners of the fish potential in new hydroelectric, irrigation, water regulation and water supply schemes is shifting fisheries development and management from the sideline into the mainstream of planning. There is also a growing awareness of the possibilities for enhancement of fisheries of existing reservoirs. Considerable thought is given today to ways of improving the existing fisheries production by way of introductions, stocking, and by intensive management techniques based on aquaculture principles. But a better understanding is still needed of interactions between the stocked and native species to enable a better prediction of the impact of introductions and stocking on the total fish production and to avoid undesirable side effects. Such knowledge is not only required for improvement of fish production in reservoirs rich in native fish stocks, but especially for those where poor, or perhaps no indigenous fish species exist. The use of cages and fishpens in reservoirs for selected fish species is seen as one of the most important aspects of considerably increasing reservoir fish yield where management of the capture fisheries has already been maximized by all other measures.

The transfer of fishery management and development experience between regions is still inadequate, although much progress has already been made among countries within regions. The meeting of the Committee for the Inland Fisheries of Africa in 1983 reviewed African reservoir fisheries with the objective of finding ways to increase benefits from African reservoir fisheries. Specific subjects were identified which would make a significant contribution to improving fishery yields, or would help in more efficient planning of new reservoir fisheries (Kapetsky and Petr, 1984). These included better and more comprehensive information on the following subjects, some of which have now been studied and summarized: pelagic fishes and fisheries (Marshall, 1984); improved approaches to reservoir-specific yield predictions including the pre-impoundment data required for such estimates (Marshall, 1984); options for dam design and reservoir operation to enhance fisheries (Bernacsek, 1984); and the relevance of terrestrial and aquatic vegetation to fish yields.

Elsewhere, a comprehensive review and analysis of reservoir fishery enhancement possibilities is still lacking, but good progress has been made in a number of subject areas such as stocking (Baluyut, 1983; Sreenivasan, 1984). Indian reservoirs have been managed for a number of years, especially with the purpose of producing table fish. In other tropical countries of Asia, with the exception of Sri Lanka, most reservoirs are new and while some of them have a relatively good fisheries management practices, many will require a better and sustained management effort for realization of their full fisheries potential. Recently, much effort has been expended on fisheries management of tropical reservoirs in Latin America, but even here data are still awaiting a thorough analysis and evaluation to be of assistance to planners and managers.

REFERENCES

1. Ali, M. El-Tahir (1984). 'Nubia (Sudan)', in *Status of African Reservoir Fisheries/État des pêcheries dans les réservoirs d'Afrique* (Eds. J.M. Kapetsky and T. Petr), *CIFA Tech.Pap./Doc.Tech.CPCA*, 10, 247–260.
2. Balon, E.K. (1974a). 'Fish production of a tropical ecosystem', in *Lake Kariba, a man-made tropical ecosystem in Central Africa* (Eds. E.K. Balon and A.G. Coche), pp. 249–573, Dr. W. Junk Publ., The Hague.
3. Balon, E.K. (1974b). 'Fishes of Lake Kariba, Africa', 144 p., T.F.M. Publications Inc., Neptune City, N.J., USA.
4. Baluyut, E.A. (1983). 'Stocking and introduction of fish in lakes and reservoirs in the ASEAN countries', *FAO Fish.Tech.Pap.*, 236, 82 p.
5. Barrow, C.J. (1981). 'Health and resettlement consequences and opportunities created as a result of river impoundment in developing countries', *Water Supply and Mngt.*, 5, 135–150.
6. Bernacsek, G.M. (1984). 'Guidelines for dam design and operation to optimize fish production in impounded river basins (based on a review of the ecological effects of large dams in Africa)', *CIFA Tech.Pap.*, 11, 98 p.
7. Bernacsek, G. and Lopes, S. (1984). 'Mozambique. Investigations into the fishery and limnology of Cahora Bassa reservoir seven years after dam closure', *FAO/GCP/MOZ/006/SWE Field Document 9*, 143 p., Rome, Italy.
8. Berra, T.M., Moore, R. and Reynolds, L.F. (1975). 'The freshwater fishes of the Laloki River system of New Guinea', *Copeia*, 2, 316–326.
9. Bhukaswan, T. (1983). 'Reservoir fishery management in Southeast Asia', in *Summary report and selected papers presented at the IPFC Workshop on inland fisheries for planners. Manila, The Philippines, 2–6 August 1982* (Ed. T. Petr), pp. 101–110, *FAO Fish.Rep.*, 288, 191 p.
10. Brown, A.W.A. and Deom, J.O. (1973). 'Summary: Health aspects of man-made lakes', in *Man-made Lakes: Their Problems and Environmental Health* (Eds. W.C. Ackermann, C.F. White and E.B. Worthington), *Geophys.monogr.*, 17, 755–764.
11. Chukajorn, T. and Pawapootanon, O. (1977). 'Annual catch statistics of freshwater fish taken from seven reservoirs in Northeastern Thailand', *Proc.IPFC*, 17(3), 206–221.
12. Clay, C.H. (1984). New reservoirs in Africa, 1980–2000. Nouveaux Reservoirs africains 1980–2000, *CIFA Occas.Pap./Doc.Occas.CPCA*, 11, 23 p.
13. COPESCAL (1982). 'Comisión de Pesca para América Latina (COPESCAL)/Commission for Inland Fisheries of Latin America (COPESCAL), Actas del Simposio sobre desarrollo y explotación de lagos artificiales. Santo Domingo, Republica Dominicana, 30 de noviembre—1 de diciembre de 1981. Proceedings of the Symposium on the development and exploitation of artificial lakes. Santo Domingo, Dominican Republic, 30 November—1 December 1981', *FAO, Inf.Pesca/FAO Fish.Rep.*, 273, 17 p.

14. Coppola, S.R. and Agadzi, K. (1977). 'Evolution of the fishing industry over time at Volta Lake 1970/76', FAO/GHA/71/533, FAO, Rome.
15. Fernando, C.H. (1977). 'Reservoir fishes in South East Asia: Past, Present and Future', in *Proc.IPFC*, 17, 3, 475–489.
16. Fernando, C.H. (1984). 'Reservoirs and lakes of Southeast Asia (Oriental Region)', in *Lakes and Reservoirs* (Ed. F.B. Taub), pp. 411–446, Elsevier, Amsterdam.
17. Ganf, G.G. (1974). 'Phytoplankton biomass and distribution in a shallow eutrophic lake (Lake George, Uganda)', Oecologia (Berl.), 16, 9–29.
18. Gliwicz, Z.M. (1984). 'Limnological study of Cahora Bassa reservoir with special regard to sardine fishery expansion', FAO/GCP/MOZ/006/SWE, Field Document 3, 71 pp.
19. Gurgel, J.J.S. (1984). 'Observations on stocking of *Sarotherodon niloticus* (Linne, 1776) into D.N.O.C.S. public reservoirs of northeast Brazil', Bamidgeh, 36(2), 53–58.
20. Henderson, H.F. and Welcomme, R.L. (1974). 'The relationship of yield to Morpho-Edaphic index and numbers of fishermen in African inland fisheries. Relation entre la production, l'indice Morho-Edaphique et le nombre de pêcheries des eaux continentales d'Afrique', *CIFA Occas.Pap.*, 1, 19 p.
21. Hye, M.A., (1983). 'Fishery potential of Kaptai lake', Agricultural Development Agencies in Bangladesh, ADAB News, 10(6), 2–6; 11.
22. Ita, E.O., (1984). 'Kainji (Nigeria)', in *Status of African reservoir fisheries. État des pêcheries dans les réservoirs d'Afrique* (Eds. J.M. Kapetsky and T. Petr), *CIFA Tech.Pap./-Doc.Tech.CPCA*, 10, 43–104.
23. Kapetsky, J.M. (1978). 'The Brokopondo reservoir: fishery yield potential, fishery research and fishery development', Report prepared for the Fisheries Division of Suriname, FAO Fisheries Department, Rome, 81 p.
24. Kapetsky, J.M. (1985). 'Twenty years of fisheries on large African reservoirs', Proceedings of the National Symposium on Managing Reservoir Fisheries—Strategies for the '80's, American Fisheries Society Special Publication.
25. Lindem, T. (1983). 'A preliminary analysis of acoustic data from Maputo Bay, Lake Niassa and Cahora Bassa', FAO, Research and Development of Inland Fisheries Project, GCP/MOZ/006/SWE, Rome, Italy.
26. Loiselle, P.V. (1972). 'Preliminary survey of inshore habitats in the Volta Lake', FAO Volta Lake Fisheries Research Project, FI/DP/GHA/67/510/2, Rome, Italy.
27. Marshall, B.E. (1982). 'The fish of Lake McIlwaine', in *Lake McIlwaine. The eutrophication and recovery of a tropical African man-made lake* (Eds. J.A. Thornton and W.K. Nduku), pp. 156–188, Dr. W. Junk Publishers, The Hague.
28. Marshall, B.E. (1984a). 'Towards predicting ecology and fish yields in African reservoirs from pre-impoundment physico-chemical data. Comment prevoir l'écologie des réservoirs africains et leur rendement en poisson a partir de données physico-chimiques réunies avant endiguement', *CIFA Tec.Pap./Doc.Tech.CPCA*, 12, 36 p.
29. Marshall, B.E. (1984b). 'Kariba (Zambia/Zimbabwe)', in *Status of African reservoir fisheries. État des pêcheries dan les réservoirs d'Afrique*, (Eds. J.M. Kapetsky and T. Petr), *CIFA Tech.Pap./Doc.Tech.CPCA*, 10, 105–154.
30. Marshall, B.E. (1984c). 'Small pelagic fishes and fisheries in African inland waters. Espèces de petits pelagiques et leurs pêcheries dans les eaux interieures de l'Afrique', *CIFA Tech.Pap./Doc.Tech.CPCA*, 14, 25 p.
31. Maciolek, J.A., (1969). 'Freshwater lakes in Hawaii', *Verh.Int.Ver.Theor.Angew.Limnol*, 17, 386–391.
32. Melack, J.M. (1976). 'Primary productivity and fish yields in tropical lakes', *Trans.Amer.Fish.Soc.*, 105, 575–580.
33. Moreau, J. (1984). 'Mantasoa (Madagascar)', in *Status of African reservoir fisheries. État des pêcheries dan les réservoirs d'Afrique* (Eds. J.M. Kapetsky and T. Petr), pp. 155–192, *CIFA Tech.Pap./Doc.Tech.CPCA*, 10, 326 p.
34. Oglesby, R.T. (1981). 'Sri Lanka. A synthesis of the reservoir fishes in Sri Lanka. A report prepared for the Project TCP/SRL/8804. Development of Fisheries in the man-made lakes and reservoirs', Field Document 2, FAO, Rome, 30 p.
35. Paiva, M.P. (1976). 'Estimava do potencial da produção de pescado em grandes represas Brasileiras', Eletrobra's, Rio de Janeiro, Brasil, 23 pp.

36. Petr, T. (1975). 'On some factors associated with the initial high fish catches in new African man-made lakes', *Arch.Hydrobiol,* 75, 32–49.
37. Petr, T. (1978). Tropical man-made lakes—their ecological impact, *Arch.Hydrobiol.*, 81:368–385.
38. Petr, T. (1984). 'Indigenous fish and stocking of lakes and reservoirs on tropical islands of the Indo-Pacific', in *Indo-Pacific Fishery Commission (IPFC), Report of the second session of the IPFC Working Party on inland fisheries. New Delhi, India 23–27 January 1984, and Report of the Joint Workshop of the IPFC Working Party on inland fisheries and the IPFC Working Party on aquaculture, on the role of stocking and introductions in the improvement of production of lakes and reservoirs, New Delhi, 24–25 January 1984, FAO Fish.Rep.*, 312, 20–26.
39. SCOPE (Scientific Committee on Problems of the Environment), (1972), 'Man-made lakes as modified ecosystems', SCOPE Report 2, International Council of Scientific Unions, Paris, 76 p.
40. Song, Z. (1980). 'Manual of small-scale fish culture', FAO, Rome, *FAO Fish.Circ.*, 727, 21 p.
41. Sreenivasan, A. (1984). 'Influence of stocking on fish production in reservoirs in India', in *Indo-Pacific Fishery Commission (IPFC), Report of the second session of the IPFC Working Party on inland fisheries. New Delhi, India 23–27 January 1984, and Report of the Joint Workshop of the IPFC Working Party on inland fisheries and the IPFC Working Party on aquaculture, in the role of stocking and introductions in the improvement of production of lakes and reservoirs. New Delhi, 24–25 January 1984, FAO Fish.Rep.*, 312, 40–52.
42. Thomi, W. (1981). 'Umsiedlungsmassnahmen und geplanter Wandel im Rahmen von Staudammprojekten in der Dritten Welt', Frankfurter Wirtschafts- und Sozialgeographische Schriften 39, Frankfurt/M.
43. Toews, D.R. and Griffith, J.S. (1979). 'Empirical estimates of potential fish yield for the Lake Bangwelo system, Zambia', *Trans. Amer. Fish. Soc.*, 108, 241–252.
44. Vanderpuye, J. (1984). 'Volta (Ghana)', in *Status of African reservoir fisheries. État des pêcheries dans les réservoirs d'Afrique* (Eds. J.M. Kapetsky and T. Petr), *CIFA Tech. Pap./Doc. Tech. CPCA*, 10, 261–320.
45. Vostradovsky, J. (1984). 'Mozambique. Fishery investigations on Cahora Bassa reservoir (March 1983–May 1984)', FAO/GCP/MOZ/006/SWE, Field Document 11, 27 p., Rome, Italy.
46. Welcomme, R.L. (comp.) (1988), 'International introductions of inland aquatic species.' *FAO Fish.Tech.Pap.* 294, 318p.
47. Woodward, J. (1974), 'Successful reproduction and distribution in Lake Kariba', in *Lake Kariba: a man-made tropical ecosystem in central Africa* (Eds. E.K. Balon and A.G. Coche), pp. 526–536, W.Junk, The Hague, The Netherlands.
48. Zaret, T.M. and Paine, R.T. (1973). 'Species introduction in a tropical lake', Science, 182, 449–455

Resource Management and Optimization
1990, Volume 7(1–4), pp. 269–281
Reprints available directly from the publisher.
Photocopying permitted by license only.
© 1990 Harwood Academic Publishers GmbH
Printed in the United States of America

AQUACULTURE IN TROPICAL ASIA

RAFAEL D. GUERRERO III
Department of Zoology, University of the Philippines College, Los Baños, Laguna, Philippines

CONTENTS

Aquaculture is a traditional practice in many countries of tropical Asia and the Pacific. The region's vast living aquatic resources from brackish- and freshwater are the main contributors to the total world production from inland waters and mariculture. The resources of coastal marine waters are also of major importance.

A variety of aquaculture production systems occur in the region. These include the culture of fishes and crustaceans in ponds, pens and cages in brackish-, fresh- and marine waters; mollusc and seaweed culture in coastal marine waters; and integrated aquaculture-agriculture farming systems in inland areas.

The potential productivity of aquaculture to supply food for the large and rapidly growing populations of Asia has assumed a greater importance with the decline of capture fisheries. Although fish culture systems are more amenable to control and efficient management than captive fisheries, many technical, economic, institutional and ecological constraints must be overcome if the potential yields of aquacultural systems are to be attained.

TABLE I
World Catches in Inland Waters (Metric Tons)

Catch Type	Catch Weight
Freshwater fishes	8,053,700
Diadromous fishes	1,103,400
Marine fishes	39,100
Crustaceans	129,500
Molluscs	254,700
	9,580,400

1. INTRODUCTION

In many parts of the world aquaculture, the farming of living aquatic resources, has long provided food, income and employment.

In 1981, the world inland fishery production exceeded 9 million t (Table I) or 12.8% of the total catch. Asia and the Pacific together contributed 84.43%, or 5.8 million t, to the world's total aquaculture output in 1980. Of this, 38.2% consisted of fishes, 33.7% of molluscs, 27.07% of seaweeds and 0.86% of crustaceans (Table II). The aquaculture resources currently utilized in the region are brackishwater ponds (Table III), freshwater ponds and cages or pens (Table IV) and marine coastal waters (Table V). In addition, more than 6 million ha of brackishwater estuaries and freshwater areas could be developed for aquaculture.

This chapter provides an assessment of the aquaculture resources and the ways in which they are utilized and managed in tropical Asia. It also discusses the biological, physical and socio-cultural factors that affect their sustainability and also act as constraints on the development of these resources.

2. AQUACULTURE PRODUCTION SYSTEMS IN TROPICAL ASIA

A variety of aquaculture production systems exist in tropical Asia. The culture of fishes and crustaceans in brackishwater ponds is the most extensive category of systems, followed by systems based on freshwater ponds, pens and cages, and mariculture in coastal areas. Integrated aquaculture-agriculture farming systems involving fish, crop and livestock production are also widespread.

2.1 Culture of Fishes and Crustaceans in Brackishwater Ponds

Fish culture in brackishwater ponds is most widely developed in Indonesia and the Philippines. Extensive fishponds also occur in Taiwan, Thailand, Malaysia, India, Bangladesh and Sri Lanka. The major species cultured are milkfish *(Chanos chanos),*

TABLE II
Aquaculture Production in Asia and the Pacific in 1980 (Metric Tons)

	Finfish	Molluscs	Crustaceans	Seaweeds	Total
Bangladesh	65,000	—	—	—	65,000
China*	941,294	1,795,467	—	1,451,997	4,188,758
Hong Kong	7,260	—	—	—	7,260
India	830,201	1,763	17,009	—	848,973
Indonesia	177,400	—	21,797	—	199,197
Malaysia	9,357	63,412	972	—	73,741
Papua New Guinea	60	—	—	—	60
Philippines	151,612	250	910	132,730	285,502
Singapore	492	—	39	—	531
Sri Lanka	17,150	—	—	—	17,150
Thailand	39,367	111,673	9,923	—	160,963

*Including Taiwan

TABLE III
Aquaculture Production of Brackishwater Ponds in Selected Countries in 1980

	Total Area (ha)	Total Production (t)	t/ha
Taiwan	19,777	37,098	1.87
Indonesia	181,792	93,644	0.52
Philippines	176,231	135,951	0.77
Singapore	118	38	0.32
Thailand	26,865	11,901	0.44

mullet *(Mugil cephalus)*, sea bass *(Lates calcarifer)*, shrimps (*Penaeus* sp.) and mud crab *(Scylla serrata)*.

The culture of milkfish is based mainly on the use of animal manures and chemical fertilizers for growing natural food (i.e. benthic algae, zooplankton and small invertebrates). Most ponds in the Philippines and Indonesia have a water depth of 20–40 cm and depend on tidal flow for water supply and drainage. Stocking densities are generally low, at 1,000–3,000 fingerlings/ha. Average annual yields are 870 kg/ha/yr in the Philippines (PCARRD 1983) and 515 kg/ha/yr in Indonesia.

Milkfish culture is more intensive in Taiwan with multi-size stocking rates of 15,000/ha, supplemental feeding and fertilization. The reported national average yield is 1.95 t/ha/yr (Kuo 1984).

The mullet is also an important food fish cultured in brackishwater ponds, particularly in Taiwan and India. The culture methods are similar to those used for milkfish. Yields range from 500–1,500 kg/ha (Nash and Shehadeh 1980).

The fry of milkfish and mullet are caught at sea during their spawning seasons and are raised to fingerling size in nursery ponds. Although the induced spawning of these

TABLE IV
Freshwater Aquaculture Production of Selected Countries in 1980

	Total Area (ha)	Total Production (t)	t/ha
Taiwan	16,771	107,922	6.33
Hong Kong	1,820[a]	7,030	3.86
Indonesia	39,785	59,359	1.49
Malaysia	13,379	1,751	0.13
Philippines	20,000[b]	149,551	7.43
Singapore	324	567	1.75
Thailand	25,192	34,634	1.37

[a]Including brackishwater production
[b]Fishpens

TABLE V
Mariculture Production of Selected Countries in 1980

	Total Area (ha)	Total Production (t)	t/ha
Taiwan	15,345	29,998	1.95
Hong Kong	18[a]	780	43.33
Indonesia	—	29,489	—
Malaysia	5,757	121,441	11.0
Philippines	427	9,022	21.13
Thailand	3,069	91,618	29.8

[a]Cages

fishes has been achieved (Nash and Shehadeh 1980; Kuo 1982), techniques for larval rearing require refinement to ensure an increasing survival rate.

The culture of shrimps in the brackishwater ponds of the region has now become a major source of foreign exchange income. The species extensively grown are *Penaeus monodon* in the Philippines, Taiwan, Thailand and Indonesia, and *P. merguiensis* and *P. indicus* in India. Yields are highest in Taiwan, at 15 tons/ha/yr. These are attained using intensive methods based on high stocking rates, artificial feeding and water quality control. Most shrimp culture operations either use intensive methods of monoculture or the extensive (low stocking and pond fertilization rates) polyculture of fish and shrimp.

Hatcheries have proliferated in the region owing to the high demand for fry from shrimp farmers coupled with the scarcity of wild spawners. In the Philippines alone there are now some 50 privately-owned hatcheries that in 1983 produced a total 85 million fry (Primavera 1984).

In Thailand, the induced spawning of sea bass and the subsequent mass production of fish seeds permitted commercial cultivation of the species, and this high-value crop is now exported to Hong Kong, Singapore and Japan, among other places. Fry are reared in concrete tanks and feed artificially. They are then grown to market size on "trash fish" in cages and ponds (Pakdee 1982; Tanomkiat 1982).

The grouper *(Epinephelus tauvina)* is another important cash crop cultured in brackishwater ponds. Juveniles from the wild are stocked in ponds and fed with "trash fish." Alternatively, they can be jointly cultivated with such fish as tilapia, on which they forage. Initial success on the induced spawning of *E. tauvina* has been achieved by Chen *et al* (1977).

The mud crab is cultured commercially in milkfish ponds in Taiwan and the Philippines. Juveniles are caught in the wild and reared in ponds, where they are fed "trash fish." At stocking rates of 5,000–10,000/m^2, yields after 4 months are 550–1,100 kg/ha/crop (Baliao 1983).

Other brackishwater fishes are of less economic value but are important contributors to regional diets. These include tilapia *(Oreochromis mossambicus)*, tarpon *(Megalops cyprinoides)* and ten pounder *(Elops hawaiensis)*. These fishes are commonly regarded as predators or competitors.

2.2 Culture of Fishes and Giant Prawn in Freshwater Ponds

Since more cultivable species are available and culture systems are highly diversified, in tropical Asia freshwater pond aquaculture is more advanced than in that conducted in brackishwater ponds.

In Thailand, some 29,500 ponds covering 6,000 ha produce over 41,000 t of 19 fish species and the giant freshwater prawn *Macrobrachium rosenbergu* (Varikul and Sritongsook 1980). According to Bhinyoying (1977), the important commercially cultured fishes in Thailand are catfish *(Clarias batrachus* and *Pangasius sutchi)*, snakehead *(Ophicephalus striatus)*, snakeskin gouramy *(Trichogaster pectoralis)*, local carp *(Puntius gonionotus)*, Nile tilapia *(Oreochromius nilotica)*, grass carp *(Ctenopharyngodon idellus)*, bighead carp *(Aristichthys nobilis)* and silver carp *(Hypopthalmicthys molitrix)*.

Intensive culture of labyrinthine air-breathing species has resulted in yields of 50–100 tons/ha/yr, with heavy stocking and feeding rates. *Clarias batrachus* and *C. macrocephalus* fingerlings are stocked at 180/m^2 and harvested after 4–5 months of culture. Feed conversion using wet "trash fish" is 6:1 (Pawapootanon 1965).

In Thailand, some 10,000 ha of marginal ricefields seasonally inundated with sea water is used to culture snakeskin gouramy. Yields in excess of 2,000 kg/ha/8 mo are obtained with application of chicken excrement and grass as manures. This aquaculture system is more lucrative than rice farming (Chang *et al* 1983).

The culture of common carp *(Cyprinus carpio)* in running water ponds or raceways is a highly productive enterprise in Indonesia. In areas of 80–100 m^2 with a flow rate of about 250 l/sec, 100g fish are stocked at a rate of 500 kg per compartment. The fish is cultured for 3 months and fed at 3% of fish biomass per day with commercial pellets containing 25 to 26% crude protein. Yields per compartment are 2–2.5 t/3 mo (dela Cruz, pers. comm.).

The culture of tilapia in freshwater ponds has developed rapidly in some countries, particularly Taiwan and the Philippines, and tilapias are regarded as the prime domesticated species for tropical fish culture in the future (Kuo and Neal 1982). The species commercially farmed are the Nile tilapia and hybrids *(O. mossambicus* × *O. niloticus* and *O. aureus* × *O. niloticus).*

In Taiwan, tilapia farming has now become the most important aquaculture enterprise (Kuo 1984). This is based on intensive feeding and efficient water quality management techniques. More than 50,000 t/yr of tilapia are produced in Taiwan from 10,000 ha of ponds. The major species cultured are all-male hybrids and red tilapia.

Tilapia is the second most important cultured fish in the Philippines, where consumer demand for it has expanded with the price increase of marine species, owing to high operating costs and declining catches (Smith and Pullin 1984). Over 500 small-scale hatcheries now use freshwater ponds to produce tilapia fingerlings for later cage and pen culture in lakes. Commercial production of Nile tilapia is done in fertilized ponds with stocking rates of 10,000–20,000 fingerlings/ha for culture periods of 4–6 mo. With marketable sizes of 100–150 g per fish, the yield/ha/crop ranges from 1–2 t (Guerrero 1983).

In West Bengal, India, city sewage is extensively used for fish culture (Jinghran 1983). Ghosh *et al* (1979) reported an equivalent yield of 9.35 t/ha/annum for *Oreochromius mossambicus* produced in a sewage-irrigated pond.

The polyculture of different species simultaneously in the same pond is widespread in the region. This type of system is ecologically sound since it utilizes fishes with complementing food habits, thus maximizing production per unit area (cf. Ruddle, this volume).

In Thailand, Malaysia and Singapore, the Chinese system is used to polyculture grass carp, bighead, silver carp and common carp. The fish are stocked in various combinations at sizes of 30–600 g and at rates of 650–1,750 individuals/ha. In Taiwan, mullet is stocked with bighead, silver carp, common carp and milkfish in addition to tilapia and sea bass. Yields (ha/yr) from well-managed ponds comprise 2–3 t of tilapia, 2.5–3 t of carps, 1 t of mullet, 0.3–0.4 t of milkfish and 0.05–0.06 t of sea bass. In India, indigenous carps, such as the surface feeding *Catla catla,* the mid-water feeding *Labeo rohita* and the bottom feeding, *Cirrhinus mrigala* and *Labeo calbasu,* are stocked in the same pond (Jinghran 1983).

Undrainable, disused tin mine pools in Malaysia are used to culture such freshwater fishes as the silver carp, bighead and the marble goby *(Oxyeleotris marmoratus).* Similar man-made ponds 0.02 to 2.5 ha in area exist in India, Pakistan and Bangladesh. The total area of such undrainable ponds in India alone has been estimated at 1.6 million ha (Anon. 1983).

Commercial culture of the giant freshwater prawn is done in Taiwan, Thailand, Malaysia, Indonesia and the Philippines. Juveniles produced in hatcheries are stocked in 0.2– 1 ha ponds at rates of 10,000–30,000/ha. With supplemental feeding, yields are 1–3 t/ha/yr (Simon and Scura 1983).

2.3 Cage Culture of Finfish in Freshwater

The Asian region has vast freshwater resources, such as lakes, rivers, reservoirs, lagoons and ricefield irrigation systems, where cage culture of fish can be undertaken. A fish cage is an enclosure with at least five sides. They are made of wood or synthetic netting and may either be floating, for use in deep waters, or fixed to the substratum in shallow waters. Water flowing through the cages, oxygenates the water in the cages and removes fish metabolites. Carnivorous fishes and certain omnivorous species stocked at high densities require supplementary feed, whereas plankton feeders and detritivores, stocked at low densities, may subsist on the natural productivity in productive waters.

The snakehead and marble goby are the freshwater fishes commercially cultured in cages in Indonesia. Cages made of hardwood, measuring 4 × 2 × 2 m are floated with round logs. The stocking rate per cage is 1,000 individuals, each of 100–150 g. Fish are fed with chopped "trash fish" at rates of 5–10 kg/cage/day, and attain a marketable size of about 600 g in 6 mo (Indra 1982).

In West Java, the culture of common carp in bamboo cages lying on the bottom of streams and rivers is practised. The cages are 1.5–3 m long, 0.9–1.5 m wide and 0.6 to 0.7 m high. Each cage is stocked with 200–400 fingerlings, each 5–12 cm in length. They feed on organic matter and organisms thriving in the sewage-laden water that flows through the cages. Yields per cage can attain 50–70 kg in 2 to 3 mo (Hickling 1971).

In Kampuchea, floating cages are made of bamboo poles and splints reinforced with wooden planks and beams. Cage sizes vary from 40–625 m^3. They are used to culture *Pangasius, Clarias* and *Oxyeleotris*. Stocking rates range from 6,000–10,000 fish/cage. The fish are fed with cooked pumpkin, banana and a combination of cooked broken rice and rice bran. Larger individuals are fed pieces of raw fish, small live fish and kitchen refuse. After about 9 months, the fish attain 1.5–2.5 kg each (Pantulu 1979).

In Malaysia, the freshwater fishes cultured in cages are the bighead, silver, grass and common carps, tilapia, marble goby, rohu and giant gouramy *(Osphronemus goramy).*

In the Philippines, the Nile tilapia is the main species commercially cultured in floating and fixed cages. In productive lakes such as Laguna de Bay, cages of 50–200 m^2 with mesh sizes of 0.5–2.5 cm are stocked with 3–4 cm fingerlings at rates of 20–25/m^2 without supplemental feeding and at 50/m^2 with supplemental feeding. The fish attain a marketable size of about 100 g each in 4–5 mo (Mane 1979).

The major freshwater fishes in Thailand commercially cultured in cages made of bamboo or wood and floated in rivers and canals are catfish, marble goby, common carp, local carp, Nile tilapia and snakehead. To culture the marble goby, wooden cages 10–15/m^2 in area and 1.5 m deep are stocked with juvenile fish (100–300 g individual) at 100/m^2. They are fed with "trash fish" at 10% of fish body weight every 2 days. After 6–8 mo fish are harvested when they weigh 0.6–1.0 kg each (Tugsin 1982).

2.4 Pen Culture of Milkfish in Freshwater

The pen culture of milkfish in a freshwater lake was first demonstrated in the Philippines in the early 1970s (Delmendo and Gedney 1976). Milkfish fingerlings are stocked at densities of 20,000–40,000/ha in enclosures (pens) made of bamboo poles and synthetic netting. With culture periods of 4–6 mo, yields of 2–4 t/ha/yr are attained. This is about 4 times that obtained from the capture fishing of milkfish.

Fish growth in pens is mainly sustained by the natural productivity of the surrounding water. Problems of pollution, the slow growth of fish because of proliferation of pens and poaching are reported by fishpen operators in the Philippines. This system of fishfarming is being used experimentally in Sri Lanka and Bangladesh, for milkfish and carps.

2.5 Cage Culture of Fishes in Estuaries and Marine Waters

The commercial production of fishes in floating cages in sheltered estuaries and marine coves is extensive in Hong Kong, Malaysia, Indonesia, Singapore and Thailand. Such cages are less expensive than land-based operations and thus provide employment opportunities to poorer, small-scale fishermen. Further, as in Hong Kong, the price of live marine fish is 4–7 times higher than that of fresh marine fish. With such an incentive, marine fish farming is a potentially profitable enterprise (Cheng 1982).

In Malaysia, the major species cultured in marine cages are the grouper *(E. tauvina),* sea bass, rabbitfish *(Siganus javus),* and the fingermark bream *(Lutjanus russeli).* When fed on "trash fish," the growth rates for the grouper and sea bass are 0.6 kg in 10–12 mo and 1.3 kg in 12 mo, respectively (Yaman 1982).

Commercial cage culture of groupers *(E. tauvina* and *Plectropomus maculatus)* and the sea bass is done in the coastal areas of Indonesia. Floating cages measuring 3 × 4 × 6 m are stocked with 800 juveniles (each weighing 150–200 g) per cage. Fish of 600–700 g are harvested after 5–6 mo. The stock of each cage is fed with 3,000 kg of low grade "trash fish" per culture period (Lanjumin 1982).

In Singapore, the grouper, sea bass and snapper (*Lutjanus* sp.) are cultured commercially in cages. Net cages ranging in size from 2 × 2 × 2 m to 5 × 5 × 3 m are floated by means of plastic or metal drums and styrofoam attached to the wooden framework, and stocked with juvenile fish weighing about 100 g each. The fish are fed daily with "trash fish" at the rate of 5–10% of fish body weight. Fish of marketable size (600 to 800 g) are harvested after about 6–8 mo (Shiew 1982).

Marine fishes such as the sea bass, grouper, snapper and rabbitfish are also cultured in cages in Thailand. For the cage culture of sea bass, floating net cages of 3 × 3 × 3 m, with a mesh size of 2–5 cm, are used. The stocking rate is 30 juveniles/m^2. "Trash fish" is fed daily at a rate of 10% of fish body weight. Fish are harvested after about one year (Tanomkiat 1982).

2.6 Mariculture in Open Waters

The culture of molluscs and seaweeds in marine coastal waters is a traditional practice in many Asian countries. The farms are located in coves and protected bays characterized by productive and pollution-free waters. Marine farming is attractive to small-scale operators since it requires minimal investment and non-skilled labour.

In the Philippines some 2,000 ha of commercial mollusc farms yield an average yield of 3.13 t/yr of green mussel *(Perna viridis)* per 400 m^2 plot, using bamboo stakes as culture material. For whole oysters (*Crassostrea* sp.) an average yield of 3.82 t/0.25 ha/yr is reported (PCARRD 1977). Pearl-oyster farming is also practised in the Philippines.

In Singapore, floating bamboo rafts with hanging ropes are used for mussel culture. A typical culture area with a density of four (4 m long) ropes per m^2 produces 100–600 kg/m^2 in 6 mo (FAO 1983b).

The cockle or blood clam *(Anadara granosa)* is cultured in the coastal waters of western peninsular Malaysia. Young clams are collected and seeded, yields attain 40–80 t/ha/yr of marketable clams (Rabanal 1983).

Experimental culture of giant clams (*Tridacna* sp.) in Papua New Guinea and Palau has shown encouraging results. *T. gigas* attain a shell length of 50 cm and flesh weight of 6 kg in 5–7.5 yr in reef areas (Munro 1983). Artificial production of giant clam seeds has been achieved in hatcheries (Heslinga and Perron 1983).

Commercial seaweed production is done in Indonesia and the Philippines. A farm of 860 m^2 in Indonesia produces 140 kg of dry seaweeds every two weeks. The red alga (*Eucheuma* sp.) in a 0.5 ha farm in the Philippines yields 2,250 kg (dry weight) in 4 mo (FAO 1983b).

2.7 Integrated Aquaculture-Agriculture Farming Systems

Farming systems involving the growing of fish with crops and livestock are age-old practices in Asia. The typical farm landholding in the region is small. With crop diversification and integrated farming, an ecological balance is maintained and a variety of products is obtained to supply the farm family's food and income (Delmendo 1980).

Three types of farming system which include fish culture occur in tropical Asia. These are the crop-fish, livestock-fish and crop-livestock-fish systems.

Of an estimated 49 million ha of irrigated ricefields in tropical Asia, less than 1%

is utilized for fish culture (Kassim *et al* 1979). The principal species cultured in the region's ricefields are common carp, tilapia, snakeskin gouramy, snakehead, climbing perch *(Anabas testudineus)*, milkfish, mullet, *Puntius* and other carps.

Rice-fish culture in Southeast Asia is perhaps the oldest integrated farming system practised in irrigated fields. The system, introduced from India some 1,500 years B.C., is one of the most efficient means of using agricultural land. It is an appropriate agro-ecosystem for small labour-intensive farming operation (Huat and Tan 1980). In Bandung (Indonesia), 47,000 ha of ricefields are used for rearing carp fry to fingerlings. The rice yield averages 7 t/ha/crop and fish production is 75–100 kg/ha (dela Cruz pers. comm.). Rice-fish farming has not advanced in the Philippines, Malaysia and Thailand because of problems in poaching, flooding or lack of water and the extensive use of pesticides (Pantulu 1980).

Livestock-fish farming is widespread in Southeast Asia. In Thailand, integrated poultry-pig-fish farming is practised. Poultry is raised above a pigsty which is over a fishpond. Using such a system production in an area of 0.24 ha is 4 t of catfish *(Pangasius pangasius)*, 8 t of pigs and 15,390 chicken eggs. The integrated raising of sheep-fish, horse-fish, duck-fish and poultry-fish is done in West Java, Indonesia.

In Hong Kong, integrated fish farms also raise ducks and geese. Ducks yield 5–6 t/ha at 2,000–3,500 ducks/ha and for 2,750–5,640 kg/ha. Pig-vegetable-fish farming in Malaysia gives higher returns than does the raising of pigs alone because of the high labour and feed costs incurred with the latter (Delmendo 1980).

3. FUTURE PROSPECTS TO AND CONSTRAINTS ON THE DEVELOPMENT OF AQUACULTURE IN TROPICAL ASIA

Aquaculture will assume a greater importance than at present in satisfying the fish demand of the rapidly growing populations of Asia because of the overfishing and decline of capture fisheries, and the enforcement of extended marine economic zones mandated by the hour of the sea (Neal and Smith 1982). As a consequence, a 5–10 fold increase in aquaculture production is projected by the year 2000 (Pillay 1979). To achieve this, the transfer of technology, massive financial investments, suitable legislation, intensive research, manpower training, and the development of institutions and other essential infrastructures will be necessary.

The constraints on aquaculture development in tropical Asia are technological, economic, institutional and ecological.

Among the major technological constraints are the inadequate supply of fry or seeds, the high cost of feeds and fertilizers, fish diseases and the scarcity of technically trained manpower.

Dependence on natural sources of fry supply inhibit a fuller utilization and intensification of aquaculture resources in the region. The problems of seed supply and quality have been stressed by (Chang *et al* 1977; Neal and Smith 1982; Cheng 1982;

Smith and Pullin 1984), among others. However, advances in the artificial spawning of various species used in aquaculture have been achieved (Kuo 1982; Varikul and Sritongsook 1980; Chen *et al* 1977). But standard techniques for hormone injection and larval rearing must be either further improved or developed.

The intensive culture in ponds and cages of fishes and crustaceans requiring high-protein diets has caused a shortage of "trash fish" or fish meal. Thus in the Philippines, for example, where these fish feed commodities are scarce and heavily imported for feeding livestock and poultry, commercial production of carnivorous fishes has been retarded.

A massive outbreak of diseases believed to be bacterial and/or viral in origin occurred in Thailand in the early 1980s and caused heavy losses to commercial fish farmers. Management and water quality problems were traced to such epizootics. Diseases have also been a major constraint in the cage culture of finfish in Hong Kong and Malaysia.

The scarcity of personnel technically trained in the research technology transfer and commercial application is a serious drawback to the development of aquaculture in many countries of the region. Traditional farming practices are still dominant. Poor farm design and inefficient management methods contribute to low productivity.

Among the economic constraints to the development of the industry are a shortage of capital, conflict in land and water uses and rights, and poorly developed marketing structures. The shortage or even lack of capital and high interest rates on credit in developing countries have discouraged the diversification and intensification of aquaculture. Competition in land use for agriculture, industrial estates and human settlements is critical in countries where land is expensive and scarce, as in Singapore, Hong Kong and Taiwan. In Laguna de Bay, the Philippines, demand for irrigation water and space for open water fishing conflicts with fishpen and cage aquaculture.

The lack or poor development of marketing structures, or even markets, for certain aquaculture products in some countries has limited the development of fish farming. For instance, whereas carps and the giant freshwater prawn have well-established markets in Thailand, Taiwan, Indonesia and Malaysia, similar markets do not exist in the Philippines.

Institutional problems besetting the industry include the failure of research institutions to plan and execute multi-disciplinary approaches to aquaculture development, the failure of funding agencies to support proposed approaches, and the shortage or lack of funding for research (Neal and Smith 1982).

With environmental deterioration being impelled by industrialization and population growth in many Asian countries, the ecological problems affecting aquaculture have escalated. Foremost among such problems are, the destructive exploitation of mangroves that threatens the natural breeding and nursery grounds of important aquatic species; the use of illegal fishing gear and methods that have extensively damaged coral reefs that serve as shelter and feeding sites for fishes, invertebrates and plants of aquacultural value; industrial pollution and agricultural pesticide contamination of natural and man-made fishing grounds; and such natural disasters as typhoons, tsunamis and "red tides."

Despite such constraints, however, the outlook for the development of aquaculture in tropical Asia is encouraging. Although still in an incipient stage in some countries, aquaculture has progressed remarkably during the last decade to become a fully-fledged industry that now provides either direct or "spin-off" economic benefits to millions of people in many countries.

REFERENCES

1. Anon., 1983. Undrainable ponds: water supply, environmental monitoring, management. Newsletter, *Network of Aquaculture Centers in Asia*, 2(2), 4–5.
2. Baliao, D D, 1983. Mudcrab production in brackishwater ponds with milkfish. Paper presented at the Seminar-Workshop on Aquabusiness Project Development and Management, SEAFDEC Aquaculture Department, Tigbauan, Iloilo, Philippines. 11 p.
3. Bhinyoying, S, 1977. Natural breeding of some important fishes in Thailand. Barcher National Inland Fisheries Institute, Department of Fisheries, Bangkok. 9 p.
4. Chang, W Y B, Diana, J S, and Chuapoehuk, W, 1983. Workshop report on strengthening of Southeast Asian aquaculture institutions. Ann Arbor: University of Michigan. 115 p.
5. Chen, F Y, Chow, M, Chao, T M, and Lim, R, 1977. Artificial spawning and larval rearing of the grouper, *Epinephelus tauvina* (Forskal) in Singapore. *Journal of Primary Industries*, 5(1), 1–21.
6. Cheng, J, 1982. Economics of marine fish farming in Hong Kong. pp 141–145 in *"Report on the Training Course on Small-Scale Pen and Cage Culture for Finfish"*, ed by R Guerrero and V Soesanto. 216 pp Manila: South China Sea Fisheries Development and Coordinating Programme.
7. Delmendo, M N and R H Gedney, 1976. Laguna de Bay fishpen aquaculture development in the Philippines, pp. 257–265. Proc. Annual Meeting of the World Mariculture Society.
8. Delmendo, M N, 1980. A review of integrated livestock-fowl-fish farming systems, pp. 59–71. *In*: Integrated agriculture-aquaculture farming systems (eds. R S V Pullin and Z H Shehadeh), ICLARM Conference Proceedings 4, Manila, Philippines.
9. Food and Agriculture Organization, 1983a. 1981 Yearbook of Fishery Statistics. Vol. 53, Food and Agriculture Organization of the United Nations, Rome.
10. Food and Agriculture Organization, 1983b. Aquaculture in Asia-Pacific Region. Regional Office for Asia and the Pacific Monograph No. 4, Bangkok, Thailand. 23 p.
11. Ghosh, S R, N G S, Rao and A N Mohanty, 1979. Studies on the relative efficiency of organic manures and inorganic fertilizer in plankton production and its relation with the water quality, pp. 107–108. *In*: Symposium on Inland Aquaculture (abstract). CIFRI: Barrackpore, India.
12. Guerrero, R D, 1983. Tilapia farming in the Philippines: practices, problems and prospects. Paper presented at the PCARRD-ICLARM Workshop on Philippine Tilapia Economics. Los Baños, Laguna, Philippine. 23 p.
13. Heslinga, G and Perron, F, 1983. Palau giant clam hatchery. *ICLARM Newsletter*, 6(1):5.
14. Hickling, C F, 1971. Fish culture. 295 pp London: Faber and Faber.
15. Huat, K K and Tan, E S P, 1980. Review of rice-fish culture in southeast Asia. pp 1–14 in *"Integrated Agriculture-Aquaculture Farming Systems,"* ed by R S V Pullin and Z H Shehadeh. 258 pp Manila: International Center for Living Aquatic Resources Management.
16. Indra, R, 1982. Fish cage culture development in east Kalimantan province. pp 163–164 in *"Report on the Training Course on Small-Scale Pen and Cage Culture for Finfish"*, ed by R Guerrero and V Soesanto. 216 pp Manila: South China Sea Fisheries Development and Coordinating Programme.
17. Jinghran, V G, 1983. Fish and fisheries of India 666 pp India: Hindustan Publishing Corp.
18. Kassim, B B, Jee, A K and Eng, T C, 1979. A review of the status of research and development activities in rice-cum-fish culture in Asia 49 pp Malaysia: Universiti Pertanian Malaysia.
19. Kuo, C-M, 1982. Progress on artificial propagation of milkfish. *ICLARM Newsletter*, 5(1): 8–10.
20. Kuo, C-M and Neal, R A, 1982. ICLARM's tilapia research. *ICLARM Newsletter*, 5(1): 11–13.
21. Kuo, C-M, 1984. The development of tilapia culture in Taiwan. *ICLARM Newsletter*, 7(1): 12–14.
22. Lanjumin, L, 1982. Development of cage culture for finfish in Riau Archipelago, Riau Province, Indonesia. pp 165–166 in *"Report on the Training Course on Small-Scale Pen and Cage Culture for*

Finfish'', ed by R Guerrero and V Soesanto. 216 pp Manila: South China Sea Fisheries Development and Coordinating Programme.
23. Mane, A M, 1979. Cage culture of tilapia in Laguna de Bay. Technical Consultation for Available Technologies in Aquaculture. SEAFDEC, Tigbauan, Iloilo, Philippines. 5 p.
24. Munro, J, 1983. Giant clams—food for the future? *ICLARM Newsletter*, 6(1): 3–4.
25. Nash, C E and Shehadeh, Z H (eds.), 1980. Review of breeding and propagation techniques for grey mullet, *Mugil cephalus* L. 87 pp Manila: International Center for Living Aquatic Resources Management.
26. Neal, R A and Smith, I R, 1982. Key problem areas in world aquacultural development. *ICLARM Newsletter*, 5(1): 3–5.
27. Pakdee, K, 1982. Hatchery and nursery of sea bass for cage culture at the Satul Fisheries Station, Satul Province, Thailand. pp 215–216 in *''Report on the Training Course on Small-Scale Pen and Cage Culture for Finfish''*, ed by R Guerrero and V Soesanto. 216 pp Manila: South China Sea Fisheries Development and Coordinating Programme.
28. Pantulu, V R, 1979. Floating cage culture of fish in the lower Mekong Basin. pp 423–427 in *''Advances in Aquaculture''*, ed by T V R Pillay and Wm A Dill. 651 pp Rome: Food and Agriculture Organization.
29. Pantulu, V R, 1980. Aquaculture in irrigation systems. pp 35–44 in *''Integrated Agriculture-Aquaculture Farming Systems''*, ed by R S V Pullin and Z H Shehadeh. 258 pp Manila: International Center for Living Aquatic Resources Management.
30. Pawapootanon, O, 1965. The cultivation of Pla Duk (*Clarias batrachus* L.) in the vicinity of Bangkok. Kasetsart University, Bangkok, Thailand (Thesis).
31. Philippine Council for Agriculture and Resources Research, 1977. Philippines recommends for mussels and oysters. PCARR, Los Baños, Laguna, Philippines. 42 p.
32. Philippine Council for Agriculture and Resources Research and Development, 1983. Philippines recommends for bangus. PCARRD, Los Baños, Laguna, Philippines. 77 p.
33. Pillay, T V R, 1979. The state of aquaculture 1976. pp 1–10 in *''Advances in Aquaculture''*, ed by T V R Pillay and Wm A Dill. 651 pp Rome: Food and Agriculture Organization.
34. Primavera, J, 1984. Seed production and the prawn industry in the Philippines. Paper presented at the National Prawn Indstury Development Workshop. SEAFDEC, Iloilo City, Philippines. 36 p.
35. Rabanal, H R, 1983. Recent aquaculture practices in the southeast Asian region. Modern Fish Farming, Philippines Federation of Aquaculturists, Inc., San Juan, Metro Manila, Philippines.
36. Shiew, L E, 1982. Cage culture of marine finfish in Singapore. pp 197–199 in *''Report on the Training Course on Small-Scale Pen and Cage Culture for Finfish''*, ed by R. Guerrero and V Soesanto. 216 pp Manila: South China Sea Fisheries Development and Coordinating Programme.
37. Simon, C M and Scura, E, 1983. Prawn aquaculture investment considerations. pp 24–29 in *INFOFISH Marketing Digest*, No.:1/83, Kuala-Lumpur, Malaysia.
38. Smith, I R and Pullin, R S V, 1984. Tilapia production booms in the Philippines. *ICLARM Newsletter*, 7(1): 7–9.
39. Tanomkiat, T, 1982. Program on cage culture at the Phang Nga Small-Scale Fisheries-Assisted Project, Phang Nga Province, Thailand. p 213 in *''Report on the Training Course on Small-Scale Pen and Cage Culture for Finfish''*, ed by R Guerrero and V Soesanto. 216 pp Manila: South China Sea Fisheries Development and Coordinating Programme.
40. Tugsin, Y, 1982. Cage culture of freshwater finfish in Thailand. pp. 205–206 in *''Report on the Training Course on Small-Scale Pen and Cage Culture for Finfish''*, ed by R Guerrero and V Soesanto. 216 pp Manila: South China Sea Fisheries Development and Coordinating Programme.
41. Varikul, V and Sritongsook, C, 1980. A review of induced breeding practices of finfish in Thailand. (abstract) in *''Report of a Workshop on Induced Fish Breeding in Southeast Asia,''* ed by F B Davy and A Chouinard. 48 pp Singapore: International Development Research Centre.
42. Yaman, A R B G, 1982. Cage culture of finfish in peninsular Malaysia. pp 173–176 in *''Report on the Training Course on Small-Scale Pen and Cage Culture for Finfish,''* ed by R Guerrero and V Soesanto. 216 pp Manila: South China Sea Fisheries Development and Coordinating Programme.

Resource Management and Optimization
1990, Volume 7(1–4), pp. 283–299
Reprints available directly from the publisher.
Photocopying permitted by license only.
© 1990 Harwood Academic Publishers GmbH
Printed in the United States of America

MARINE REGULATED AREAS: AN EXPANDED APPROACH FOR THE TROPICS

NICHOLAS V.C. POLUNIN
Department of Biology, University of Newcastle, Newcastle-Upon-Tyne NE1 7RU, England

CONTENTS

Protected areas such as national parks and nature reserves combat environmental problems only in so far as they preserve designated sites from further damage. They scarcely solve such underlying problems as mangrove destruction, the dynamiting of coral reefs, or plain over-fishing.

In a broader view regulations related to areas should nevertheless be a major mode of conservation. Various types of regulation other than strict protection may in any case be more appropriate to the coastal zones of many tropical countries. Preservation there may be an economic luxury, whereas social pressures dictate that exploitation should continue. A realistically designed regulated area should be planned with particular objectives in mind, and deal most of all with problems such as those of over-fishing, population replenishment, and conflicts between fisheries and between various uses of single ecosystems.

However, there exist limitations to implementing this approach. Poor knowledge of larval recruitment patterns as yet precludes satisfactory design of replenishment areas. The multi-species complexity of tropical fisheries makes simple management measures inadequate in many ways. There is limited knowledge of how certain beneficial ecosystem functions are maintained naturally; for example, this lack inhibits satisfactory planning of mangrove areas sufficient to sustain various coastal fisheries. Such knowledge gaps will not be filled quickly. Interim measures may nevertheless be feasible. Thus existing regulated areas may offer a focus for exploring the design of potential replenishment zones. Attempts to model whole ecosystems for which some basic data are available will help to clarify some of the major ecological processes involved. There is little evidence yet that traditional marine reserves can contribute to modern management to any great extent.

1. INTRODUCTION—A QUESTION OF EMPHASIS

Regulation is often thought desirable where human activities adversely affect useful natural processes and products. The feasibility of regulation depends on such factors as the value of the resource and the cost of the measures proposed. Management may take many forms: one approach involves the designation of delimited areas within which certain rules prevail.

In this context prominence has been given to national parks and nature reserves, concepts which have been applied more recently to marine conservation problems than to terrestrial ones. Why is this so? In part the answer is that the recognition of human impacts and sophisticated scientific investigation of them have tended to come later to the sea than to the land. This, of course, reflects the long-held belief of the land-dwelling majority that the oceans and their resources are too vast to be damageable. Evidently, such an attitude persists to this day, even in marine protected areas of the southeastern USA (Davis, 1981).

Are regulatory measures considered 'successful'? Results are none too obvious, although a sense of frustration is evinced in some cases (Castañeda & Miclat, 1981; Rapson, 1983). This might, however, be the consequence of a forward planning which will bear fruit only in future. Alternatively, given that the value of such regulated areas has yet to be demonstrated, it could equally be that there is something inherently inappropriate in the idea, a mis-match for example, between the concept with its terrestrial precedent and the marine context to which it is now being directly transposed. Evidence for the latter is provided by Salvat (1981), who presents cogent reasons for coral reef management and gives prominence to reserves. He does not, however, explain how such protected areas can solve the kinds of problem which he identifies. Similarly Salm and Clark (1984) assess threats to the reefs of the Seribu Islands of the western Java Sea, but with statements such as "the small wooded islands . . . have enormous potential for the development of watersports (notably snorkeling, SCUBA, diving, swimming and picknicking) yet enjoyed by few in Indonesia, and are an exciting natural history laboratory for education and research," it is evident that what they have in mind is a national park for elite holiday-makers and scientists rather than for the impoverished majority of Javanese.

The case for such a measure being the objective response to a deteriorating human condition is far from convincing, frequently made though it is. In such examples the protected area appears, at best, to be a political expression on behalf of untramelled nature or aesthetics, at worst it may be mere, if unintentional, reinforcement for commercial enterprise. Most countries can scarcely award themselves the intangible benefits of the "holding strategy" proposed by Bradbury and Reichelt (1981) for the Great Barrier Reef, and elsewhere more than lip service must be paid to the oft-discussed meeting-ground of environmental conservation and economic development. How, in the present context, might this lack be addressed?

"Conservation" can not do everything, but one of its difficulties may still be in professing to do too much, and thus failing to be decisive. A single area is unlikely to be designed to successfully fullfil several major aims simultaneously. Appropriate design is more likely to follow from an accurate statement of purpose. Just as there is a range of aims in conservation, so too among regulated areas is there a number of potential types with which to respond to given demands. Processes of preservation have often been associated with these, but areas other than national parks and nature reserves may yet prove more relevant, more realistic and more workable.

To explore how this might be so it is important to appreciate the kinds of human activities that require regulation, as well as some of the ecological characteristics of the marine regions involved. It will then be possible to assess how design might be adapted to deal with sample problems.

2. THE SETTING

2.1 Populations

Tropical marine animals tend to produce more planktotrophic larvae than do those at higher latitudes (Mileikovsky, 1971). The tendency, therefore, is for these populations to be all the more dispersed by means of copious larvae capable of sustaining themselves for long periods. Lecithotrophic larvae are given a fixed yolk supply at birth and may live more briefly in the plankton. Many tropical populations are thus widely distributed.

Nevertheless, the duration of larval life and current regimes of the oceans may vary, becoming more or less conducive to the geographical isolation of some populations (Scheltema, 1971). These geographical isolates may be subject to extinction, especially where they are economically valuable, susceptible to exploitation, and population growth occurs only slowly.

Genetic variability in populations is apparently unusually high in coral reef communities, though not in tropical marine ecosystems as a whole (Somero & Soulé, 1974; Redfield *et al.*, 1980). Thus in genetic terms each species there would tend to be more intrinsically valuable there than in the other communities.

Founder events, whereby any new recruits reaching a new locality are likely to be different from their parent population because of the high mortality of larvae and their original genetic variability, are more probable. This may enhance possibilities for geographical differentiation, and in turn rates of actual speciation. Such processes, aided by a profusion of land and current barriers, may help to explain the large number of marine species in mid-Indo-West-Pacific waters (Valentine & Jablonski, 1983).

2.2 Communities, Ecosystems and Provinces

The communities of the tropics are highly diverse compared to those at greater latitudes. The number of species on coral reefs is especially large, whereas other communities, such as mangroves, are not so obviously rich. The high diversity, with more species consequently packed into the same area of habitat, means that interactions between species are especially important. On coral reefs symbioses are prominent, and as a consequence perturbation in one part of the ecosystem is highly likely to have ramifications elsewhere. Although much biological accommodation may occur in such communities, the stability and environmental constancy earlier inferred for such systems has been exaggerated. Seasonal and perennial change may, in fact, be substantial.

The high primary productivity of such shallow-water benthic ecosystems as those dominated by sea-grasses, reef corals and mangrove trees is a conspicuous feature of the tropics. This biological activity is generally thought to contrast with processes further offshore. Pathways of inorganic nutrient supply are considered to be important. Flow of nutrients to surface layers of the unproductive ocean water is evidently slow because the thermocline, below which nutrient-rich water occurs, rarely breaks down, although it does so seasonally at higher latitudes. Although it may be periodically enhanced by upwelling, river outflow and animal inputs, the supply of nutrients to coral reefs is thought also to be slow, but to be partly compensated for by recycling mechanisms within the ecosystem (Larkum, 1983). Predation on reefs is probably high at all trophic levels, and this efficient dissipation of primary production makes for high secondary production of consumers (Grigg, Polovina & Atkinson, 1984).

The high localised productivity of certain communities makes them something of biological oases in the surrounding tropical waters. The high biomass of fish on reefs and their associated slopes is thus prominent, as is the association of certain productive fisheries, such as those for penaeid prawns, with mangroves. Nutrient, organic matter and other flows in and out of ecosystems emphasise that although the biota and environments of such systems may be individually distinctive, each ecosystem is not independent of its neighbours. The ''interconnectedness'' of shallow-water tropical marine ecosystems should, in any case, be obvious from their close association in many places (Fig. 1).

Where barriers to dispersal occur and an environment becomes so different that the species cannot survive in it as they can elsewhere, distinct communities are seen to replace each other in analogous habitats. Various schemes for these biogeographic provinces have been put forward (Briggs, 1974; Schopf, 1980; Hayden, Ray & Dolan, 1984). Several provinces can be delimited within the tropical zone. The richest is that covering the Indo-West Pacific. Within it there is a central region of maximal diversity, while in outlying margins of the province endemicity may be locally high, even if overall species richness is reduced there.

2.3 Human Conditions

The high biotic diversity of tropical marine ecosystems may reduce the susceptibility of most species to extinction, because there are more species to choose from. Never-

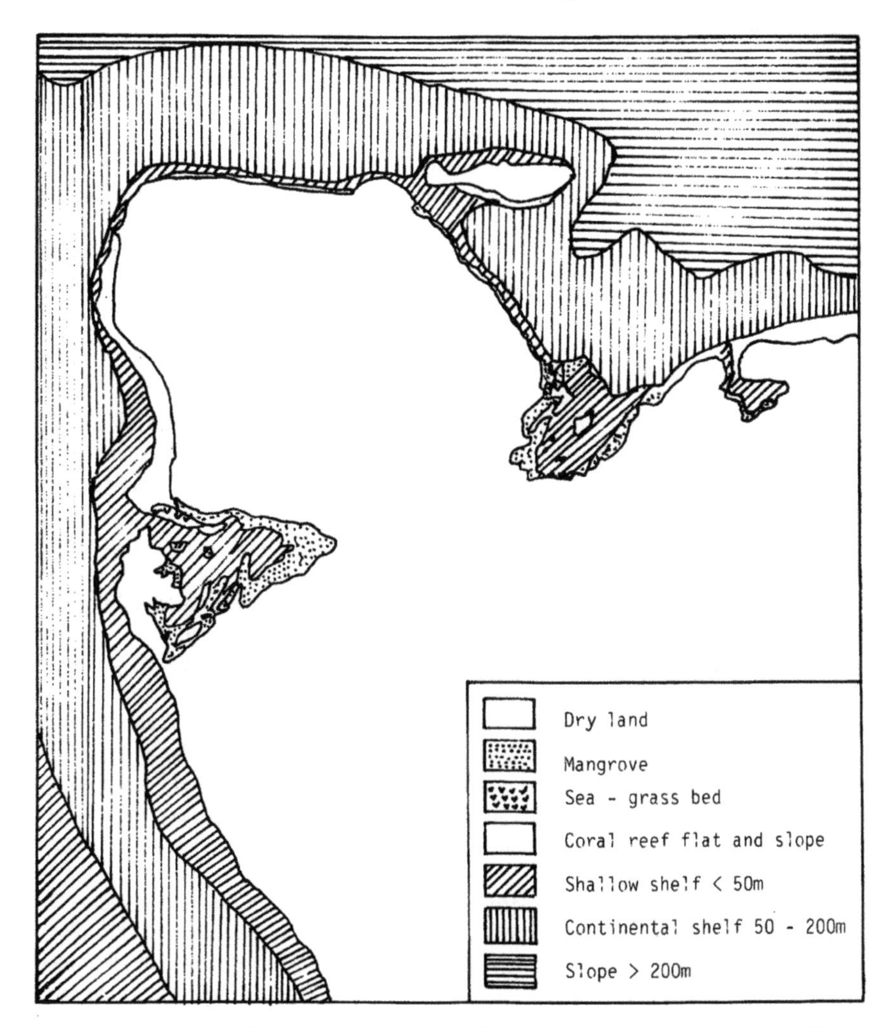

FIGURE 1. The distribution of Preservation, Scientific Research, Marine National Park 'A' and 'B', and General Use 'A' and 'B' Zones, and of Replenishment Areas within the Capricornia Section of the Great Barrier Reef Marine Park

theless, the likelihood of depletion exists where such indiscriminate fishing-gears as trawls are used, and where the species aimed for are large, valuable and accessible. This probability of over-exploitation is greatly increased where the marine-dependent human population is growing rapidly, or where it is concentrated in coastal areas; both are common conditions on small islands.

Many tropical fisheries have been rapidly developed in the last twenty years, mainly through capital-intensive operations. On the other hand the vast majority of small-scale fishermen have benefited little. Trawlers can operate offshore, but inshore fishing-grounds are often the most profitable. Yet it is to these inshore zones that subsistence and small-scale fishermen are confined for technical and economic reasons, and sometimes too by social-cultural constraints. Competition between the two fisheries sectors has thus become commonplace. In some instances it has culminated in physical conflict (eg. Polunin, 1983).

3. OBJECTIVES OF REGULATION

3.1 Human Impacts

The activities of man have various impacts on tropical marine ecosystems. Some are beneficial, such as the construction of intertidal fish-ponds which often provide a high, predictable and sustainable yield of protein. At low levels, increased loading with limiting inorganic nutrients may well be beneficial to primary production in many places. But human interventions can also have adverse effects, and the principal concern is that these are steadily increasing in extent and magnitude.

The major impacts of human activities on the tropical seas can be grouped broadly into three categories: over-exploitation of populations, problems of habitat degradation and pollution. A decline in the yield of a fishery is a tragedy: a useful, potentially sustainable resource is then no longer as productive and valuable as it might have been. The destruction of habitat can have many adverse consequences, such as the loss of the productive basis of a fishery, or alteration of the hydrological environment leading to marked shoreline changes. Pollution can include that by man-made chemicals, by accelerated releases of such elements as heavy metals, and by resuspension of fine silts that dramatically affects processes in the water column and on the bottom.

As a basis for regulating human activities, delimited areas can function to reduce all three types of impact. Except on a short-term basis, it would be impractical to physically cordon off areas as a barrier against pollution. But if broad zones are delimited within which unnatural releases to the environment are severely controlled, the incidence of pollution can be locally reduced. Regulated areas, however, are more obviously appropriate to problems of over-exploitation and habitat destruction.

In the long term, at least, few human impacts can be viewed in isolation. Oil pollution may destroy mangroves and secondarily lead to declines in fisheries. Careful regulation of a fishery may be of little use if water quality steadily deteriorates.

3.2 Aims of Control

The purpose of instigating controls on human activities is not blandly to inhibit them. Any process of economic and social development should be viewed to some extent as

TABLE I
Some Categories of Regulated Area, With a Possible Priority Ranking of Their Major Objectives

	Biosphere Reserve	National Park	Natural/Cultural Monument	Nature Reserve	Resource Reserve	Sustained Yield Harvest Zone	Water Quality Control Zone	Coastal Management Zone
	A	B	C	D	E	F	G	I
Protect ecosystem processes	1	2	2	2	1	2	1	2
Maintain yield of biological resources	1	2	3	2	1	1	2	2
Maintain biotic diversity	1	1	2	1	2	3	2	2
Research/Environmental Monitoring	1	2	3	1	3	3	2	3
Education, protect aesthetic/cultural sites	3	2	1	2	3	3	2	3
Stimulate rational use and development	3	3	3	3	2	1	1	1

an exploration in the right direction, but one which has many scarcely-predictable consequences. Some consequences may greatly enhance the benefits of any development, whereas others may have adverse effects.

The broad objective in iniating regulation is thus to mitigate the negative side-effects, thereby giving more scope to those that are beneficial. Yet in many cases knowledge is minimal, and on scientific grounds it would be better not to tamper at all with the populations or ecosystems. Unfortunately, also on practical grounds, the growing demands from burgeoning human populations and rising aspirations require some compromise with this ideal situation. Were that not the case, much of the world's fishing would have been halted long ago, on the grounds that scientific understanding was and still is too poor for designing and monitoring management plans.

3.3 Types of Area

Table I indicates how some major objectives of marine conservation can be met by various types of regulated area. This is not intended to include all possible objectives or areas, since both are subject to local interpretation and a more comprehensive treatment is provided by Salm and Clark (1984). Rather, Table I illustrates a potential range of regulatory states. Where a particular area is allotted priority objective rating

"1", this indicates that such an objective is the kind which the area should address. A rating of "3" does not mean that the objective involved may not be attained, but rather that in the design of a conservation plan emphasis should be placed on other aims. These areas need not be mutually exclusive.

Area types A–D (Biosphere Reserve, National Park, Natural/Cultural Monument, and Nature Reserve) should be already familiar, since, to greater or lesser degree, they have already been established in several parts of the world. A "Resource Reserve" is one which aims to maintain particular useful processes of natural ecosystems. An example might be the protection of mangrove green-belts to support fisheries or to maintain the coastal *status quo*. In a similar way forested watersheds on land can be protected to safeguard water supplies. A "Sustained Yield Harvest Zone" is one in which particular fisheries are managed in order to maintain productivity at a high level. Such regulated areas can be said to exist already in many countries. A "Water Quality Control Zone" is one in which all effluents and activities which may adversely affect water quality would be managed. This might be in an area of intensive mollusc culture. Few controls can be viewed in isolation, and a "Coastal Management Zone" is one which might incorporate many different types of regulated area. An existing tropical example of this is the Great Barrier Reef Marine Park of Northeastern Australia (Fig. 2).

It is not intended to imply by Table I that there is a fixed number of types of regulated area, each with its own inflexible criteria and predetermined design, since in reality a near-infinite array of types of area should be possible given the different limitations, opportunities and other circumstances of each site.

4. DESIGN

4.1 Over-Fishing

Reducing human exploitation can lead to some recovery in a hard-pressed fishery. This was inadvertently demonstrated by the reduction of fishing in the European North Sea during both World Wars (Cushing, 1975). In this case the recovery in catch per unit fishing effort was dramatic, but then the reduction in fishing pressure itself was draconian. What proportion of a total fishing-ground would need to be so protected to produce a beneficial effect in a fully-operational fishery? There remain too many unknowns to answer this question, even for the handful of well-studied fish populations. It is possible, nonetheless, to consider some factors which might influence exploration in this direction.

To begin with, recruitment to a fish population is not a simple process. At least in some species the relationship of recruitment to stock size may be convex (dome-shaped): maximal recruitment thus occurs at intermediate stock densities. This means that the mere act of depleting a fishery is not necessarily a process which will be reversed by preventing fishing in some areas, and thus supposedly permitting more

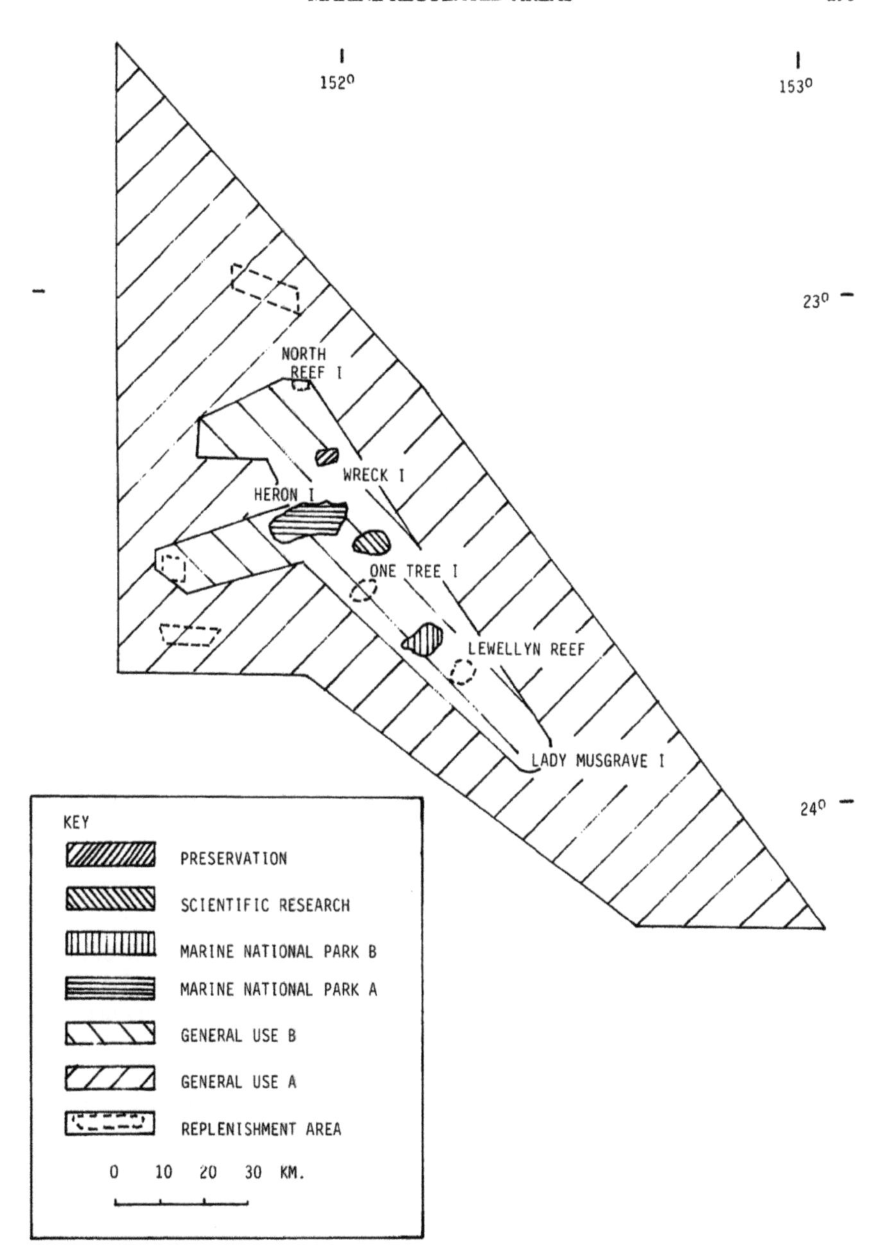

FIGURE 2. The major marine environments of the Bali Barat National Park, Indonesia (After Polunin, Halim & Kvalvågnaes, 1983)

adults to reproduce. If the recruitment-to-stock relationships elucidated by Cushing (1975) are widely applicable, it is only in cases of severe depletion that controls could have a beneficial effect.

How might appropriate replenishment areas be established in an instance of heavy over-fishing? Since most reef fishes migrate only locally if at all, it is useful to consider such populations first. These sedentary species might be relatively easy to restore: the greater the proportion of habitat protected from fishing, the swifter should be the recovery. Given the larval nature of dispersal in most species, however, such areas may have to be substantial. If area "A" receives most of its recruits from area "B" when the latter continues to be fished while the first does not, little recovery may occur in the replenishment area. As fishing reduces the average size of individuals and in turn this reduces reproductive output because larger fish are more fecund, then a regulated area might increase recruitment to a fished area. The regulated area itself, however, may continue to be starved of recruits. If vast areas that are recruitment self-contained cannot be set aside, then the possibility arises of establishing a patch-work of such areas. Here again, though, replenishment would be inefficient and thus slow. It follows that some basic biological and oceanographic knowledge is necessary for the successful design of a replenishment area.

Migratory species may create even greater problems. A well-known localised nesting/spawning ground, as in the case of most sea-turtles, provides a focus for control, but if the killing of a nesting female represents an obvious waste of potential recruits, the fishing of a female preparing for reproduction on the feeding-grounds is little less so. Again, regulated areas over extensive parts of breeding, nursery and feeding grounds will presumably be necessary for any speedy replenishment. This should be possible where the basic biology of the animal is known, where enforcement of the regulated areas is effective, and where interference by other fisheries can be minimised.

4.2 Competition Between Fisheries

If the dynamics of single-species fisheries are poorly understood, then those of multi-species ones typical of tropical waters remain virtually unintelligible. Goeden (1982) has shown how some serranid fish may be susceptible to over-fishing and that such depletion may have many consequences for ecologically related species. The best that can be done is to consider a simple two-species case. There are several possible permutations and one is as follows. Capture of species "A" is prohibited by trawling gear as it is the object of a directed trap fishery. Trawlers cannot avoid catching species "A", but they must return it to the sea when caught. The trawl fishery itself is directed at another principal species, species "B". Though not in the tropics, such a case has been analysed by Somerton and June (1984). Species "A"is the red king crab *(Paralithodescamtschatica)* and species "B" is the yellowfin sole *(Limanda aspera)*. The crab is a high-value species and the trawl fishery for the sole is developing.

Somerton and June (1984) suggest that trawling in any one area yields a gain to the sole fishery and a loss to the crab fishery. The loss occurs because crabs would have been harvested by the trap fishermen had they not been taken by the trawlers. If these gains and losses are measured in terms of gross revenue from the catch (ex-vessel prices), then it may be relatively simple to assess the net value of trawling in any one area. One complication is that the value of a crab catch has to be assessed not only in terms of value in a current year, but also of expected value that the crabs would have contributed in future years had they not been caught. This can be estimated from simple equations if data are available on such parameters as age-specific natural and fishery mortalities, the age/length relationship, ex-vessel price and the economic discount rate.

If the results are standardised to unit fishing effort, the value of each area is no longer a function of the fishing conducted. Areas can thus be compared on the basis of exploratory fishing results. In this case it was proposed that areas with positive (sole fishery value exceeds crab fishing loss) or zero value remain in the trawling zone, whereas those with negative value entered the zone protected for the trap fishery. The possibility exists, therefore, of redesigning the protected area on the basis of each year's fishing results. But a static regulated zone based on results from several years could be more easily implemented.

That example is simple in several respects. It would become more complicated were fisheries mutually interfering, were more than two dominant species involved, and were a wasteful by-catch also considered. Wider application of the analysis requires, among other things, that such economic decision-making will solve disputes. Where more "social" values are at stake, as in many densely-populated tropical coastal regions, decisions as to regulated zones may have to become more economically arbitrary. In its desire to resolve the serious conflict between small-scale and modern trawlermen, the Indonesian Government opted first for coastal trawler-exclusion zones, the width of which varied as a function of trawler size, but later decided to close completely the coastal fishing-grounds to trawlers.

4.3 Community Genetic Resources

A long-term aim of conservation should be the maintenance of high biological diversity, since species-rich communities may represent an untapped source of useful compounds for structural or pharmaceutical purposes. With the increasing importance of aquaculture it can be expected that eventually wild genetic resources will be actively sought to maintain the genetic vigour of captive populations. Given the broad knowledge of variations in biological richness at the level of provinces and ecosystems, it should be possible to incorporate such genetic objectives into reserve design on a worldwide basis. Protected areas should accordingly be planned for the major biogeographic provinces and representative ecosystems within them (Polunin, 1982).

The question remains as to the size of such reserves. Goeden (1979) has demonstrated for the tropical marine context how species-area curves can be used to predict what area of habitat may be necessary to maintain maximal diversity. In practice, the process of larval recruitment again precludes establishing any such zone in isolation from adjacent areas. Further, with coastal ecosystems being closely associated, provisions must be made for some much broader area of regulation if high-diversity core zones are to be properly protected. It is thus likely that only extensive areas, such as those proposed for Biosphere Reserves, could be widely applicable in this regard.

4.4 Multiple Use of an Ecosystem

A single area is commonly subject to more than one major use. Multiple-use provides opportunities for conflict which, without regulation, are to the detriment of the uses involved. Regulated areas can mitigate this problem by delimiting zones to which only compatible uses are confined. What are the compatible sets of uses, and what areas would be required to support each set?

As an example the case of an island with a limited area under mangrove forest can be taken. An increasing human population is threatening this ecosystem, although it is useful in many ways. The first step in an analysis would be to describe these major uses (Table II), and then to assess levels of compatibility among them. This allows an evaluation of how many different types of zones might ultimately be involved. Already a new set of questions arises. What spatial scales are involved? Some compatibility of subsistence gathering might, for example, be consistent with fish-pond conversion if the area is sufficiently well planned and were some retention of forest useful for fish-pond operations (e.g., provision of plant fodder, or poles for bank support). How would exclusive uses be weighted so that the areas allocated to them would be proportional to measures of their value? Clearly such additional answers are possible only for each local situation.

Initial criteria could, however, be established for the selection of areas of mangrove appropriate to each set of compatible uses. Thus fish-pond conversion requires extensive flats not susceptible to intensive erosion and not underlain by acid-sulphate soils. Mangroves designated for subsistence gathering must be located near human settlements dependent on them for their livelihood. Mangroves critical to shoreline protection would be those where a narrow band of forest parallels developed land, since even minimal over-cutting could be deleterious. Using such guidelines different sets of compatible uses can be allocated to suitable areas of the available mangrove.

4.5 Use Conflicts in a Reserve

The creation of a reserve may itself cause conflicts, since not uncommonly it precludes those prior uses deemed incompatible with the reserve's objectives. It will also introduce new uses, some of which may require management. By zoning areas for different uses, some conflicts can be avoided.

TABLE II
Conceivable Uses and Use-Compatibility Matrix for a Hypothetical Mangrove Area

i) Major uses:
- A. Wood production
- B. Firewood and other subsistence gathering
- C. Fish-pond conversion
- D. Salt-farm conversion
- E. Land reclamation for agriculture/forestry/building
- F. Capture fisheries
- G. Aesthetics/preservation/education
- H. Prevention of erosion
- I. Sewage processing/water quality control

ii) Compatibility matrix*:

	A	B	C	D	E	F	G	H	I
A	—	2	1	0	0	2	2	2	2
B		—	0	0	0	2	1	2	1
C			—	1	1	0	0	0	0
D				—	1	0	0	0	0
E					—	0	0	0	0
F						—	2	3	2
G							—	3	2
H								—	3
I									—

*Key:
0 no compatibility
1 low compatibility
2 some potential compatibility
3 high compatibility possible

One such potential problem is between the reserve itself and conventional fishing, where this is not obviously destructive. Certainly, if the overall boundaries of the reserve are to have any meaning, such activities must be monitored in populated areas. But it should also be accepted that excluding them outright can be damaging in the long term, if local opposition is needlessly aroused. In such cases zones should be established where particular types of traditional fishing are allowed.

Another problem is that created by increased visitor pressure on particular sites. Some damage is inevitable, but its effects can be limited by designating intensive visitor-use sites and preparing the ground accordingly. One major such facility in coral reef areas is anchor-buoys. Both the above types of zone were incorporated in the marine areas of the Bali Barat National Park, in Indonesia (Polunin *et al.*, 1983).

A third type of problem may arise when a reserve is the focus of intensive, scientific research, which entails long-term and disruption-free monitoring and experimentation. This objective may be accomplished by delimiting particular areas for specialised study. A scheme for zoning the Great Barrier Reef Marine Park for these and other

TABLE III
Major Types of Regulated Area and Their Use Rules in the Great Barrier Reef Marine Park (After Cocks, 1984)

i) Zone types
PR = Preservation
SR = Scientific Research
MNP(B) = Marine National Park (B)
MNP(A) = Marine National Park (A)
GU(B) = General Use (B)
GU(A) = General Use (A)

ii) Activities allowed, forbidden and subject to consent in the various zones

	Zone types#					
Activities	PR	SR	MNP (B)	MNP (A)	GU (B)	GU (A)
Non-manipulative research	P	P	P	P	P	*
Research station activities	X	P	P	P	P	P
Manipulative research	X	P	P	P	P	*
Private power boats and day cruise vessels	X	X	*	*	*	*
Recreation activities (not fishing)	X	X	*	*	*	*
Tourist ships	X	X	X	P	P	*
Observatory structure/erection	X	X	X	P	P	P
Netting	X	X	X	P	*	*
Recreational line fishing	X	X	X	*	*	*
Other line fishing	X	X	X	X	*	*
Trolling	X	X	X	X	*	*
Spear fishing	X	X	X	X	*	*
Collecting coral, shells, fish	X	X	X	X	X	P
Non-tourist ships	X	X	X	X	X	*
Trawl fishing	X	X	X	X	X	*

# Controls * = Allowed
X = Forbidden
P = Subject to consent and permit

potentially-conflicting uses has been outlined by Cocks (1984). Use regulations pertaining to the various zones are shown in Table III, and a resultant plan for part of the Great Barrier Reef in Fig. 2.

5. FUTURE STUDIES

5.1 Traditional or Modern?

Thus far I have taken the approach that there should be no substitute for objective appraisal of actual problems and the application of appropriate regulations. This assumes, however, that such measures can be fully implemented as designed. Alternatives have been sought by those who recognise that in many countries regulations cannot be properly enforced, however well-designed. It is thus important to recognise that in many parts of the tropics marine regulated areas of sorts are old-established. But are these traditional regulated areas appropriate to modern problems?

Appreciating the rapid acculturation of traditional peoples and that much lack of enforcement is a result of local opposition to regulations, some observers have suggested that local small-scale users might be incorporated into the system by recognising their patterns of traditional ownership. From his work in the oceanic Pacific Johannes (1978) proposed that restraint of exploitation at the level of the community, and thus for the conservation of limiting resources, was one reason why such marine tenure had developed. Although it remains to be shown that the traditional system did indeed function in that manner, there is also doubt as to the wide applicability of the concept. In Indonesia and New Guinea, for example, marine coastal tenure appears to have come into being as a means of resolving disputes. Conflicts could evidently arise for more than one reason, including competition over valuable marine resources, but many other issues could also be influential (Polunin, 1984). Because of its uncertain derivation, and because it is by no means established that existing patterns of tenure relate to some equilibrium between human populations and marine resources, it is thus difficult to see how traditional ownership could contribute widely to regulation of a fishery by limiting access to it. In some areas the tenure system may inhibit resource development, but this is usually undesirable.

Even if these traditional controlled areas were deemed to be sustainable for a local population, it is not certain that they would be able to cope with such pervasive problems as those pertaining to migratory species and rapidly growing human populations (Polunin, 1984). The whole topic, however, is neglected and deserves thorough investigation.

5.2 Replenishment

Although the concept of protected areas acting as means of replenishment to overexploited fisheries is often discussed, it should now be clear that the actual practice of such regulation has been little explored. To approach the matter from a purely biological perspective—an understanding of the life-cycles and population dyanmics involved—would no doubt take impossibly long in the tropics, although great progress is being made in areas such as the analysis of growth and reproduction of reef fishes. Perhaps a better, if temporary, approach is to use available protected areas to explore the subject experimentally. Among these are the replenishment areas now incorporated into the Great Barrier Reef Marine Park. Similar areas, if properly managed, should also provide foci for research into the recovery of heavily-fished stocks, both within regulated zones as well as at varying distances from them. The methods of monitoring recovery need not be complicated: a useful review for coral reef fishes is provided by Russell *et al.* (1978).

5.3 Ecosystem Processes

Although much has been inferred about the fisheries potential of coral reefs (Smith, 1978) and about the trophic support of fisheries afforded by sea-grass beds and mangroves (Hamilton & Snedaker, 1984), little is known about the processes involved. Much more research is needed on the trophic and other support mechanisms, and thus on estimating the areas and characteristics of ecosystems required to sustain a given fishery. With such information, the management of the ecosystems involved could be placed on a sounder basis than is possible at present. The attempt by Grigg, Polovina and Atkinson (1984) to model a Hawaiian reef ecosystem and draw some conclusions for management therefrom is an important step forward. The indication that natural predation may be intense on reefs suggests that populations may not be able to sustain a greatly increased mortality through fishing. Conversely, if large carnivores were depleted, the yield to fisheries might be considerably enhanced.

ACKNOWLEDGEMENTS

I thank Sue Wells and John Munro for their comments on a draft of this paper, many ideas for which have arisen from my work in Indonesia, as part of a IUCN/WWF programme, and in Papua New Guinea, while on the staff of the University of Papua New Guinea. I am grateful to both institutions for their support. I am grateful also to Crane, Russack and Co., Inc., the Great Barrier Reef Marine Park Authority and Applied Science Publishers Ltd. for permission to reproduce Table 3, Figure 1 and Figure 2, respectively.

REFERENCES

1. Bradbury, R.H. & Reichelt, R. (1981). 'The reef and man: rationalizing management through ecological theory', *Proc. 4th Int. Coral Reef Symp.*, 1, 219–223.
2. Briggs, J.C. (1974). *Marine Zoogeography*, McGraw-Hill, New York.
3. Castañeda, P.G. and Miclat, R.I. (1981). 'The municipal coral reef park in the Philippines', *Proc. 4th Int. Coral Reef Symp.*, 1, 283–285.
4. Cocks, K.D. (1984). 'A systematic method of public use zoning of the Great Barrier Reef Marine Park, Australia', *Coastal Zone Management Journal*, 12, 359–383.
5. Cushing, D.H. (1975). *Marine Ecology and Fisheries*, Cambridge University, London.
6. Davis, G.E. (1981). 'On the role of underwater parks and sanctuaries in the management of coastal resources in the southeastern United States', *Environ. Conserv.*, 8, 67–70.
7. Goeden, G.B. (1979). 'Biogeographic theory as a management tool', *Environ. Conserv.*, 6, 27–32.
8. Goeden, G.B. (1982). 'Intensive fishing and a 'keystone' predator species: Ingredients for community instability', *Biol. Conserv.*, 22, 273–281.

9. Grigg, R.W., Polovina, J.J. and Atkinson, M.J. (1984). 'Model of a coral reef ecosystem. III. Resource limitation, community regulation, fisheries yield and resource management', *Coral Reefs,* 3, 23–37.
10. Hamilton, L.S. and Snedaker, S.C. (Eds) (1984). *Handbook for Mangrove Area Management,* UNESCO, Paris.
11. Hayden, B.P., Ray, G.C. and Dolan, R. (1984). 'Classification of coastal and marine environments', *Environ. Conserv.,* 11, 199–207.
12. Johannes, R.E. (1978). 'Traditional marine conservation methods in Oceania and their demise', *Ann. Rev. Ecol. Systematics,* 9, 349–364.
13. Larkum, A.W.D. (1983). 'The primary productivity of plant communities on coral reefs', In *Perspectives on Coral Reefs* (Ed. D.J. Barnes), pp. 221–230, Clouston, Canberra.
14. Mileikovsky, S.A. (1971). 'Types of larval development in marine bottom invertebrates, their distribution and ecological significance: a re-evaluation, *Mar. Biol.,* 10, 193–213.
15. Polunin, N.V.C. (1982). 'Marine genetic resources and the potential role of protected areas in conserving them', *Environ. Conserv.,* 10, 31–41.
16. Polunin, N.V.C. (1983). 'The marine resources of Indonesia', *Oceanogr. Mar. Biol. Ann. Rev.,* 21, 455–531.
17. Polunin, N.V.C. (1984). 'Do traditional marine 'reserves' conserve? A view of Indonesian and New Guinean evidence', In *Maritime Institutions in the Western Pacific,* Senri Ethnological Studies 17 (Eds. K. Ruddle and T. Akimichi), 267–283.
18. Polunin, N.V.C., Halim, M.K. and Kvalvågnaes, K. (1983). 'Bali Barat: an Indonesian marine protected area and its resources', *Biol. Conserv.,* 25, 171–191.
19. Rapson, A.M. (1983). 'Economic management of lagoons', *Ocean Management,* 8, 297–304.
20. Redfield, J.A., Hedgecock, D., Nelson, K. and Salini, J.P. (1980). 'Low heterozygosity in tropical marine crustaceans of Australia and the trophic stability hypothesis', *Mar. Biol. Letters,* 1, 303–313.
21. Russell, B.C., Talbot, F.H., Anderson, G.R.V. and Goldman, B. (1979). 'Collecting and sampling of reef fishes', In *Coral Reefs: Research Methods* (Eds. D.R. Stoddart and R.E. Johannes), pp. 329–345, UNESCO, Paris.
22. Salm, R.V. and Clark, J.R. (1984). *Marine and Coastal Protected Areas: A Guide for Planners and Managers,* International Union for Conservation of Nature and Natural Resources, Gland, Switzerland.
23. Salvat, B. (1981). 'Preservation of coral reefs: scientific whim or economic necessity? Past, present and future', *Proc. 4th Int. Coral Reef Symp.,* 1, 225–229.
24. Scheltema, R.S. (1971). Larval dispersal as a means of genetic exchange between geographically separated populations of shallow-water benthic marine gastropods', *Biol. Bull.,* 140, 284–327.
25. Schopf, T.J.M. (1980). *Paleoceanography,* Harvard University, Cambridge, Mass.
26. Smith, S.V. (1978). 'Coral-reef area and the contributions of reefs to processes and resources of the world's oceans', *Nature,* 273, 225–226.
27. Somero, G.N. and Soulé, M. (1974). 'Genetic variation in marine fishes as a test of the niche-variation hypothesis', *Nature,* 249, 670–672.
28. Somerton, D.A. and June, J. (1984). 'A cost-benefit method for determining optimum closed fishing areas to reduce the trawl catch of prohibited species', *Can. J. Fish. Aquat. Sci.,* 41, 93–98.
29. Valentine, J.W. and Jablonski, D. (1983). 'Speciation in the shallow sea: general patterns and biogeographic controls', In *Evolution, Time and Space: The Emergence of the Biosphere* (Eds. P.W. Sims, J.H. Price and P.E.S. Whalley), pp. 201–226, Academic, New York.

AUTHOR INDEX—VOLUME 7

SUBJECT INDEX—VOLUME 7

For Product Safety Concerns and Information please contact our EU
representative GPSR@taylorandfrancis.com
Taylor & Francis Verlag GmbH, Kaufingerstraße 24, 80331 München, Germany

www.ingramcontent.com/pod-product-compliance
Ingram Content Group UK Ltd.
Pitfield, Milton Keynes, MK11 3LW, UK
UKHW021443080625
459435UK00011B/357

* 9 7 8 0 3 6 7 3 5 3 5 6 8 *